BROOKLANDS BOOKS

CW01654803

Alfa Romeo
GIULIETTA
Gold Portfolio
1954~1965

Compiled by
R.M. Clarke

ISBN 1 85520 066X

Distributed by
Brooklands Book Distribution Ltd.
'Holmerise', Seven Hills Road,
Cobham, Surrey, England
Printed in Hong Kong

BROOKLANDS BOOKS

BROOKLANDS ROAD TEST SERIES

AC Ace & Aceca 1953-1983
Alfa Romeo Alfasud 1972-1984
Alfa Romeo Alfetta Coupes GT. GTV. GTV6 1974-1987
Alfa Romeo Giulia Berlinas 1962-1976
Alfa Romeo Giulia Coupes 1963-1976
Alfa Romeo Giulietta Gold Portfolio 1954-1965
Alfa Romeo Spider 1966-1990
Allard Gold Portfolio 1937-1958
Alvis Gold Portfolio 1919-1969
American Motors Muscle Cars 1966-1970
Armstrong Siddeley Gold Portfolio 1945-1960
Aston Martin Gold Portfolio 1972-1985
Austin Seven 1922-1982
Austin A30 & A35 1951-1962
Austin Healey 100 & 100/6 Gold Portfolio 1952-1959
Austin Healey 3000 Gold Portfolio 1959-1967
Austin Healey 'Frogeye' Sprite Col No.1 1958-1961
Austin Healey Sprite 1958-1971
Avanti 1962-1983
BMW Six Cylinder Coupes 1969-1975
BMW 1600 Col. 1 1966-1981
BMW 2002 1968-1976
Bristol Cars Gold Portfolio 1946-1985
Buick Automobiles 1947-1960
Buick Muscle Cars 1965-1970
Buick Riviera 1963-1978
Cadillac Automobiles 1949-1959
Cadillac Automobiles 1960-1969
Cadillac Eldorado 1967-1978
High Performance Capris Gold Portfolio 1969-1987
Chevrolet Camaro SS & Z28 1966-1973
Chevrolet Camaro & Z-28 1973-1981
High Performance Camaros 1982-1988
Camaro Muscle Cars 1966-1972
Chevrolet 1955-1957
Chevrolet Corvair 1959-1969
Chevrolet Impala & SS 1958-1971
Chevrolet Muscle Cars 1966-1971
Chevelle and SS 1964-1972
Chevy Blazer 1969-1981
Chevy EL Camino & SS 1959-1987
Chevy II Nova & SS 1962-1973
Chrysler 300 1955-1970
Citroen Traction Avant Gold Portfolio 1934-1957
Citroen DS & ID 1955-1975
Citroen SM 1970-1975
Citroen 2CV 1949-1988
Shelby Cobra Gold Portfolio 1962-1969
Cobras & Replicas 1962-1983
Chevrolet Corvette Gold Portfolio 1953 1962
Corvette Stingray Gold Portfolio 1963-1967
High Performance Corvettes 1983-1989
Daimler SP250 Sport & V-8250 Saloon Gold Portfolio 1959-1969
Datsun 240Z 1970-1973
Datsun 280Z & ZX 1975-1983
De Tomaso Collection No.1 1962-1981
Dodge Charger 1966-1974
Dodge Muscle Cars 1967-1970
Excalibur Collection No.1 1952-1981
Facel Vega 1954-1964
Ferrari Cars 1946-1956
Ferrari Cars 1973-1977
Ferrari Dino 1965-1974
Ferrari Dino 308 1974-1979
Ferrari 308 & Mondial 1980-1984
Ferrari Collection No.1 1960-1970
Fiat-Bertone X1/9 1973-1988
Fiat Pininfarina 124 + 2000 Spider 1968-1985
Ford Automobiles 1949-1959
Ford Bronco 1966-1977
Ford Bronco 1978-1988
Ford Consul, Zephyr Zodiac MkI & II 1950-1962
Ford Cortina 1600E & GT 1967-1970
Ford Fairlane 1955-1970
Ford Falcon 1960-1970
Ford GT40 Gold Portfolio 1964-1987
Ford RS Escorts 1968-1980
Ford Zephyr Zodiac Executive MkIII & MkIV 1962-1971
High Performance Escorts Mk1 1968-1974
High Performance Escorts Mk II 1975-1980
High Performance Mustangs 1982-1988
Holden 1948-1962
Honda CRX 1983-1987
Hudson & Railton 1936-1940
Jaguar and SS Gold Portfolio 1931-1951
Jaguar Cars 1957-1961
Jaguar Cars 1961-1964
Jaguar Mk2 1959-1969
Jaguar E-Type Gold Portfolio 1961-1971
Jaguar E-Type 1966-1971
Jaguar E-Type V-12 1971-1975
Jaguar XKE Collection No.1 1961-1974
Jaguar XJ6 1968-1972
Jaguar XJ6 Series II 1973-1979
Jaguar XJ6 & XJ12 Series III 1979-1985
Jaguar XJ12 1972-1980
Jaguar XJS Gold Portfolio 1975-1988
Jaguar XK120.XK140.XK150 Gold Portfolio 1948-1960
Jeep CJ5 & CJ6 1960-1976
Jeep CJ5 & CJ7 1976-1986
Jensen Cars 1946-1967
Jensen Cars 1967-1979
Jensen Interceptor Gold Portfolio 1966-1986
Jensen Healey 1972-1976
Lamborghini Cars 1964-1970
Lamborghini Cars 1970-1975
Lamborghini Countach Col No.1 1971-1982
Lamborghini Countach & Urraco 1974-1980
Lamborghini Countach & Jalpa 1980-1985
Lancia Stratos 1972-1985
Land Rover 1948-1973 - A Collection
Land Rover Series II & IIa 1958-1971
Land Rover Series III 1971-1985
Land Rover 90 & 110 1983-1989
Lincoln Gold Portfolio 1949-1960
Lincoln Continental 1961-1969
Lotus and Caterham Seven Gold Portfolio 1957-1989
Lotus Cortina Gold Portfolio 1963-1970
Lotus Elan Gold Portfolio 1962-1974
Lotus Elan Collection No.2 1963-1972
Lotus Elite 1957-1964
Lotus Elite & Eclat 1974-1982
Lotus Turbo Esprit 1980-1986
Lotus Europa 1966-1975
Lotus Europa Collection No.1 1966-1974
Lotus Seven Collection No.1 1957-1982
Marcos Cars 1960-1988
Maserati 1965-1970
Maserati 1970-1975
Mazda RX-7 Collection No.1 1978-1981
Mercedes 190 & 300SL 1954-1963
Mercedes 230/250/280SL 1963-1971
Mercedes Benz SLs & SLCs Gold Portfolio 1971-1989
Mercedes Bens Cars 1949-1954
Mercedes Bens Cars 1954-1957
Mercedes Bens Cars 1957-1961
Mercedes Bens Competion Cars 1950-1957
Mercury Muscle Cars 1966-1971
Metropolitan 1954-1962
MG TC 1945-1949
MG TD 1949-1953
MG TF 1953-1955
MG Cars 1959-1962
MGA Roadsters 1955-1962
MGA Collection No.1 1955-1982
MGB Roadsters 1962-1980
MGB GT 1965-1980
MG Midget 1961-1980
Mini Cooper Gold Portfolio 1961-1971
Mini Moke 1964-1989
Mini Muscle Cars 1961-1979
Mopar Muscle Cars 1964-1967
Mopar Muscle Cars 1968-1971
Morgan Three-Wheeler Gold Portfolio 1910-1952
Morgan Cars 1960-1970
Morgan Cars Gold Portfolio 1968-1989
Morris Minor Collection No.1
Mustang Muscle Cars 1967-1971
Oldsmobile Automobiles 1955-1963
Old's Cutlass & 4-4-2 1964-1972
Oldsmobile Muscle Cars 1964-1971
Oldsmobile Toronado 1966-1978
Opel GT 1968-1973
Packard Gold Portfolio 1946-1958
Pantera Gold Portfolio 1970-1989
Plymouth Barracuda 1964-1974
Plymouth Muscle Cars 1966-1971
Pontiac Tempest & GTO 1961-1965
Pontiac GTO 1964-1970
Pontiac Firebird 1967-1973
Pontiac Firebird and Trans-Am 1973-1981
High Performance Firebirds 1982-1988
Pontiac Fiero 1984-1988
Pontiac Muscle Cars 1966-1972
Porsche 356 1952-1965
Porsche Cars in the 60's
Porsche Cars 1960-1964
Porsche Cars 1964-1968
Porsche Cars 1968-1972
Porsche Cars 1972-1975
Porsche Turbo Collection No.1 1975-1980
Porsche 911 1965-1969
Porsche 911 1970-1972
Porsche 911 1973-1977
Porsche 911 Carrera 1973-1977
Porsche 911 Turbo 1975-1984
Porsche 911 SC 1978-1983
Porsche 914 Gold Portfolio 1969-1976
Porsche 914 Collection No.1 1969-1983
Porsche 924 Gold Portfolio 1975-1988
Porsche 928 1977-1989
Porsche 944 1981-1985
Range Rover Gold Portfolio 1970-1988
Reliant Scimitar 1964-1986
Riley 11/2 & 21/2 Litre Gold Portfolio 1945-1955
Rolls Royce Silver Cloud 1955-1965
Rolls Royce Silver Shadow 1965-1981
Rover P4 1949-1959
Rover P4 1955-1964
Rover 3 & 3.5 Litre 1958-1973
Rover 2000 + 2200 1963-1977
Rover 3500 1968-1977
Rover 3500 & Vitesse 1976-1986
Saab Sonett Collection No.1 1966-1974
Saab Turbo 1976-1983
Shelby Mustang Muscle Cars 1965-1970
Stubebaker Gold Portfolio 1947-1966
Stubebaker Hawks & Larks 1956-1963
Sunbeam Tiger & Alpine Gold Portfolio 1959-1967
Thunderbird 1955-1957
Thunderbird 1958-1963
Thunderbird 1964-1976
Toyota Land Cruiser 1956-1984
Toyota MR2 1984-1988
Triumph 2000. 2.5. 2500 1963-1977
Triumph GT6 1966-1974
Triumph Spitfire 1962-1980
Triumph Spitfire Col No.1 1962-1982
Triumph Stag 1970-1980
Triumph Stag Collection No.1 1970-1984
Triumph TR2 & TR3 1952-60
Triumph TR4-TR5-TR250 1961-1968
Triumph TR6 1969-1976
Triumph TR6 Collection No.1 1969-1983
Triumph TR7 & TR8 1975-1982
Triumph Herald 1959-1971
Triumph Vitesse 1962-1971
TVR Gold Portfolio 1959-1988
Volkswagen Cars 1936-1956
VW Beetle Collection No.1 1970-1982
VW Golf GTi 1976-1986
VW Karmann Ghia 1955-1982
VW Kubelwagen 1940-1975
VW Scirocco 1974-1981
VW Bus. Camper. Van 1954-1967
VW Bus. Camper. Van 1968-1979
VW Bus. Camper. Van 1979-1989
Volvo 120 1956-1970
Volvo 1800 1960-1973

BROOKLANDS ROAD & TRACK SERIES

Road & Track on Alfa Romeo 1949-1963
Road & Track on Alfa Romeo 1964-1970
Road & Track on Alfa Romeo 1971-1976
Road & Track on Alfa Romeo 1977-1989
Road & Track on Aston Martin 1962-1990
Road & Track on Auburn Cord and Duesenburg 1952-1984
Road & Track on Audi & Auto Union 1952-1980
Road & Track on Audi 1980-1986
Road & Track on Austin Healey 1953-1970
Road & Track on BMW Cars 1966-1974
Road & Track on BMW Cars 1975-1978
Road & Track on BMW Cars 1979-1983
Road & Track on Cobra, Shelby & GT40 1962-1983
Road & Track on Corvette 1953-1967
Road & Track on Corvette 1968-1982
Road & Track on Corvette 1982-1986
Road & Track on Datsun Z 1970-1983
Road & Track on Ferrari 1950-1968
Road & Track on Ferrari 1968-1974
Road & Track on Ferrari 1975-1981
Road & Track on Ferrari 1981-1984
Road & Track on Fiat Sports Cars 1968-1987
Road & Track on Jaguar 1950-1960
Road & Track on Jaguar 1961-1968
Road & Track on Jaguar 1968-1974
Road & Track on Jaguar 1974-1982
Road & Track on Jaguar 1983-1989
Road & Track on Lamborghini 1964-1985
Road & Track on Lotus 1972-1981
Road & Track on Maserati 1952-1974
Road & Track on Maserati 1975-1983
Road & Track on Mazda RX7 1978-1986
Road & Track on Mercedes 1952-1962
Road & Track on Mercedes 1963-1970
Road & Track on Mercedes 1971-1979
Road & Track on Mercedes 1980-1987
Road & Track on MG Sports Cars 1949-1961
Road & Track on MG Sprots Cars 1962-1980
Road & Track on Mustang 1964-1977
Road & Track on Nissan 300-ZX & Turbo 1984-1989
Road & Track on Peugeot 1955-1986
Road & Track on Pontiac 1960-1983
Road & Track on Porsche 1961-1967
Road & Track on Porsche 1968-1971
Road & Track on Porsche 1972-1975
Road & Track on Porsche 1975-1978
Road & Track on Porsche 1979-1982
Road & Track on Porsche 1982-1985
Road & Track on Porsche 1985-1988
Road & Track on Rolls Royce & B'ley 1950-1965
Road & Track on Rolls Royce & B'ley 1966-1984
Road & Track on Saab 1955-1985
Road & Track on Toyota Sports & GT Cars 1966-1984
Road & Track on Triumph Sports Cars 1953-1967
Road & Track on Triumph Sports Cars 1967-1974
Road & Track on Triumph Sports Cars 1974-1982
Road & Track on Volkswagen 1951-1968
Road & Track on Volkswagen 1968-1978
Road & Track on Volkswagen 1978-1985
Road & Track on Volvo 1957-1974
Road & Track on Volvo 1975-1985
Road & Track - Henry Manney at Large and Abroad

BROOKLANDS CAR AND DRIVER SERIES

Car and Driver on BMW 1955-1977
Car and Driver on BMW 1977-1985
Car and Driver on Cobra, Shelby & Ford GT 40 1963-1984
Car and Driver on Corvette 1956-1967
Car and Driver on Corvette 1968-1977
Car and Driver on Corvette 1978-1982
Car and Driver on Corvette 1983-1988
Car and Driver on Datsun Z 1600 & 2000 1966-1984
Car and Driver on Ferrari 1955-1962
Car and Driver on Ferrari 1963-1975
Car and Driver on Ferrari 1976-1983
Car and Driver on Mopar 1956-1967
Car and Driver on Mopar 1968-1975
Car and Driver on Mustang 1964-1972
Car and Driver on Pontiac 1961-1975
Car and Driver on Porsche 1955-1962
Car and Driver on Porsche 1963-1970
Car and Driver on Porsche 1970-1976
Car and Driver on Porsche 1977-1981
Car and Driver on Porsche 1982-1986
Car and Driver on Saab 1956-1985
Car and Driver on Volvo 1955-1986

BROOKLANDS PRACTICAL CLASSICS SERIES

PC on Austin A40 Restoration
PC on Land Rover Restoration
PC on Metalworking in Restoration
PC on Midget/Sprite Restoration
PC on Mini Cooper Restoration
PC on MGB Restoration
PC on Morris Minor Restoration
PC on Sunbeam Rapier Restoration
PC on Triumph Herald/Vitesse
PC on Triumph Spitfire Restoration
PC on VW Beetle Restoration
PC on 1930s Car Restoration

BROOKLANDS MOTOR & THOROGHBRED & CLASSIC CAR SERIES

Motor & T & CC on Ferrari 1966-1976
Motor & T & CC on Ferrari 1976-1984
Motor & T & CC on Lotus 1979-1983

BROOKLANDS MILITARY VEHICLES SERIES

Allied Mil. Vehicles No.1 1942-1945
Allied Mil. Vehicles No.2 1941-1946
Dodge Mil. Vehicles Col. 1 1940-1945
Military Jeeps 1941-1945
Off Road Jeeps 1944-1971
Hail to the Jeep
US Military Vehicles 1941-1945
US Army Military Vehicles WW2-TM9-2800

BROOKLANDS HOT ROD RESTORATION SERIES

Auto Restoration Tips & Techniques
Basic Bodywork Tips & Techniques
Basic Painting Tips & Techniques
Camaro Restoration Tips & Techniques
Custom Painting Tips & Techniques
Engine Swapping Tips & Techniques
How to Build a Street Rod
Mustang Restoration Tips & Techniques
Performance Tuning - Chevrolets of the '60s
Performance Tuning - Ford of the '60s
Performance Tuning - Mopars of the '60s
Performance Tuning - Pontiacs of the '60s

BROOKLANDS BOOKS

CONTENTS

BROOKLANDS BOOKS

ACKNOWLEDGEMENTS

Brooklands Books owe their existence to the understanding and generosity of the world's leading motoring journals which, for many years, have allowed the reproduction of their road tests and other important features. In this instance, our thanks go specifically to Alfa Romeo, Auto Age, Autocar, Automobile Topics, Auto Parade, Autosport, Car (South Africa), Car and Driver, Car Collector, Cars Illustrated, Cars of the World, Classic and Sportscar, Foreign Cars Illustrated, Motor, Motor Trend, Road & Track, Sports Car Graphic, Sports Car Reports, Sports Car, Sports Cars Illustrated, Sports Car World, Thoroughbred & Classic Cars, Wheels and the World Car Catalogue.

No less important in the case of this book has been the help given to us by the Alfa Romeo Owners' Club and, in the particular, by Michael Lindsay, who gave us valuable guidance on its content. It was at his suggestion that we included stories on the early Giulia 1600 Sprint Speciale and the Giulia 1600 Spider, which had bodies similar to the Giuliettas which are the main subject of this book. Lastly, our thanks also go to motoring writer and enthusiast James Taylor for agreeing to set the scene with the introductory paragraphs which follows.

R.M. Clarke

Although the 1900 was the first mass-produced Alfa Romeo, the model which really popularised the marque was its smaller companion, the Giulietta. This car had an extraordinarily difficult birth for, despite the company's determination to produce a second model, funds were not available and had to be raised by means of a lottery. The prizes, naturally enough, were to be examples of the new Alfa Romeo.

Unfortunately, development of this new model proved slow. Somewhat unwisely. Alfa Romeo announced the names of the prizewinners before there were any cars to give away. So, when the press got hold of the story and turned it into a scandal, they had to act fast.

Mechanical components, most notably a new 80bhp, 1290cc, twin-cam engine, were already in production. So the coachbuilder Bertone was commissioned to draw up a 2+2 coupé body to clothe them in time for the Turin Motor Show in spring 1954. The result was called the Giulietta Sprint, and examples were given to the lottery prizewinners. But the design was so widely acclaimed that what had started as an order for a limited production run turned into a mass-production exercise.

Soon afterwards, the factory-styled saloon which had originally been planned as the first Giulietta came on stream and, by mid-1955, there was also an open two-seat Spider version with coachwork by Pinin Farina. Despite their disparate styling origins, however, the Sprint, Spider and saloon all bore a family resemblance.

The model range became even more complicated in 1956, when twin-carburettor 90bhp Veloce versions of all three models were introduced, and further tuning later took the power up to 115bhp for two special coupé models. These were the Bertone-bodies Sprint Speciale, made between 1957 and 1962, and the delightful Sprint Zagato (or SZ), with Zagato-styled body made from 1959 to 1961. Production of the mainstream variants ended in 1965.

The Giulietta established the pattern for future models from Alfa Romeo: its responsive, high-revving, small-capacity engine was allied to nimble road behaviour and sensitive but predictable handling, making for a car which was real fun to drive. Today, even the humble saloons attract enthusiast interest, and the pretty coupés and spiders remain an eloquent testimony to the pre-eminence of Italian stylists in the 1950s. This book is a mine of information on the story of these cars, and will appeal to owners and dreamers alike.

James Taylor

Woodcote, Oxon.,
September 1990

PUBLIFOTO

Departing from our rule of no "cheesecake", this view of the Alfa Romeo Giulietta shows its petite size, mentioned in article at right.

THIS modern Juliet is a wench to win the heart of any motor-minded Romeo before she even gets out of second gear! The lovely creature of whom I speak is none other than the new Alfa Romeo "Giulietta" to whom I lost my heart at first meeting. It is without question the best and most exciting small car of our day, the Italian concept of a sort of Modern Super MG.

I made the acquaintance of this little Italian beauty in the forecourt of the Alfa Romeo factory in Milan on a bright sunny morning and was surprised to see how tiny she was (photographs taken at the recent Motor Show did not do her justice), but garbed in the exciting traditional Alfa Red her petite lines were set off to the best advantage.

This girl, however, does not depend upon her looks alone. She has looks *and* ability—the ability to go very rapidly (100 mph); to make rough roads feel like boulevards (like a Lancia): to corner like a super sports car (Ferrari); and yet to behave like a lady in traffic.

Truly, the road-holding and comfort of this amazing little car is streets ahead of anything now offered the public. It will stay put on its course around a fast bend despite crossing tram lines diagonally or encountering a nasty patch of pavement while in a controlled slide. Never a trace of the front or back wanting to patter outwards or to alter the line already taken. The steering is perfect, the four-speed, all-syncromesh gearbox a delight, the driving position superb, and the brakes fantastic.

The tach is red-lined at 6500 and the incredibly smooth 4-cylinder twin-cam engine

NEW FROM ITALY

by Roger Barlow

A favorite of P. Farina appears again on a Cadillac chassis.

An Abarth modified Alfa Romeo 2000 has a special frame. The twin-cam engine develops 135 bhp at 6000 rpm from 1975 cc.

Collaboration by Abarth and Ghia resulted in this distinctive Alfa Romeo 2000. Weight is given as 1960 lbs, top speed 124 mph.

will reach and pass this point not only with ease but downright enthusiasm. It gives the feeling that it is not an engine made up of lots of separate bits and pieces but rather of having somehow been formed miraculously out of a solid block of metal. There is simply no change in the mechanical feel or sound (and absolutely no vibration) as the pointer on the rev counter sweeps round the dial from 5,000 to 6,000, to 6,500 to 7,000. I know that the 12-cylinder Ferraris have much this same willingness to rev, but they give one the worrisome feeling of a great deal of machinery whirling around at high speed. The little 4-cylinder "Giulietta" revs so happily that one is convinced it will never be strained, and that, therefore, one can use these high revs with complete freedom. This is not to suggest that it is an engine which performs only at high rpm. Not at all—this 1290 cc unit will pull away from 20 mph in top gear quite as though it were about 2 litres in capacity. The makers claim only 65 bhp, and for once I believe the Italians have made an understatement! Quite likely this engine will receive additional attention for competition and be made to develop considerably more power in the future.

This remarkably low and compact little coupe has an unexpected amount of room for both luggage and passengers. Headroom and leg rooom are more than adequate for a 180-lb 6-footer, but for my taste this car should be made as an open seater. There is definite indication that the builders are seriously considering this and apparently planning to make more of an effort to market their product in the USA in the future. I think there is little doubt that they would find a large number of small car enthusiasts quite happy to pay about $3250 for this delightful motor car. Actually this figure is only indicative of the bracket into which this model would probably fall and is not official. This combination of one of the greatest names in the world of motoring, a racing background, advanced yet simple mechanical specification, beauty, comfort and performance should put Alfas right into the thick of the competition for a fair share of the world's sports car market. Since they plan shortly to offer this model as a 4-seater sedan, it will also compete in the more sedate category of a "family" type transport!

Alfa Romeo Giulietta Specifications

Wheelbase	93.7 in.
Tread	50.0 in.
Tire Size (mm)	155 x 380
Dry weight	1790 lbs.
Engine	4 cyl, dohc
Bore & stroke	2.91 x 2.95 in.
Displacement	1290 cc
Max. bhp	65 at 6000
Axle ratio	4.55
No. of speeds forward	4
Electrical system	12 v.
Top speed	99 mph
Price	not available

PRETTY GIULIETTA

Leslie Brooke Describes a Brief Encounter with Alfa Romeo's Latest "1300"

It was after the Monte Carlo Rally, while thinking about going home to England, that it occurred to me to call and see my old friend Guidotti, the racing manager of Alfa Romeo. I offered to let him drive our little Standard Ten to see what he thought about it, and in exchange, to my surprise, he suggested I might like to try the new Alfa Romeo Giulietta. Naturally, I was delighted.

ATTRACTIVE: A pleasing contrast in lines by two Alfa Romeo "models". The low overall height and general compactness of the "Giulietta" is apparent.

Undoubtedly the designer had the old Latin saying *e fortis dolce* well in mind when this very beautiful little car was put down upon the drawing board. Grace is in every line of the Giulietta's coupé bodywork, yet within it is surprisingly roomy, despite a kerb weight of about 750 kg. Having taken in the car's lines, the next thing to strike one forcibly is the gleam of fluted aluminium brakes, of quite colossal dimensions, clearly visible through the apertures in the wheels. Upon opening the bonnet, one is confronted by the shapely all-aluminium four-cylinder engine; its head is hemispherical, the valves operate Jaguar-fashion directly from overhead camshafts, mounted in two remarkably compact bulges in the head casting; carburetters are two; electrics are by Lucas, Morelli or Bosch, to choice. Six thousand r.p.m. is the approved maximum for this delightful unit, although Guidotti confided to me that he has had over 120 m.p.h. from a Giulietta!

In the driving seat, moving smoothly and incredibly swiftly off, one immediately feels "this is an Alfa Romeo" in its deepest sense; to take it through the gears is sheer joy; its handling is impeccable and performance seemingly effortless at any range. As to braking, believe me, there is no trouble in stopping one of these cars from any speed in a highly satisfactory manner. The helical spring suspension gave rather a sharp motion, but the feeling at all times was of complete stability.

Later, when following behind the Alfa, I could watch how stable it was on bends of all kinds, the body and occupants riding smoothly while the wheels danced up and down over the bumps. Never before, in my opinion, has there been a 1,300 c.c. car with such pleasing appearance, exceptional acceleration and enviable maximum speed as in the new Alfa Romeo Giulietta.

Brief Specification

Engine: 4-cyl., o.h.c.; 74 x 75 min. (1,290 c.c.); 65 b.h.p. at 6,000 r.p.m.

Transmission: 4-speed gearbox; single dry-plate clutch; hypoid final drive: ratio, 4.55 to 1.

Suspension: I.f.s. by helical springs, rear by helical springs.

General: 155 x 380 mm. tyres; wheelbase, 7 ft. 9¾ ins.; track, 4 ft. 2 ins.; length, 12 ft. 9¼ ins.; width, 5 ft. 0¾ in.; dry weight, 1,760 lb.; hydraulic brakes.

NON-STOP FROM LAND'S END TO JOHN O' GROAT'S IN 1902

Following upon recent non-stop long distance runs that have been undertaken for publicity purposes, the Renault Company claim to have been the first car manufacturers to have staged such a run in this country—as far back as 1902. In that year a two-cylinder Renault was driven from Land's End to John o' Groat's, remaining in motion the whole time.

Cans of petrol were picked up from the roadside and the tank replenished while on the move, the cans subsequently being dumped overboard. The whole journey was completed in three days and four nights without stopping either the engine or the road wheels, and the resultant publicity boosted sales of Renaults in this country by more than 10 times in that and the following year.

STYLE + SPEED = *Alfa Romeo*

OUR cover car this month is the pretty little Alfa-Romeo "Juliet," released last year, but so far not seen in Australia. It has a four-cylinder, twin overhead camshaft engine which gives 65 b.h.p. from 1,290 c.c.

Other features are coil spring and wishbone independent front suspension and a rear axle located by trailing arms and suspended by coil springs.

Although Alfa builds the chassis, two of Italy's best-known coachbuilders, Bertone and Ghia, construct the bodywork.

The body seats two in luxury and the car will give a speed approaching 100 m.p.h. All-up weight is only 1,760 lb.

Other Alfa-Romeos are equally glamorous. The bigger 1900 Super Sprint saloon pictured here has similar bodywork, but is a bigger car, with a 1975 c.c. engine tuned to 115 h.p. on 8 to 1 compression.

Alfa also dominated post-war Grand Prix racing. For the last few years they have concentrated on such futuristic-looking but completely realistic sports cars as the Disco Volante, or flying saucer, with a powerful three-litre engine. Rumor has it that they will return to G.P. racing to compete against Mercedes, Lancia, Ferrari, Gordini and Maserati.

- Above: Alfa's sporting activities are concentrated at present on F.I.A. sports cars such as the Disco Volante (Flying Saucer).
- Left: Low-swept and graceful frontal styling for the 1900 Sprint.
- Below: The car on the cover—the "Juliet".

GRAND TOURER: The Alfa Romeo Giulietta taking a fast open bend in the Wharfedale district of Yorkshire.

As readers of AUTOSPORT know, John Bolster normally road-tests cars for this magazine, but when Jack Walton offered to lend me his new Alfa Romeo Giulietta, I was quite glad that other commitments prevented J. V. B. from doing this test. During my week's "ownership" of the Alfa, I covered a considerable mileage, in what must be regarded as probably the finest small-capacity high-performance machine yet to be marketed.

I collected the car from Walton's own garage, Riffa at Pool-in-Wharfedale. It was immaculate, but Jack's racing manager ruefully pointed out a long score on the polished red panels, caused by a female in a Ford. Among the reception committee was Peter Bolton,

Giulietta — Lovely Giulietta

Alfa Romeo's 1.3-litre Sprint Model is Revealed as an 100 m.p.h. Machine of Great Charm and Superb All-round Performance

director of Bolton's of Leeds who supplied the Alfa, first Giulietta Sprint to be delivered in Great Britain.

One could rave all day about the beauty of this little machine. The lines are so perfect that it would be impossible to find fault with it. Even dyed-in-the-wool vintage types admitted that the proportions could not be bettered, and that the Giulietta set a completely new standard in appearance for a Grand Touring vehicle. It is not, of course, designed for the six-foot and over brigade, but it is astonishingly roomy nevertheless. Nominally a two-seater coupé, there is plenty of space in the rear for a couple of extra seats, wells being provided under a removable luggage-shelf for that purpose.

Equipment is lavish, and there is that meticulous attention to detail which springs from a competitions background. For example, the headlight dipping arrangement and the operation of the traffic "blinkers" are controlled by flick levers mounted on the steering column. On the wheel boss a horn-type push-switch is wired up direct to the headlights. This follows normal Italian practice as headlights are being increasingly used to warn approaching traffic—a very sound system. The instruments

BY GREGOR GRANT

comprise speedometer, revolution counter, ammeter, oil pressure, oil and water temperature gauges. These are contained in a neat hooded structure, illuminated at night by a carefully diffused and non-dazzling panel-light system.

Visibility is remarkable by means of the curved windscreen and sloping bonnet line. The large "wrap-around" rear window gives the interior the airiness of an open model. Non-draught ventilation is achieved by the use of rear quarter-lights which can swivel outwards slightly. The ultra-efficient heater unit supplies fresh air for hot-weather motoring via a high-speed, fan-type blower.

The seats are delightfully comfortable, and have a generous amount of adjustable travel. On the passenger's side a grab-handle is thoughtfully provided on the roof—doubtless for use during rally motoring in the Alps. Finish throughout is superb, Walton's car being painted in real Alfa red. As on all Giulietta models left-hand drive is fitted, but, with the superb visibility, this is no disadvantage. In fact, the writer has a distinct preference for l.h.d. high-performance cars—particularly when mist and fog are experienced in this country.

Treated as a coupé, luggage space is vast. With the rear seats installed and occupied, there is not much room for baggage in the neat rear locker, which is quite normal for cars of *Gran Turismo* pedigree. There is no danger of the bonnet top springing open, as it is hinged at the front, a feature which should be made compulsory on all fast cars with top-opening bonnets. The underside of the Giulietta's bonnet-top is treated for anti-drumming by fitting a fabric overlay, which appears to retain sound-proofing "goo". The makers have gone to endless trouble to make the car soundproof and as rattle-free as possible, liberal quantities of the tarry-looking substance being found in every concealed spot—even at the back of the luggage locker.

A body-cum-chassis method of construction has been adopted. Front suspension is by double wishbones of unequal length, controlled by helical springs which enclose Girling telescopic

WIDE-ANGLE view to the rear is supplied by the large window. Laminated safety glass is used throughout.

EXQUISITE in appearance and proportion the Giulietta is a triumph of the Italian coach-builders' art, and sets the standard for modern, small-capacity "Gran Turismo" machines.

dampers. Worm-and-roller steering is used. Large vertical springs are used at the rear, also enclosing Girling dampers. Radius arms are used on each side to locate the rigid axle, an anti-roll bar arrangement being fitted on the driver's side. The drive is taken through a dry-plate clutch to a four-speed constant-mesh gearbox and hypoid final drive. The Girling brakes have 10½ ins. drums at the front, and 10 ins. at the rear, being developed from those used on later Type 159 formula cars.

The four-cylinder engine has two overhead camshafts operated by dual chains, the four sparking plugs being located centrally in the hemispherical cylinder head combustion spaces. Makers claim 65 b.h.p. at 5,500 r.p.m. with an 8 to 1 compression ratio, but in view of the performance figures obtained on this particular car, this would appear to be a modest estimate. Carburation is supplied by a double-choke Solex Type 32 PAIAT instrument which is provided with an enormous air-cleaner-cum-silencer.

A large number of British-made components are used, including Lucas starter, dynamo, coil and distributor, Hardy-Spicer propeller shaft, Vandervell engine bearings, Valves Ltd. of Coventry valves, Fram oil filter, Girling dampers and brakes.

On the road the Giulietta's manners are well-nigh perfect. She is smooth, and, apart from a healthy exhaust note at high r.p.m., remarkably silent. Even the gears are reasonably quiet, and the axle is completely free from any suspicion of hum. The brakes are just about the most powerful I have experienced on any car, and do not appear to have any "fade" characteristics whatsoever. The headlights make for safe fast-driving, but the "peep-peep" horn is definitely not up to Continental standards.

Whether the Giulietta under-steers or over-steers I couldn't care less. To my way of thinking the steering is just about ideal, and the car seems to prefer conducting itself without any necessity for wheel-winding on any type of bend. It has the feel of a genuine race-bred machine, with road-holding to match. The suspension cannot be faulted. Roads which I know to be rough and bumpy can be treated with disdain, with a complete absence of bucking and jumping which is often experienced with light, short-chassis vehicles.

Acceleration is extraordinarily good for a 1.3-litre closed car, and it was only after the speedometer had been carefully checked for accuracy that some of the performance figures were accepted. The instrument, curiously enough, was dead accurate, being marked in kilometres over which Jack Walton had pasted a ring giving m.p.h. reading. Standstill to 50 m.p.h. in 9.8 secs., and a standing quarter-mile in 18.9 secs. are figures which speak for themselves.

The maximum speed (mean of two directions) was 101.5 m.p.h., the best one way being 103.5 m.p.h. Even at this speed the engine was dead-smooth and there was not the least sign of vibration. Ideal cruising speed is 75-80 m.p.h., when the power-unit appears to be able to go on for ever. Fuel consumption is astonishingly light. Driving really hard, it does not drop below 26 m.p.g. As an experiment, a Solex test-tank was fitted, and, circling round an airfield circuit at a steady 40 m.p.h., 42 m.p.g. was obtained.

Nevertheless this otherwise matchless small Alfa Romeo comes in for two main points of criticism. The steering

OFFICE: Sensible grouping of the instruments inside a non-dazzle hood is a feature of the Alfa interior. Space is also provided for a radio set. Visibility is remarkably good.

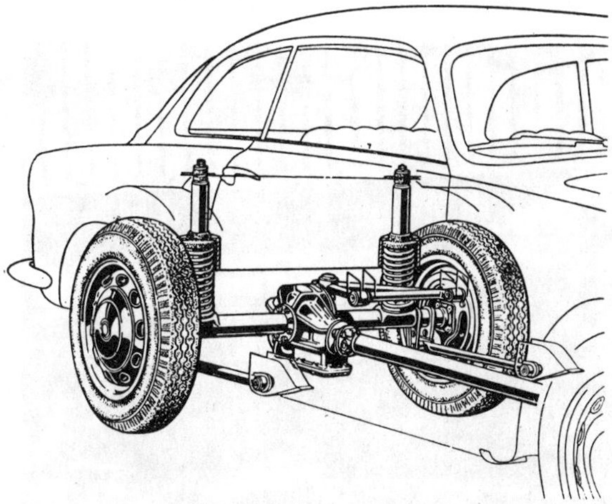

SOSPENSIONE POSTERIORE: (*Above*) *Layout of the rear suspension of the Giulietta, showing the radius arms and anti-roll bar arrangement.*

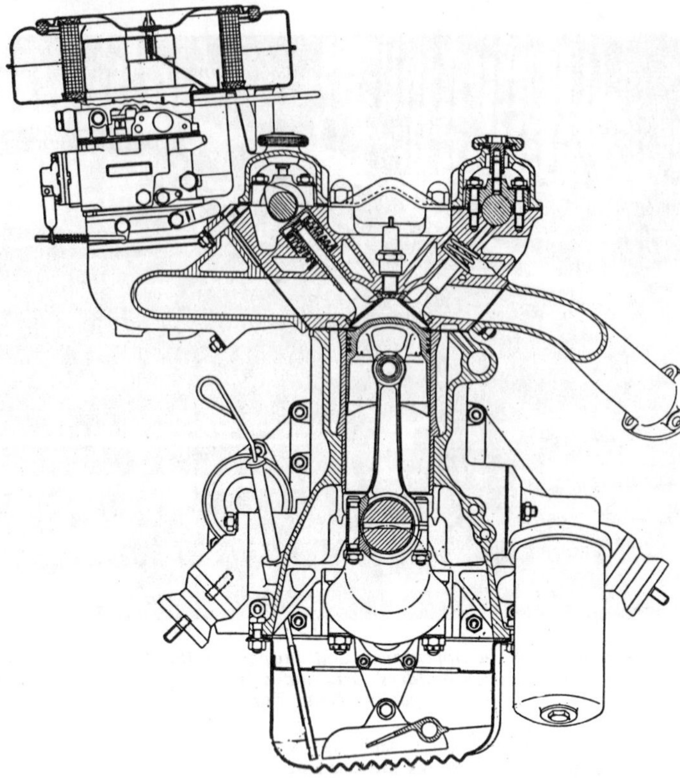

MOTORE: (*Above*) *Transverse section of the ultra-efficient Giulietta twin-o.h.c. engine. Valve operation is via "jam-pots".*

★

(*Left*) *The power-unit fits snugly under the forward-hinged bonnet.*

column gearchange, although a good example of its type, is not the sort of thing one wishes to find on an Alfa. Reverse gear is sometimes difficult to engage, while a really quick change from second to third is not only hard to achieve, but often impossible. A short-travel, floor-located lever wou' make all the difference to this car. Also, the pistol-type handbrake seems to be a somewhat cheap fitting for such a quality high-performance machine. Oil consumption should also be mentioned. While this may be peculiar to the car tested, a pint of lubricant every 150 miles seems to be excessive for a car with just over 2,000 miles on the clock.

It was a wrench having to part with this lovely little vehicle. During the time it was in my hands I thoroughly enjoyed every minute at the wheel, and, to judge by the reaction of the general public wherever the car went, its beauty of line stamps it as a real thoroughbred. One of these days British manufacturers may realize the appeal in this type of motor-car, and that there is an ever-growing section of the buying public which would "go" for a Grand Tourer of roughly the same specification. The Giulietta should have a ready sale in U.S.A. In Italy it is not particularly expensive as Italian quality machines are reckoned, and it is only British purchase tax and import duties which bring its total cost in the U.K. to just over £2,260.

Jack Walton has a car of which he can indeed be proud, and there is little doubt that this red Alfa Romeo will cause countless numbers of people to break the Tenth Commandment!

Specification and Performance Data

Car Tested: Alfa Romeo Giulietta Sprint.

Engine: Four cylinders 74 mm. x 75 mm. (1,290 c.c.). Two overhead camshafts (chain-operated). 8 to 1 compression ratio. 65 b.h.p. at 5,500 r.p.m. Solex 32 PAIAT d/c carburetter. Lucas coil and distributor.

Transmission: Single dry-plate clutch. 4-speed constant-mesh gearbox with steering column control. Ratios, 4.555, 6.172, 9.565, and 16.327 to 1. Reverse, 15.327 to 1. Hardy-Spicer open propeller shaft. Hypoid rear axle.

Chassis: Integral construction. Independent front suspension with double wishbones and helical springs. Rear axle rigid with vertical helical springs, radius arms and anti-roll bars. Girling hydraulic telescopic dampers. Bolt-on 15 ins. wheels with 155 x 15 ins. tyres. Girling hydraulic brakes with 10½ ins. drums (front) and 10 ins. (rear). Pistol-type handbrake.

Equipment: 12-volt electrical equipment, Lucas starter and dynamo, Marelli lamps and battery. Speedometer, rev. counter, oil pressure, water, oil temperature gauges. Flashing direction indicators. 14-gallon rear petrol tank.

Weight, with oil, water and spare wheel, 17 cwt. 87 lb.

Leading Dimensions: Overall length, 13 ft. Overall height, 4 ft. 4 ins. Overall width, 5 ft. Track (front and rear), 4 ft. 2 ins. Floor to roof, 42 ins. Steering wheel to seat back, 18½ ins. max., 14 ins. min. Seatback to clutch pedal, 23 ins. max., 18½ ins. min. Floor to edge of seat, 9½ ins. Width of boot, 39½ ins. Capacity, 10 cubic ft.

Performance: Maximum speed, 101.5 m.p.h. Speeds in gears, 1st 32 m.p.h., 2nd 50 m.p.h., 3rd 75 m.p.h. Standing quarter-mile, 18.9 secs. Acceleration 0-30 m.p.h. 3.6 secs., 0-40 m.p.h. 7 secs., 0-50 m.p.h. 9.8 secs., 0-60 m.p.h. 13 secs., 0-70 m.p.h. 20.1 secs.

Fuel Consumption: Driven hard, 26 m.p.g.

Price: £1,595 plus £565 14*s*. 2d. Purchase Tax.

Makers: Alfa Romeo, Milano. Concessionaires G.B., Thompson & Taylor, Ltd., Cobham, Surrey. Supplied by Bolton's of Leeds, Ltd., 14 Harrison Street, Briggate, Leeds 1.

ACCELERATION GRAPH

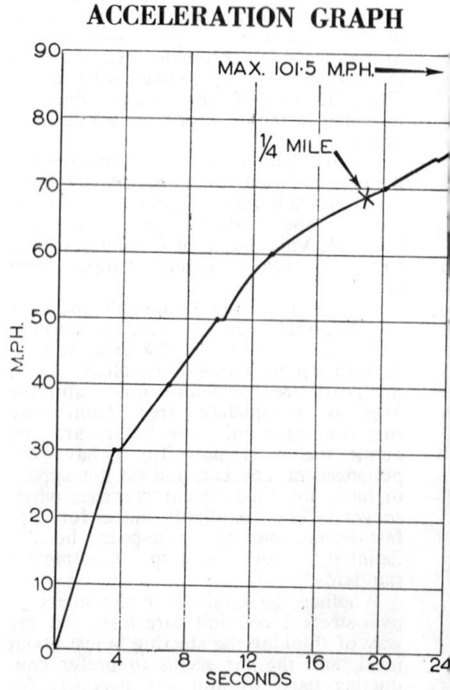

ROAD TEST:
Alfa Romeo Giulietta Spyder

The interior views, below, show the roomy seating, the now obsolete column shift lever and the neat crackle-finished dohc engine.

So well proportioned is the Giulietta that a photo gives no clue to its size.

The Spyder-type convertible body gives a slight loss in top speed compared to the hard-top.

photography: Poole

PROOF that the Italian automotive industry is at last serious about entering the American market is offered by the long-awaited "$3000" Alfa Romeo. Many of us have heard, read about, and occasionally seen one of the sleek, expensive Italian imports, but of the 1% of our population who could readily afford such a car only a few hundred have been willing to give up the luxurious floating ride and six passenger spaciousness of their $5000 "prestige" automobile.

Now we have the "Giulietta" or more simply, the Alfa Romeo 1300, indicating the engine capacity in cubic centimeters (actually 1290 cc). The 1300 is available as a 4-door sedan with a unit frame and body structure and a "dreamy" double overhead camshaft engine developing 50 bhp. This car sells in the U. S. A. for about $2500, but we are concerned here with the more specialized version of the 1300. This is a tiny car with a long name, the Alfa-Romeo-Giulietta-Sprint, available either as a 2-4 seater hard top coupe, or a 2-seater Spyder, the latter being actually a convertible coupe type body. The Sprint, or "1300-S," versions are, of course, smaller and lighter than the standard sedan, and minor engine refinements give an output of 65 bhp. The principal differences between the hard top Sprint Coupe and the convertible Spyder are the price (the coupe is $1200 higher!) and the wheelbase, the convertible being 7 inches shorter.

Without a doubt the 1300-S Spyder is the most fascinating small sports car we have ever driven. With an engine of only 78.7 cubic inches displacement it provides a performance which approaches the modern American behemoth so closely that one wonders if claims of 300 bhp are not closer to an honest 150. An acceleration time of under 15 seconds for 0 to 60 mph, and an honest top speed of 100 mph were impossible less than 4 years ago, from our lowest priced "big 3" makes.

But performance data in this case is only a part of the story. The greatest charm of the Giulietta Sprint is its handling qualities. Much has already been written about the delightful road manners of this new car including a report on a prototype which appeared in R & T over a year ago (Nov. 1954, page 32). Every word of these glowing reports is true—the adhesion to the road is absolutely uncanny. In cornering there is moderate understeer, very light caster-pull on the steering wheel rim, no noticeable body roll and no squeal from the Pirelli tires until the absolute limit of adhesion has been exceeded. Even then the instant steering response (3.2 turns lock-to-lock) and a very small turning circle (30 ft) give easy and confident control of slides when a tight turn is taken too fast. This is strictly a fun car, either for the new-to-sports-car man or the I've-tried-them-all type.

Second only to the impeccable handling and roadability of the Giulietta must come the brakes which are close to if not absolutely the best we have ever experienced. Pedal pressure is light without being touchy and when applied hard the feel is very solid and deceleration approaches the magic one-*g* figure. The bi-metallic drums have a most impressive look too, with 72 massive cooling fins per drum placed at an angle for positive air circulation. There's no fade here.

Riding in the car it is almost impossible to believe that the wheelbase is only 87 inches, so good is the ride. Credit for the smooth ride under any condition must be shared by the 4-coil suspension system and low tire pressures allowed by the 155 millimeter (5.99") tires. The only criticism which could possibly be made of the suspension was a tendency to judder when making "all-out" starts during the acceleration tests and this was not serious—only unexpected of a conventional rear axle well supplied with control arms, a pair of short radius rods on each side and a wishbone above, slightly to the left of center.

The engine is both a delight to the eye and a joy to drive. It sings along at any desired rate of revolutions very quietly and with no trace of vibration at any speed. The tachometer is redlined at 6200 rpm, but it reached and held 6600 rpm during the timed high speed runs with no sign of stress or strain. We used the recommended 6200 rpm rev limit during the acceleration checks, but there is reason to believe that 7000 rpm in each gear could be used safely, returning slightly better figures. However, this was the only Spyder west of the Mississippi and it had only 2400 kilometers (about 1500 miles) on the clock when we were handed the keys by Clem Atwater Motor Sales at their San Fernando Valley home. Needless to say, we returned the car in perfect condition and, incidentally, recorded 27.6 mpg overall including the performance testing, 100 miles of highway driving and considerable traffic work.

ROAD AND TRACK ROAD TEST NO. F-4-56

ALFA ROMEO GIULIETTA SPYDER

SPECIFICATIONS

List Price	$2995
Wheelbase, in.	87 in.
Tread, front	50 in.
rear	50 in.
Tire size	155-15
Curb weight, lbs.	1940 lbs.
distribution	56/44
Test weight	2220 lbs.
Engine	4 cyl.
Valves	dohc
Bore & stroke	2.91 x 2.95 in.
Displacement	1290 cc
Compression ratio	8.00
Horsepower	65
peaking speed	5500
equivalent mph	82.5
Torque, ft/lbs	79.5
peaking speed	4000
equivalent mph	60
Mph per 1000 rpm	15.0
Mph at 2500 fpm	76
Gear ratios (overall)	
4th	4.55
3rd	6.17
2nd	9.57
1st	16.3
R & T high gear performance factor	47.4

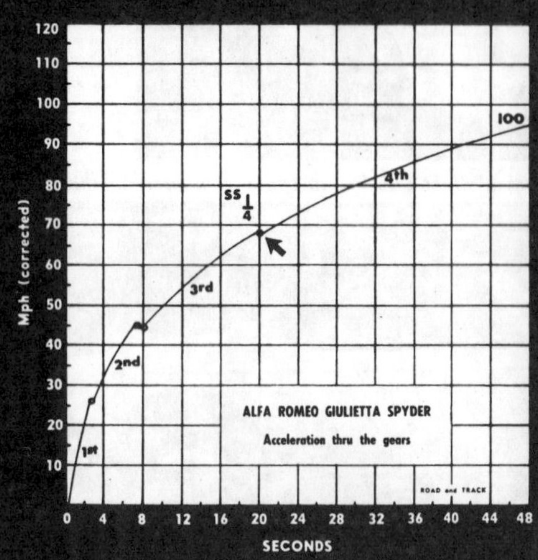

PERFORMANCE

Timed top speed	100
Max speeds in gears—	
3rd (6600)	74
2nd (6600)	47
1st (6700)	28
Shift points from—	
3rd (6250)	69
2nd (6300)	45
1st (6200)	26
Mileage range	27/33 mpg.

ACCELERATION

0-30 mph	3.8 secs.
0-40 mph	6.0 secs.
0-50 mph	10.0 secs.
0-60 mph	14.8 secs.
0-70 mph	21.9 secs.
0-80 mph	29.4 secs.
0-90 mph	41.4 secs.
Standing ¼ mile	20.0 secs.

TAPLEY READINGS

Gear	Lbs/ton	Mph	Grade
1st	560	at 21	29%
2nd	430	at 30	22%
3rd	310	at 40	16%
4th	230	at 45	12%

Total drag at 60 mph, 78 lbs.

SPEEDO ERROR

Indicated mph	kph	Actual mph
30	48.3	25.0
40	64.4	35.2
50	80.5	45.5
60	96.6	55.1
70	113	64.3
80	129	74.0
90	145	83.2
107	172	100.0

ALFA ROMEO

This particular car was originally shown at the Fall Show in Paris. The body was built by Pinin Farina, and production models (also by Farina) show no important changes. The curb weight of 1940 lbs. was higher than expected (the unit frame and body coupes are said to weigh 1980 lbs.), but large box section body sills are an added necessity on a "roofless" chassis—which adds weight aplenty. The chassis felt adequately rigid and had none of the cowl shake so often encountered in open roadster type bodies.

On comfort, the excellent ride has already been mentioned. If we can accept the interesting theory that the T-Bird outsold the Corvette because 15 Americans want wind-up windows to 1 who prefers side curtains, then the Alfa Romeo Spyder's success is assured, for it alone in its price class has glass windows. On this car the cranking operation took an inordinate amount of work, but the windows are nice to have. The seating position is excellent, the seats themselves comfortable. However, one criticism must be made. The interior appears to have been designed only for drivers between the heights of 5'4" and 6' approximately. Anyone shorter than the lower limit will find himself (or herself) peering under the steering wheel rim, and our 6'4" staff member found that when the top was in place it fitted rather like a hat.

The sealed cover over the gasoline filler cap is not being used on production models.

The less said about the column-mounted shift lever, the better. It was next to the worst we have ever encountered, especially on the shift from 2nd to 3rd. The travel on this shift was too short, and there was insufficient leverage to make the change quickly and positively. We estimate that all the acceleration figures from 0 to 50 and up could be reduced by one full second with the new floor-mounted shift lever, which fortunately is being supplied on all production cars. We examined this new feature on a car just off the dock but were not able to try it out. It appears to be a very serviceable and usable stick and is distinguished by a very large knob which feels just right to grasp.

For the production model with floor shift, our highest tribute lies in the triviality of the only criticism left to level at the car—in the roar of U.S. traffic, the Giulietta's feeble "beep-beep" horn is like a mouse squeaking among lions. Now, the question remains: when is the factory going to come through with enough Spyders to fill orders? ●

Lightweight Alfa-Romeo "Juliet"

REVEALED to the public for the first time at the Turin Automobile Show was a new Alfa Romeo sports coupe called the Guilietta 1300 Sprint with a 1.3-liter engine rated at 65 h.p. English version of the name is Juliet. This is a new field for Alfa Romeo, which has heretofore been confined to cars in the medium-size class, such as the 1900.

The Juliet coupe is shown as a two passenger model, but is reported to be the forerunner of a sedan version. The new 4-cylinder engine closely resembles the "1900" in design, having twin camshaft overhead valves and a light alloy head. Bore and stroke are 74 x 75 mm. (2.91 x 2.95 in.) and 1290 c.c. displacement (79 cu. in.). The intake manifold couples the two center and two outer cylinders. In unit with the engine is a dry-plate clutch and four-speed transmission. Rated horsepower of 65 is reached at 6000 r.p.m.

A two-piece propeller shaft transmits power to a conventional rear axle with a 4.55:1 ratio. The front independent suspension is by coil springs and wishbones. The king pin pivots in ball joints at top and bottom. Brakes are hydraulic with two leading shoes on the front wheels. Large bi-metal drums are finned tranvsersely to aid in heat dissapation. The disc wheels are perforated also to aid in brake cooling.

Externally the car retains the look of an Alfa, including a smaller version of the familiar shield-shaped grille. The aerodynamic body is by Bertone. It is a 2-pass. model with luggage space behind the seats and a package shelf beneath the rear window. Additional luggage space is provided in the sloping tail. The entire tail panel including the king-size rear window opens up on a side hinge. Opening of the tail section is controlled by a handle in the trim panel on the right-hand side, so luggage is safe whenever the car doors are locked.

Visibility is excellent due to a large curved windshield and the exceptionally large curved rear window. Door windows and quarter windows are also good size. Instruments, including tachometer, are grouped in three dials under a single cowl to prevent reflection.

Although maximum power is reported at only 65 b.h.p. the Juliet is said to be capable of 100 m.p.h. A strong contributing factor undoubtedly is the low weight of about 1760 lbs. which should also guarantee brisk acceleration.

Wheelbase of the Alfa Romeo Juliet is 94 in. and tread is 50 in. both front and rear. Overall length is only 153 in. and 60½ in. is the overall width. ★

Alfa-Romeo's

Racing experiences have taught Alfa-Romeo many lessons in how to build cars that perform and last. Here's one of their new beauties.

Powered through sharp corner at high speed, Giulietta demonstrates substantial lean but has almost no tendency whatsoever to break away.

Four-cylinder, 1,290-cc, twin-ohc Giulietta engine used in "Sprint" has modified carburetion, compression ratio upped from standard 8:1.

Giulietta SPRINT

ROAD AUTO AGE TEST

By D. M. BARTLEY

ONE OF THE NICER things for an auto editor about making a trip to Europe is the opportunity to drive some of the latest European cars on their own home ground, so to speak. One gains an entirely different impression of a car driving it on terrain and in the traffic for which it was designed rather than on alien ground.

One of the cars which I drove last summer in Italy was Alfa-Romeo's Giulietta "Sprint"—an extraordinarily beautiful little automobile in all aspects. As a matter of fact, the Giulietta, in our opinion, was *so* special that we decided not to tempt anyone with details about it until it was available in America. Now it is, so read on and be tempted!

After a very pleasant morning spent at the Alfa-Romeo office in Milan, discussing their plans for exporting their cars to America, some fairly guarded questions and answers about the "possibility" of racing cars, etc., we picked up a lovely light blue Giulietta "Sprint" at the garage. With a factory driver at the wheel, we headed toward the outskirts of Milan and while a passenger, noted with pleasure a variety of details. The bucket-type seat was exceedingly comfortable, the ride, even over rough cobblestones, very smooth, and even when the driver cornered sharply at considerable speed, there was only a slight sense of being "thrown" by the centrifugal force. Although the two-door coupe model of the "Sprint" which we were testing has a sort of banquette seat in back, this is primarily a two-passenger car.

From the driver's seat, a whole new batch of likable features made themselves known. Though the seating was a little low for me, it seemed fine for the factory man who is of average height. The instruments were positioned under a hemispherical hood and were large, clearly calibrated and easy to read. All controls, both hand and foot, were placed for natural, easy reach.

Still in the city traffic—so characterized in Italy by abrupt, spectacular and completely unexpected maneuvers on the part of most of the other drivers as to make rush hour in any American city seem like restful driving by comparison—the Giulietta was plenty responsive enough in the shift, acceleration and brake departments to permit us to compete satisfactorily. As the car was not thoroughly broken in, we found the steering column shift lever a little stiff yet, but were able to make surprisingly fast shifts in spite of this. (And how nice it is to have such a nice smooth four-speed synchromesh gearbox to play around with!)

The Giulietta "Sprint" utilizes the exceedingly well-designed 1,290-cc, (74 x 75 mm), 8:1 compression ratio, twin-ohc, four-cylinder Alfa engine—with a few changes. Carburetion is modified somewhat, compression is increased slightly, and output is upped from 50 bhp at 5,200 rpm to 67 bhp at 6,000 rpm.

When it comes to accelera- (*Continued on page* 23)

Final adjustments are made on a nearly-completed Giulietta. Notice enormous brakes which are the same used on the 1,900-cc Alfa-Romeo.

Spyder model of Giulietta "Sprint" has same engine, brakes, suspension, etc., as closed coupe but has a slightly shortened wheelbase.

SCI
tests the Alfa Romeo Giulietta

Giulietta accelerating in SCI road test. High torque gives car excellent pulling power in top gear even on long, steep grades.

Alfa Giulietta: a champagne car at a beer price.

WHEN the men who run Alfa Romeo named their new 1.3 liter model "Giulietta" – Juliet, in English – they were thinking of character as much as of an obvious play on names. Shakespeare's Juliet was a tender child whose passionate nature belied her demure specifications. Alfa's Giulietta is much the same.

Its dimensions are toy-like. Its displacement and horsepower are minute. Its top speed will give no one gray hairs and the fierceness of its acceleration is as much a matter of chassis feel as it is of gravities or feet per second squared. But the Giulietta is *all* machine and it's one of the most exciting, eagerly responsive cars I've ever driven.

The little Alfa has a vital personality, and it mirrors your own mood. If you're feeling dreamy you can drive it as somnolently as you would a heavy Detroit lugger. But if you are in a mood to *go*, it can turn you on like a jug of adrenalin. I've never driven a safer car. Its brakes, steering and roadholding are sheer perfection. In most cars your chances of worming your way out of a real road emergency depend largely on luck; it's something like riding a loco

The Giulietta displays usual simple Alfa Romeo grille. Tires are Pirelli Cinturato type with unique tread pattern which give car fine adhesive qualities on wet or dry surfaces.

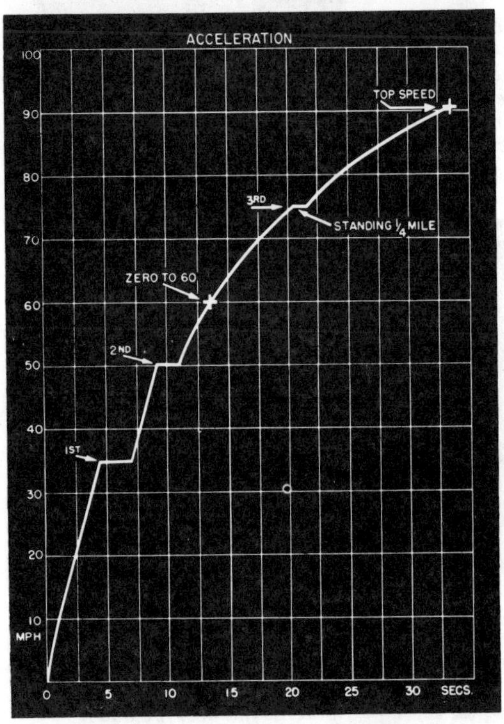

Acceleration chart shows Giulietta's high torque performance. Long horizontal lines at each shift point were due to column shift linkage being out of adjustment. With floor shift, there is no doubt that the car would have reached top speed in about 25 seconds.

PERFORMANCE
ALFA ROMEO GIULIETTA SPYDER SPRINT

TOP SPEED:

Two-way average ...... 90.4 mph
Fastest one-way run (one mile approach to traps) ...... 93.8 mph

ACCELERATION FROM ZERO TO:

20 mph ...... 2.4 secs.
30 mph ...... 3.7 secs.
40 mph ...... 7.7 secs.
50 mph ...... 9.2 secs.
60 mph ...... 13.7 secs.
70 mph ...... 17.5 secs.
80 mph ...... 24.6 secs.
Standing ¼ mile ...... 20.7 secs.

SHIFT POINTS:

First ...... 35 mph
Second ...... 50 mph
Third ...... 75 mph

SPEEDOMETER CORRECTION:

Approximately 4 mph fast throughout speed range.

FUEL CONSUMPTION:

Hardest possible driving, actual ...... 28 mpg

SPECIFICATIONS

POWER UNIT:

Type ...... 4 cyl., in line
Max. bhp ...... 65 bhp @ 550 rpm
Piston displacement .. 78 cu. ins. – 1290 cc
Bore x stroke ...... 2.91 x 2.95 ins. – 74 x 75 mm
Bore: Stroke Ratio .. 1:1.01
Compression Ratio .. 8.0 to 1
Valve arrangement .... Vee-inclined, dohc

DRIVE TRAIN:

Clutch type ...... Single-plate, dry
Transmission Ratios:
1st 3.59
2nd 2.10
3rd 1.36
4th 1.00
Final drive ratio 4.55

CHASSIS:

Suspension, front ...... Inclined coil springs, unequal length wishbones.
Suspension, rear ...... Solid axle, coil spring, radius rods.
Shock absorbers ...... Girling tubular, F & R
Steering type ...... ZF worm and roller
Steering wheel turns . . . 3 from lock to lock
Steering turning diameter ...... 30 ft.
Brake type ...... Bimetallic drums, 10.5 in. diam. F, 10-in. diam. rear. Two leading shoes at F.
Brake lining area ...... 264 sq. in.
Wheel studs ...... (15/32") 12 mm, 4½-in. circle diam.
Tire size ...... 15.5 x 15
Rim width ...... 5.75 ins.
Wheelbase ...... 94 ins.
Tread ...... 50 ins., F & R

GENERAL:

Length ...... 156 ins.
Width ...... 60 ins.
Height ...... 50 ins.
Weight, ready to go.. 1910 lbs.

RATING FACTORS:

Bhp per cu. in. ...... .834
Bhp per sq. in. piston area 2.45
Piston speed in ft. per. min @ 60 mph.... 1950 fpm
Piston speed in ft. per min. @ max. bhp.. 2708 fpm
Brake lining area per ton ...... 293 sq. ins.

Uncovered engine shows dohc set-up and chain drive. Head and block are light alloy. Cylinder liners are wet. Carburetor is two-throat Solex. Acorn nuts holding cover are chrome finished.

Rounding a sharp bend at 40 mph, the small Alfa tenaciously grips road without trace of slide or dip. Author felt Giulietta could take curve at 50 mph nicely. There is no excessive dip at either end of the car. Steering is almost perfectly neutral.

Three quarter front view shows the little Alfa's aerodynamic-type body. High beam is ultra-high, used in lieu of horn in Italy. This may be good for Quiet Week but illegal in U.S.

Profile of new Giulietta. Convertible top folds completely out of sight behind seat backs. Stowing top is no problem for average person because of spring loaded framework.

elephant in a jungle stampede — you can't control him but maybe his beef will get you through. But the Giulietta is like an extension of your nervous system. It responds as quickly as your own reflexes. Although it pretends to be no more than a sports-touring car it gives you that feeling of oneness with the machine that is a quality of the racing thoroughbreds.

This character makes the Giulietta a perfect expression of Alfa Romeo's unique traditions. From the date of its first touring car production, 1910, Alfa has built racing cars. Furthermore, Alfa has always staked its reputation on its racing record — a precarious practice avoided by nearly all car manufacturers. Down through the years, Alfas sold to the public often have been no more than detuned racing cars and nearly every Alfa chassis has been patterned after great and successful racing machinery. The company was one of the most vigorous pioneers of the fine, small-displacement, high-revving engine at a time when nearly all car manufacturers associated quality with lumbering road locomotives. In the Twenties the company faced up to the inherent superiority of the dual overhead camshaft layout and has stuck with it ever since, in spite of its higher cost. And of all high-performance cars, none has a better name for reliability than Alfa.

This is part of the inheritance that is expertly invested in the Giulietta. Cars of comparable quality — the 1750 cc Alfa of the early Thirties, for example — sold for around $5000 when dollars were tough to come by. But now, for the first time in its history, the company has converted to real mass production and you don't have to be one of the wealthy few to own a pure bred Alfa. The Giulietta comes in three forms. Cheapest is the four-door Berlina at about $2700, followed by the Spyder Sprint convertible two-seater at $3200 and the Coupe Sprint at $4200. All three forms have the same running gear and, except for minor modifications, the same engine.

View shows width of rear brake drum which, in part, accounts for superb braking. Note rubber cushion to damp axle under extreme compression; strap halts rebound.

ENGINE

The heart of the Giuletta is a four-cylinder in-line power unit with chain driven overhead cams, nearly equal bore and stroke, and a displacement of 1.3 liters or 79 cubic inches. It cannot compete with its prewar forebears in terms of raising gooseflesh on lovers of straight-cut gears; the chain drive is quiet and so are the tappets. Nevertheless, the 1.3 does have the buzz-saw whine you would expect from an engine made to wind well beyond 6000 revs.

Alfa Romeo was one of the world's great pioneers in light alloy metallurgy. The 1.3 reflects this experience and only those parts that must be made of steel are. Block, head, sump, cam cover, intake manifold and fan are among the light alloy parts and the cylinder barrels are wet liners; steel sleeves in contact with the water jacketing. These are ideally cooled and so are the valves, spark plugs and combustion chambers.

The porting and manifolding of the 1.3 have the straight-through cross section that is one of the goals of the d.o. layout. The valves are operated by cup or piston-type cam followers which fit over double nested valve springs. There's a sign of economizing here. The classic Alfa tappet adjustment by means of serrated discs on the cups has been replaced by the now almost-universal but less convenient "button" adjustment. To change tappet clearances by this method it is necessary to remove the camshafts, remove the cups, and install a coin-shaped steel "button" of the desired thickness between the end of the valve stem and the cup.

The 1.3 has five main bearings — the most you can put in a four cylinder engine — massively supported in an unusually deep and stiff crankcase. The bearings and bearing caps for crankshaft and camshafts are clearly numbered and there are many other signs of the designer's consideration for the mechanic, including excellent overall accessibility.

The only visible difference between the Berlina and Sprint engines is in their carburetors. In place of the standard model's single-throat carb the Sprint has a dual-throat Solex in which the throttle plate in the secondary throat begins to open after the primary throttle is about half open. Fuel economy and performance with this linkage both are amazingly good. But the car can be made drastically hotter by altering the linkage so that both throttles open simultaneously. I experimented briefly with the Spyder Sprint test car and obtained these results:

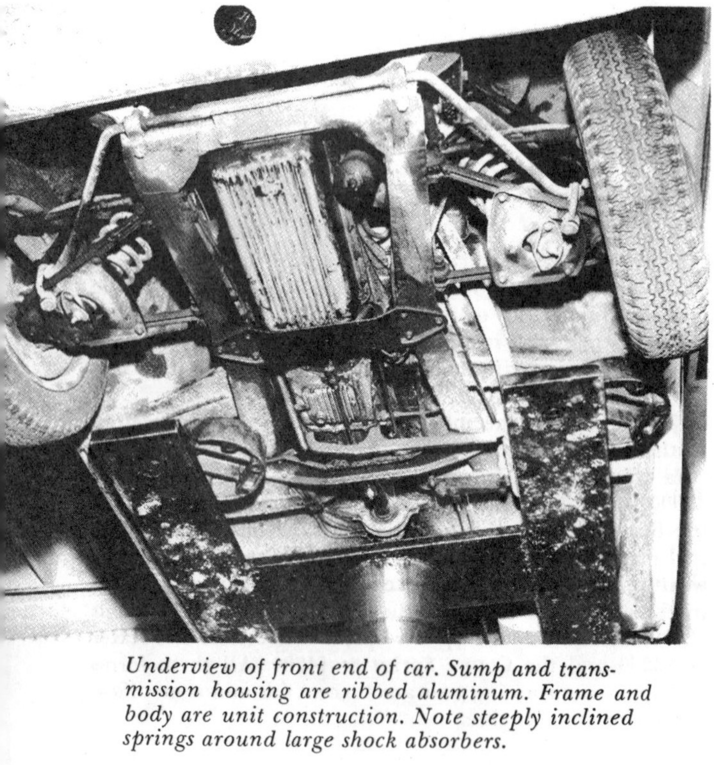

Underview of front end of car. Sump and transmission housing are ribbed aluminum. Frame and body are unit construction. Note steeply inclined springs around large shock absorbers.

Housing of final drive assembly is of light alloy — axle tubes are of steel. Exhaust system has two small, straight-through mufflers. Top of springs are anchored by tubular pillars extending through to top of rear deck.

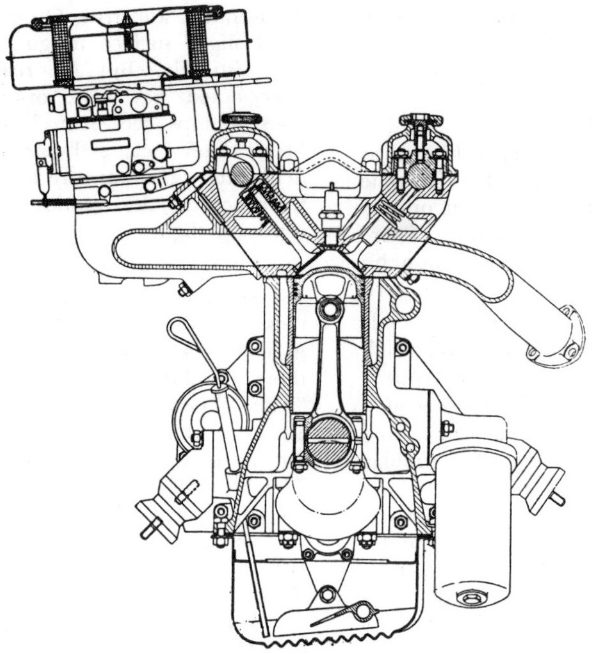

Diagram of engine cutaway shows typical Alfa dohc set-up. High torque characteristics allow engine to deliver 85% of its bhp to the rear wheels.

Measuring rim width. Outside width is 5.75 inches. Between each pair of holes, in wheel disc, is welded a small stud on which balance weights can be mounted.

Aceleration from	Stock Linkage	Modified Linkage
0 to 30 mph	3.7 secs.	3.7 secs.
0 to 40 mph	7.7 secs.	6.6 secs.
0 to 50 mph	9.2 secs.	8.2 secs.

As for internal differences between the 50 bhp standard engine and the 65 bhp Sprint, there is a large lack of this sort of information in the U. S., to which the Giulietta is brand new. But it's obvious that the Sprint engine has very different camshafts and its compression ratio may be higher than the nominal eight to one.

The performance of this tiny power unit is uncanny. When I first took the test car over and headed for open country I was astonished by its unexpected torque. The car would accelerate smoothly in top gear from about 17 mph and long grades had scarcely any slowing effect on the car's top-gear pulling power. Later I put the Giulietta on a chassis dynamoter and witnessed the road horsepower readings. Between no load and full load the muscular little engine dropped only 500 rpm. A popular 3½ liter d.o. engine, for example, can be counted on to drop about 1500 revs under the same conditions. And according to the dyno, 55 of the 1.3's 65 horses were reaching the rear wheels!

CLUTCH AND TRANSMISSION

The early Giuliettas had a pronounced fierceness of the clutch which vintage fanatics clucked over contentedly. But

Attractively finished engine is exceptionally accessible in roomy compartment. Camshaft cover is done in black crackle-paint. Fuel line fittings are black enamel, and acorn nuts are chrome. Note air-ducts à la American machines.

Alfa—Giulietta

in its current form the Giulietta's clutch is soft enough to be quite comparable with clutches that survive in American practice. You can let it out slowly while bringing the engine revs up gently, or you can let it out fast and use higher revs. Smooth first-gear starts can be made with engine speeds as low as about 1800 rpm.

The choice of transmission ratios is excellent, first being useful up to about 35 mph, second to 50 and third to 75. Torque is so ample that fourth is a perfectly acceptable gear for traffic but if you want to be aggressive, third will really get the job done. The gears emit a faint, pleasant whine.

But all is not perfection in the cog box. It is described as having "four forward synchromesh speeds" but there is no synchro on bottom gear. This is the first discovery and it's quickly followed by others. My test car was well broken in but the column-shift linkage remained stiff, much too stiff. Getting into second from first was often noisy. The long zig-zag shift from second to third sometimes failed — third was simply a closed door, so back to neutral and then back for another try to get in. On these repeat shifts the synchro would be inoperative, a fact announced by the noise of grinding gears. This also occurred during downshifts but could be avoided by double-clutching and speeding up the engine.

As far as our test car was concerned, some of these malfunctions must have occurred because the column-shift linkage was out of adjustment. Which, all by itself, is a sufficiently lethal indictment against the column shift. No manufacturer has ever built finer floor-shift transmissions than Alfa Romeo — Theirs are legendary. It's good to hear that the very latest Giulietta Sprint models *are* arriving in the U. S. equipped with floor shift.

STEERING

Giuliettas are fitted with ZF worm-and-roller steering. There are more expensive ways of building a steering mechanism but I've handled nothing, including worm-and-wheel, that is more perfectly precise. There is no lost motion in it and it's as quick and direct as a bicycle's steering. How this is accomplished with three turns of the wheel from lock to lock I'm not prepared to explain.

This steering has all the feel you could ask for or need, yet it's light. Little effort is required in parking and at 10, 40, or 80 mph you can hold the rim of the wheel between the tips of thumb and one finger and steer the car through a maze.

SUSPENSION

The Giulietta was first exhibited at the Turin Show of '54 and at that time its rear suspension was based on traditional quarter-elliptic springs. There may or may not be a connection between that fact and some early reports from Italy criticizing these cars for body roll. At any rate, the rear suspension of the mass-produced Giuliettas is a very different matter. It consists of a solid axle with a vertical coil spring at each end. The coils enclose tubular shock absorbers and are mounted on vertical pillars that extend to the top of the unit-frame structure, just under the rear deck. No torque tube is used, but axle torque is controlled by a leading radius rod at each extremity of the axle. Large rubber pads are provided to keep the axle from striking the frame during extreme compression of the springs and straps are used to limit downward axle travel during rebound.

Lateral movement of the rear axle is controlled by a triangle of steel tubing, two corners of which are anchored to the frame-body structure while the third is attached to the final-drive housing by means of a ball joint. This housing, incidentally, is a jewel-like little structure of deeply-ribbed light alloy and it would do honor to a fine midget race car.

The front suspension is more orthodox and consists of long, unequal length wishbones and steeply inclined coil springs that enclose tubular shock absorbers. A safety cable, attached to each lower wishbone and to the frame, limits rebound travel of the front suspension.

The Giulietta's ride and roadholding qualities are additional Alfa achievements that can be described only in superlatives. In the corners it is absolutely glued, has the feel of a Ferrari or a Kurtis. I am not an exponent of the controlled slide and cannot evaluate the little Alfa in those terms, but in terms of a straightforward race car adhesion, this car has it. Also like a good race car, the body and frame do not oscillate softly above the suspension while cornering, as nearly all touring cars do. With the Giulietta, body, frame, and running gear all behave as one well-knit unit.

On the straightaway the car tracks dead-straight and true. On good pavement slight surface irregularities are felt distinctly. But on really foul, chuck-holed pavement or dirt, the worst bumps feel scarcely different. Many times I charged into deep ruts deliberately and was sure I had missed them because the reaction was so deceptively slight.

BRAKES

The Giulietta's brakes are just like those on the bigger, faster, costlier Alfa 1900 and are closely related to those developed on the Type 159 "Alfetta" Grand Prix cars. Here again, and here above all, mere superlatives are not enough. Look at it this way. The Mercedes-Benz 300 SL has magnificent brakes. It has 257 sq. ins. of friction area for its 2240 lbs. of dry weight. The Giulietta is claimed to have 264 sq. ins. for its 1800 lbs. of dry weight. In terms of brake lining area per ton, this gives the little Alfa 293 sq. ins. against the 300 SL's 230! You can be sure that the performance of these brakes is not easily forgotten.

The front brake drums are huge light alloy castings containing ferrous liners and having deeply machined diagonal ribs which serve as air-pumping vanes and increase the heat-dissipating surface. The rear drums have conventional circumferential ribs and are of similar bimetallic construction.

I have never experienced brakes of such superb quality. The Giulietta's stopping distances are fractions of those of Detroit cars and of many sports cars. In spite of this immense deceleration the car's occupants are not flung forward nor does the car nose down. You can perform braking feats with ease in the little Alfa that would be desperately reckless in another car. It stops in a straight line. You do not have to churn wildly at the steering wheel in an effort to keep the car on the pavement during emergency type stops. Finally, the brakes are inexhaustibly fade proof. And beyond that, the transmission ratios are so chosen that you can always find just the right gear for steep mountain descents and hardly need to touch the brakes at all.

BODYWORK

The low-priced Berlina has a mass-production body with lines that are pleasing but not breathtaking. Seats are of the bench type and provide comfortable accommodation for four passengers. Interior appointments, including instruments, are confined to the austere necessities.

The medium-priced Spyder (a name once used for wiry, light-weight carriages) has a beautifully contoured body by Pinin Farina. The appointments are *de luxe* throughout for it and the Coupe. Instruments include a large rev counter immediately over the steering column, a metric speedometer, gauges for water and oil temperature, oil pressure, fuel capacity, and a bevy of warning lights. Bucket seats with folding backs give excellent lateral support and there is ample head and leg room for occupants six feet tall.

The convertible top has the advantage of folding completely out of sight behind the seat backs. But unlike the rest of the coachwork, this top deserves a good deal of criticism. Its metal framework is heavily spring-loaded and stowing the top can be an annoying job for two persons. There is no positive locking device to hold the stowed top in place against its eager springs and it occasionally tries to unfurl itself, which results in grinding the top fabric against the seat backs. The top has no headliner, rattles like a snare drum in a strong wind and is insecurely fastened to the rear deck.

Actually, these are minor defects that can be corrected by the lucky owner for a very few dollars and it's to be expected that a firm with Alfa's high standards will be quick to check them at their source.

The Coupe I had hopes of testing was sold after I had fondled it for less than 15 minutes. I had seen photos of the car many times before but not until I was in its physical presence was I able to appreciate it fully. The body is by Bertone and it's my own feeling that of all the thousands of unusual and beautiful postwar Italian bodies, hardly any can equal its simple, sculptured beauty.

The Coupe's visibility is splendid. Its interior appointments are comparable with the Spyder's, and it is relatively spacious. The area behind the front bucket seats can be used for great stacks of luggage, for seating children on long trips or for adults on an "occasional" basis. But *the* virtue of the Coupe is its exquisite style.

A car that's a treasured collector's item today is the 1.5 liter, four-barrel Brescia Bugatti. It sold new in the early Twenties for about $2500. Another hand-built treasure was the 1.5 liter Alfa of the late Twenties which sold for around $5000. Times have changed and small-scale volume production has penetrated to the purebred field. The Giulietta is today's improved equivalent of the baby Bugs and Alfas and it's to be had for a range of prices from $2700 to $4200. It's a legitimate jewel, bargain-priced for those who recognize its worth. #

Alfa-Romeo's Giulietta "Sprint"

Continued from page 15

tion tests, the driver is well aware of these additional horses. Doing several runs for each figure (and all of them under the handicap of having to watch for an occasional misplaced piece of traffic), zero to 40 mph was accomplished in what averaged out to be about 6.2 seconds; 0 to 60 took an even 12.5 seconds, and 0 to 75 a very respectable 20.5—all of them comparing most favorably to times for a substantial number of true sports cars.

While we were photographing the car, I was surprised to notice the amount of lean which was apparent when I watched it taking some sharp corners at a good clip. As a passenger and driver, I had not been aware of excessive lean at all, but as a "sidewalk" observer, it seemed to me that the "Sprint" heeled over somewhat farther than most European cars (though nowhere nearly as much as most Detroit cars).

I must admit that we had practically no opportunity to drive the car in "serious" type mountains, but the little time we could spend in non-level terrain, the Giulietta fully came up to expectations. Again, as on flat corners, the use of the gears and the fine brakes, plus the "Sprint's" natural tendency to "track" rather than break away, combined to make for exceptional roadability and to impart a feeling of confidence not often so quickly acquired in an unfamiliar touring car.

Probably for the average driver this confidence is inspired in a very large part by the exceptional brakes of the Giulietta. Alfa-Romeo uses the same very large Al-fin brakes which were originally designed for the larger and more powerful "1900 Super" fast touring car.

After we brought the car back to the factory, we took a little time to examine it more closely. Though it is true of most European-made cars, it is always a particular pleasure to re-discover the beautiful attention to minute detail which characterizes them. The clean, simple but modern lines of the Giulietta itself become even more attractive when you run a finger under the edge of the fender, for instance. It's so beautifully beaded that it is as smooth as the body itself. Each chrome section of the grille fits snugly into its niche and joins evenly with the other sections. All body joints are smoothly made and flawlessly finished off. The interior appointments, dashboard, upholstery, flooring, door and window handles, etc., are equally well-executed.

Even so brief a tour of the Alfa-Romeo plant as we made the following day was enough to show why the Giulietta has so few flaws. In every operation we watched—engine assembly, upholstering, painting, glass fitting, insulating and some 25 or 30 others—the workmen handled their materials carefully, took time to smooth off the last rough edge, re-measure, re-check, re-test.

In the interests of impartiality, I will mention that our European correspondent, Johnny Lurani, disagrees with me about the almost-perfection of the Giulietta "Sprint." Briefly, the areas of disagreement center around the gears—he dislikes the steering column lever and maintains that *really* fast shifts are impossible—and the suspension. In his opinion, the car leans far too much.

From the viewpoint of an exceptionally fine racing driver, which Johnny is, these are legitimate criticisms. No one who is accustomed to a short, very sturdy stick shift will ever truly like a column type, and the Giulietta certainly can't claim the almost absolutely level cornering qualities of some other fast touring cars (such as the Lancia Grand Turismo). I tend to think that these criticisms would hold true for what you might call an "absolute enthusiast"—which Johnny also is—but not be entertained by a more "average" type like me.

A few last facts which you might like to have: the closed two-seater Giulietta "Sprint" has a top speed of around 105 mph; it consumes approximately one gallon of gasoline every 27½ miles, and it sells in America for about $3,000. ●

GIULIETTA "SPRINT" SPECIFICATIONS

ENGINE: 4 cyl., twin overhead camshafts; bore, 2.91 in.; stroke, 2.97 in.; total displacement, 78.65 cu. in.; developed hp, 65 bhp at 6,000 rpm; torque, not available; compression ratio, 8.0+ to 1.
TRANSMISSION: 4-speed synchromesh (over-all ratios 4.5, 6.1, 9.5 and 16.3 to 1); steering column control.
REAR AXLE RATIO: not available.
SUSPENSION: front, independent coil springs and wishbones; rear, coil springs.
BRAKES: Al-fin 4-wheel hydraulic; 2 cylinders per wheel at front; floating brake shoes; large dimension bi-metal drums.
DIMENSIONS: Wheelbase, 94 in.; rear track, 50 in.; width, 60 in.; height, 52 in.; over-all length, 153 in.; turning circle, 34 ft.; ground clearance, 5.5 in.; weight (approx) 1,763 lbs.

PERFORMANCE

ACCELERATION:
Zero to 40: 6.2 seconds.
Zero to 60: 12.5 seconds.
Zero to 75: 20.5 seconds.
TOP SPEED: 100-105 mph.

The MG's side curtains snap on to give protection against storms or cold weather, but no window ventilation adjustment is possible

Sports cars, both imported and domestic, are achieving increasing popularity in the United States, and for good reason. As the family sedan grows more bulbous and less maneuverable, more and more drivers look for a car that's fun to drive, a vehicle that provides sprightly transportation with a minimum of metal and machinery. And, of course, as second—or third—cars become increasingly a part of the American family pattern, the sports car offers a tempting combination of virtues, while its deficiencies are largely offset by the availability of more conventional family vehicles.

The British *MG*, a sports car which has been seen in this country for many years, played a pioneering role in acquainting the American public with the sports-car type. This year, for the first time since its introduction, it has been completely redesigned and streamlined, and thus finally shorn of its old-fashioned "vintage" look.

Among soft-top sports cars, the Alfa is unusual in having roll-up glass windows. With top up, inside mirror gives poor rear view

For the current test report, CU's auto consultants have paired the new *MG "A"* with the Italian *Alfa-Romeo Giulietta Spyder*—to give the little vehicle its full and impressive title. The *Alfa*, about the same size as the *MG*, costs nearly a thousand dollars more.

Both cars have small four-cylinder engines, four-speed hand-shifted transmissions, and seats for two; both have curb weights just edging a ton; both have folding tops. But where the *MG* has side curtains, the *Alfa* has windows which crank up and down. Both cars have trunks too: the *Alfa's* is reasonable in size and quite useful, that of the *MG* (see photo page 484) houses the spare tire and has room for very little else.

Both cars are really small and light—four feet shorter overall and more than a foot narrower than a *Plymouth*, for instance, and three-quarters of a ton lighter in weight.

At its port-of-entry price of $2195, the *MG* is the lowest-

the MG "A" and Alfa-Romeo

The responsiveness of these two imported sports cars will win the admiration of discerning drivers

priced genuine sports car readily obtainable in the U. S.—and one of the handsomest at any price. It lacks many of the detail refinements of the *Alfa-Romeo* which, however, costs $900 more.

The *Alfa* (though it competes in a smaller-displacement category), is just as much a sports car as the *MG*; it handles and takes corners as well if not better; but it is much less austere in appointments, offers much more comfort, is of more advanced mechanical design.

As a result of CU's appraisal and testing of the two cars, its auto consultants feel that while the *MG* will appeal chiefly to sports car addicts, the *Alfa* will, in addition, please many motorists who want and can afford a small, high quality, personal pleasure car that takes up little room, handles like a dream, rides acceptably, and transforms into a snug and well-heated convertible.

Road behavior

The list of the "sports car virtues" is not a long one, but on it are items that every U. S. motorist ought to learn about—preferably at first hand. Stability on the road is one of these virtues, and oversized brakes (by American standards) constitute another (see photo page 484).

More important still, and virtually unobtainable in ordinary cars, is the sports car's extreme maneuverability. This is compounded of a great reduction in size coupled with exceedingly quick and precise steering, "eager" response to the steering on the car's part because of its light—and well-distributed—weight plus a tight-to-the-road feeling due to the sports car's low center of gravity, low seating, freedom from lean on curves, and relatively stiff springing.

The net result of these differences from conventional cars is, a feeling of release for the driver—like discarding his heavy winter clothing in the Spring. Governing the behavior of the car in relation to the road becomes a simple pleasure rather than a nose-to-the-grindstone chore.

Sports cars do, of course, differ widely in the degree to which they approach the ideal in their handling qualities, but both the *MG "A"* and the *Alfa-Romeo Giulietta Spyder* lie well toward the top of the list in these characteristics. Steering is easier, more sensitive, and slightly quicker on the *Alfa*, and the *MG's* tires squealed noticeably on corners, even at normal speeds, at the recommended air pressure.

Both cars had excellent brakes, and had no trouble with fade or in giving straight, quick stops on CU's downhill coasting test. Aside from grabbing when damp, the *MG* brakes behaved well, but the *Alfa* brakes rate special mention. (Italian cars in general have outstanding brakes, and one look at a relief map of mountainous Italy will show why.) The *Alfa* brakes are enormous for a one-ton car. The drums are of light alloy, with iron liners; the front brakes, which on all cars do more work than the rear, have drums with machined helical cooling fins, the rear drums have cast-in ribs. Brake pedal pressure is low, the brakes take hold gently and provide seemingly effortless stopping.

Neither car showed any lack of bounce control—as do the majority of U. S. sedans—when taken over CU's rough road loop at the standard test speed.

Performance

All that can be said of the acceleration, flexibility, and speed of the *MG "A"* and the *Alfa-Romeo Giulietta Spyder* is that both do very well considering the size (piston displacement) of their engines—91 and 79 cubic inches respectively. A *Chevrolet 6* will give them a hard time accelerationwise, while most any U. S. *V-8* will out-accelerate them with one hand tied behind its back.

Moreover, the maximum performance of the *MG* and the *Alfa* (in which, as the Facts and Figures show, the two cars are roughly equal) depends on frequent gearshifting and use of the high revolutions per minute of which both

Continued on next page

MG "A" AND ALFA-ROMEO continued

engines are capable. High engine speeds entail engine noise, of which the *Alfa* makes less than the *MG*.

There is a contrast in gear-shift operation between the two cars. Both have four forward speeds, and floor-mounted gear-shift levers. But the *MG's* gear lever is short, stiff to operate, and permits quick shifts. The *Alfa* has a longer, heavier lever that shifts the gears more easily but requires more movement. The *MG's* low gear is not synchronized; all *Alfa's* four forward speeds are.

Of the two cars, the *MG* is considerably more flexible, with more pulling power at moderate speeds in high gear, but the *Alfa* is the smoother-running powerplant—as indeed it ought to be, in the light of its construction. Compared with the *MG* engine, which is a "hotted-up" production four-cylinder job which, like most U.S. engines, has push-rod-operated overhead valves, the *Alfa* is an expensive-to-build piece of automotive jewelry.

The *Alfa* engine utilizes chain-driven double overhead camshafts, and has racing-type hemispherical combustion chambers. The bore and stroke are nearly equal—approximately 3 x 3 inches. Cylinder block, cylinder head, crankcase and transmission housing are of light alloy. Cylinder bores, as in many heavy-duty vehicles, are inserted cast-iron sleeves in direct contact with the cooling water. Nearly everything possible has been done to produce an engine capable of sustained high horsepower output per cubic inch and per pound of weight.

Fuel economy

In addition to the contribution made by a small, lightweight engine to good car handling characteristics, the high cost of fuel in both Great Britain and Italy in part dictates the small engine displacement of both the *MG* and the *Alfa-Romeo*. Both give excellent gas mileage; the *Alfa* being the better of the two in about the proportion one would expect from its lesser cubic-feet-per-mile displacement. At 50 and 60 mph, however, the *Alfa's* superiority over the *MG* increased; in fact, at the latter speed, it gave better mileage than any other car CU has tested except one —the 1954 *Porsche* coupe.

To the American buyer of an *Alfa-Romeo*, in view of its price, any margin in gas mileage over a cheaper car is academic as far as overall costs are concerned, but outstanding gas mileage serves as an indicator of efficiency in design.

Comfort

While good handling should be paramount in a sports car, as much comfort should be built into it as is consistent with its small size, light weight and precise handling. The great majority of sports cars, in fact, are used as personal rather than sports cars most of the time, and their owners have every right to expect a reasonable degree of comfort. ***Continued on page 28***

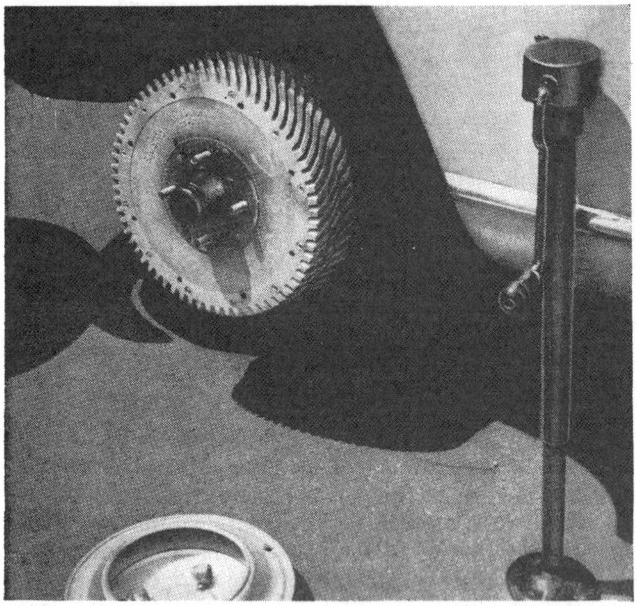

The cooling fins on the Alfa's big front brake drum help in preventing brake "fade." Alfa's fine jack is at right

Golf shoes show the scale. MG's tiny trunk, loaded with spare tire and tools, leaves little space for additional luggage.

With top up, hand signals from the MG—or opening doors from outside—means fighting a stiff, springloaded flap

FACTS AND FIGURES

	Alfa Romeo	MG-A
PRICE		
AT COASTAL PORT OF ENTRY	$3230*	$2195
DIMENSIONS		
WHEELBASE	87 in.	94 in.
OVERALL LENGTH	152 in.	156 in.
OVERALL WIDTH	61 in.	58 in.
OVERALL HEIGHT, TOP UP	51 in.	50 in.
OVERALL HEIGHT, TOP DOWN	47 in.	46 in.
ROAD CLEARANCE	6.1 in.	5.4 in.
TURNING CIRCLE DIAMETER (wall-to-wall)	35 ft.	33 ft.
STEERING FACTOR	0.57	0.59
WEIGHT AND TIRES		
CURB WEIGHT	1955 lb.	2000 lb.
% WEIGHT ON REAR WHEELS	45%	49%
TIRE SIZE	6.00x15 in. (approx.)	5.60x15 in.
THEORETICAL TIRE CARRYING CAPACITY (above curb weight)	1095 lb. (approx.)	740 lb.
ENGINE		
TYPE	Double overhead camshaft 4	Overhead-valve 4
BORE AND STROKE	2.91x2.95 in.	2.88x3.50 in.
PISTON DISPLACEMENT	79 cu. in.	91 cu. in.
COMPRESSION RATIO	8.0	8.3
MAXIMUM ADVERTISED HORSEPOWER	65@6000rpm	68@5500rpm
ENGINE SPEEDS		
AXLE RATIO	4.55	4.30
ENGINE REVS PER MILE, HIGH GEAR	3772	3523
PISTON TRAVEL PER MILE, HIGH GEAR	1855 ft.	2055 ft.
PERFORMANCE		
LEVEL ACCELERATION		
0 to 60 mph	17.5 sec.	17.3 sec.
covered ¼ mile, standing start	21.3 sec.	21.8 sec.
45 to 65 mph (third gear)	10.9 sec.	10.6 sec.
ACCELERATION ON 9% GRADE (third gear)		
from 30 to 40 mph	15.8 sec.	13.7 sec.
from 30 to 50 mph	41.0 sec.	30.5 sec.
top speed attainable on grade	60 mph	62 mph
ECONOMY		
CONSTANT-SPEED GASOLINE MILEAGE		
at steady 30 mph	40.4 mpg	37.6 mpg
at steady 50 mph	39.3 mpg	35.3 mpg
at steady 40 mph	37.0 mpg	31.6 mpg
at steady 60 mph	34.4 mpg	29.1 mpg
TRAFFIC GAS MILEAGE (simulated traffic test)	23.7 mpg	21.2 mpg
HIGHWAY GAS MILEAGE	33.8 mpg	31.4 mpg
OVERALL GAS MILEAGE	26.4 mpg	24.2 mpg
OIL CONSUMPTION RATE AFTER BREAK-IN	for 2550 mi. 2100 mi./qt.	for 2600 mi. 725 mi./qt.

**Includes built-in fresh-air heater.*

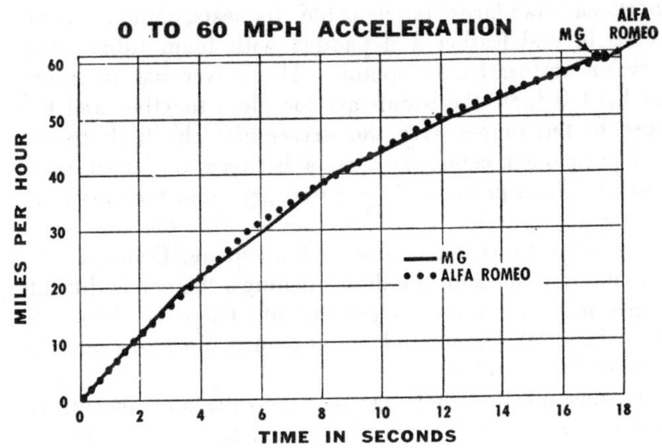

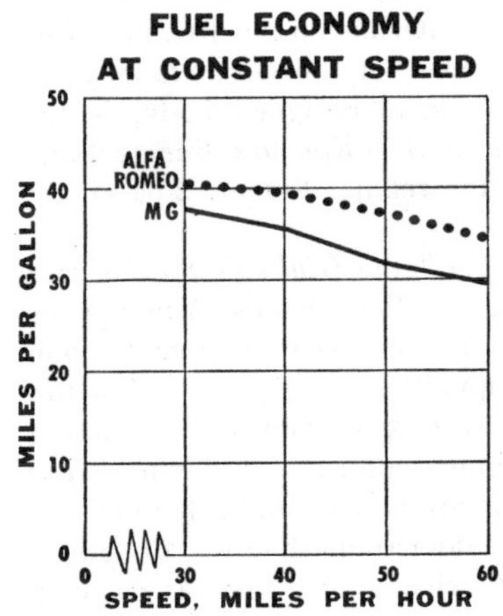

Mileage per gallon of fuel is recorded at various constant speeds for CU's test cars. The acceleration is measured at wide open throttle with the car moving from stand-still to 60 mph

A GUIDE TO THE FACTS AND FIGURES

PRICE AT COASTAL PORT OF ENTRY. Includes Federal tax, but not local taxes, freight, optional extras, accessories, or conditioning charges.

ROAD CLEARANCE. Distance from road surface to lowest part of car likely to hit high spots in the road.

TURNING CIRCLE. Path traced by outtermost tip of front bumper with wheels all the way to left, as for a tight U-turn. Some cars turn shorter to the right.

STEERING FACTOR. Number of turns of the steering wheel required to make a "right-angle" turn of 30-foot radius.

CURB WEIGHT. Weight of equipped car, full of gas, oil, and water, ready (at the curb) for occupants.

TIRE CAPACITY. Official but conservative load-carrying rating for the four tires, minus curb weight of car. Low capacity predisposes to more rapid tire wear.

ENGINE DATA. From factory figures.

PERFORMANCE. 0-60 mph, ¼ mile runs with all gears used to maximum advantage; 45-65 mph and 9% grade runs with floored accelerator but no manual shifting.

ECONOMY. Constant speed tests offer controlled comparison between cars. In normal driving at comparable average speeds mileage will be much lower. Traffic pattern involves moderate acceleration, 35 mph maximum, average speed for course of about 21 mph. Highway gas mileage is measured on a parkway, cruising at 55 mph, averaging about 51 mph. Mileage for most drivers will fall between Highway and Traffic Gas Mileage, depending on individual driving patterns.

MG "A" AND ALFA-ROMEO continued

Here, it seems to CU, is where the *MG "A"* falls down badly; its extremely hard riding qualities are, even by sports car standards, indefensible. Its seats, though upholstered in real leather and padded with foam rubber, are very uncomfortable in contour. The driver has no room for his left foot; the pedals are too close together and too close to the tunnel over the driveshaft; the high tunnel and permanent center arm rests between the front seats also prohibit squeezing three passengers into the car in an emergency. Interior ventilation, even with the top down and side curtains not in use, is inadequate. Owing to the size and placement of the door openings, the car is difficult to get into and worse to get out of. There is almost no room inside the trunk and no lock on it, no glove box, no outside door handles.

To sum up, the whole car is a too-snug and uncomfortable fit for two, as though it had been designed for streamlined appearance first and passenger accommodation second. The paint job and trim materials on the *MG*, however, are superior in quality to those of the *Giulietta Spyder*, which, notwithstanding its Farina-designed body, is the specially low-priced model in *Alfa-Romeo's* line—selling, in fact, for $900 less than the regular 2-door, 2-passenger *Giulietta Sprint* coupe.

Compared with the *MG "A,"* the *Giulietta Spyder* is a masterpiece of comfort and practical details. Although its seats are very low to the floor they are exceptionally comfortable. The *Alfa's* riding qualities, aided by very low tire pressures, are reasonably soft and comfortable without much sway or any loss of bounce control on big bumps. The wide doors make getting in and out easy for such a low car, and inside there is more shoulder room than the *MG* offers, and plenty of room for the driver's feet. The car has conventional door handles inside and out. The top attaches more easily to the windshield than the *MG* top and for some reason the fabric drums less at high speed. The *Alfa* has an effective built-in fresh-air heater and two ventilating ducts. Trunk space, while not large by American standards, is usable and efficient. Both car doors and trunk can be locked.

The little Alfa has faults, however. Maximum leg room is less than on the *MG*, and the distance between steering wheel rim and brake pedal is below average, which may bother the long-legged driver. Instrument panel chrome reflects in the windshield. The glove box has no door. Though located in the trunk, the battery is hard to service, and in CU's car, despite its being covered, it regurgitated acid solution over the trunk floor.

Summing up

To one who wants a solid and thoroughgoing sports car at a reasonable price, and is genuinely indifferent to major and minor discomforts in his vehicle, the *MG "A"* is a good buy. It is a handsome car, built for the most part of well-tried and simple components, which can be serviced with comparative ease, and, except for its seemingly springless ride, has no serious vices in its road behavior.

With the *Alfa-Romeo Giulietta Spyder*, to give it its full name for the last time, the owner will have more trouble vis-a-vis the *MG* getting service, but is much better off in the matter of comfort, and has a vehicle built, without penny-pinching, to very exacting standards in the matter of advanced mechanical design and road behavior.

The *Spyder* ("Spyder" is an obsolescent word for a "spidery," lightweight carriage) is, at present, the only imported sports car at its price that has roll-up windows, but it can lay no other claim to being a bargain. As a personal car it is an almost completely ingratiating connoisseur's item, and, although the sporting Italian drivers for whom the *Giulietta* was designed prefer the lighter, smoother-lined *Sprint* coupe to the convertible *Spyder*, it is hard to see how any knowledgeable motorist who gets acquainted with the *Alfa* can fail to respond to the facility with which the husky little plaything does what it was designed to do.

ALFA ROMEO

IMMEDIATELY after World War II competition sports cars carrying the Alfa Romeo nameplate were a sensation in the sporting world simply because Alfa executives had the foresight to store their racing machinery safely away from their plants that were completely destroyed during the war. Today the Alfa Romeo factory is one of the more solid Italian firms producing cars which are designed to cope with the rather rugged driving requirements of Italian traffic conditions and mountain roads.

Basically, Alfa Romeo cars are highly desirable pieces of precision machinery whose engineering has evolved from the firm's extremely successful racing experience. Newest in the line that should particularly appeal to the enthusiasts is the Giulietta 1300 series which comes in a sedate four-door sedan body and is equipped with a small, 50-horsepower (79 cubic inch displacement) engine that's capable of giving the lightweight car a top speed of about 86 miles per hour. In the still more desirable Sprint form, the engine is boosted to 65 bhp at 6000 rpm and top speed is slightly over 100 mph. Two Giulietta Sprint bodies are offered, a two-passenger convertible and a streamlined coupe that should have immediate styling appeal to most enthusiasts. The coupe holds two passengers and has enough space for a considerable amount of luggage behind the seat or for carrying an additional two passengers.

Alfa Romeo cars feature unit body and frame construction and have a suspension system engineered for good roadability and handling on curves. This consists of independent coil spring front suspension and coil springs over a rigid axle at the rear. The four-speed, manual transmission is operated by a shift lever mounted on the steering column. Instrumentation includes usual units plus a tachometer, all contained in a hooded panel immediately in front of the driver. Huge 10-inch brakes have helical cooling fins.

ALFA ROMEO GIULIETTA SPRINT

Alfa Romeo also produces a 1900 Super and Super Sprint car, both of which have more powerful engines (with 90 and 115 horsepower ratings respectively) and are slightly larger in body size. So far, few of these latter models have reached the U.S. The Giulietta cars, however, are imported and sold by established foreign car dealers. Owners of these aerodynamic small cars are enthusiastic in their praise of the machines, claiming excellent agility in traffic, competition-type cornering, and gas mileage of at least 25 miles per gallon. The coupe sells for $4085 at port of entry; the convertible is priced at $3175. ■

driving around with walt woron

in the new 1300-cc Alfa Romeo Giulietta

photos by Bob D'Olivo

GIVE THE ALFA ROMEO GIULIETTA the right place to be driven hard, where it can be pushed thru corners as fast as *you* dare to go, and it will get you there quicker and safer than most cars with 4 times the horsepower. To appreciate this Italian engineering masterpiece from Milan, you would want to drive it hard on curving byways, and not pussyfoot your way among the behemoths on Main Street, U.S.A. It's only where you can stuff your foot in it, revving the engine between 4000 and its maximum of 6000, that you can wring out its best performance. And tho it's extremely light to the touch, making it a cinch for traffic driving, it's the wide open spaces that this Shakespearean heroine yearns for—there is where she really asserts herself.

The Giulietta is a fitting car for the Alfa Romeo firm to be producing in their 50th anniversary year. In 1906 Alessandro Darracq was producing a combination taxi-car, but for 3 years he encountered one problem after another. In 1909 his garage was absorbed by A.L.F.A. (Anonima Lombarda Fabbrica Automobili), which in turn became the property of engineer Nicola Romeo in 1911. Seven years later the company was reorganized and its name was again changed, this time to its present one, Alfa Romeo, S.P.A.

Since that time there has been a long procession of Alfa Romeo cars, starting with the pre-war P-1, P-2, Tipo 312, and 158, and the postwar 1900, Flying Saucer, T.I. Super, Giulietta Berlina, 1900 Super and 2000 Gran Turismo Sportiva. Racing has been a part of the Alfa Romeo firm since 1923, so the thorobred strain of the Giulietta is not hard to understand.

The Giulietta Sprint that I picked up from the importer, Hoffman Motors, Western Division, Inc., is what Alfa calls a 2-2 coupe. It has the same running gear and powerplant as the Spider (roadster), but its wheelbase is 7½ inches longer to accommodate the smallish jump seats. The determining factor

All controls and instruments are grouped conveniently for driver. Comfortable seats are separate semi-buckets

A set of 4 pieces of fitted luggage can be bougnt and strapped compactly into place on this rear bench

Trunk capacity is reduced by the spare tire and tools. Trunk lid projects far enough to bump heads.

Sprint, a small luxury coupe with the accent on quality . . .

The Alfa Romeo Giulietta Sprint, an example of how well a car can be built

for selecting one car over the other would probably lie in the need for jump seats, weatherproofing vs. open air, and price. The Sprint delivers at port of entry for $4070, while the Spider goes for $3135. Performance-wise, the 2 cars would be pretty close to a toss-up. Handling-wise, they're both superb.

Corners taken at average speeds will make you wonder if this 4-coil-sprung car is like its American-built counterparts that sway all over the road in a like condition. Body lean is there, doubly noticeable to your passenger, and you wonder if the car is going to stick. Maybe you even overcorrect a little, for this is easy to do with a car that has a steering ratio as quick as this one: 1½ turns of the small wheel take you from a dead-ahead position to full lock either way. Then, as you begin to pick up speed and boot it thru the corner you begin to drift, sliding all 4 wheels, but you are far from acutely aware of it. Then you're coming out of the corner and if you haven't judged it right, you find that you might have to shift down, for you won't come out fast enough unless you have your revs up.

As for its ability to keep its nose pointed straight ahead when you want it, it's no wonder I seldom corrected the wheel when on an arrow-like road. Traveling at any speed you can whip the wheel to either side, and instead of diving toward a ditch, it returns to dead ahead like a gyroscope that's been knocked off its normal axis and is now seeking its original level. The only objection I found came in strong, gusty winds, where you *do* have to correct because of the buffeting you get. At its top speed of just under 100 mph it feels as sure-footed as it does at 40-50 mph; the sensation of speed (aside from the landscape slipping past) comes from the added engine and wind noise instead of a lurking uneasiness.

If you were to entirely discount the fact that the Giulietta Sprint is powered by a 78-cubic-inch, 4-cylinder engine, you would probably be disappointed in the fact that performance is not better. Taking it into account, you would no doubt marvel at its ability to be pushed thru the gears from standstill to 30 mph in 5.6 seconds, to 60 mph in 16.6 seconds, and to flash by the ¼-mile marker in 20.7 seconds, doing 67.4 mph. You would also be pretty much impressed with the fact that it

CONTINUED ON PAGE 39

Dual A-frames, coil springs, telescoping shocks, and a stabilizer bar form the sway-proof front suspension

Rigid rear axle is suspended by a coil spring, co-axial shock absorber, and trailing link combination

The Giulietta's a jewel of fine workmanship, as is evinced by the design and finish of its tiny engine

SCI Technical Report:

Low, wide and handsome is this frontal view of the Veloce. Body and chassis of the Veloce is practically identical to the Giulietta. Bumpers have been removed to cut down turbulence for SCI test runs.

The Alfa Romeo Veloce Spyder

BY RUSS KELLY

ONE of the biggest surprises of the automotive year was the slow realization that Alfa Romeo, of Milano, Italy was making, for them, a mass invasion of the American light car market. The semi-mass production of a car aimed at a foreign sales program is really new for Alfa and is sure to prove enlightening to both the Italian manufacturer and the American consumer.

The natural decision of Alfa to publicize this new product —the 1290 cc Giulietta, on the race course has caused a couple of super hot models to put in their appearance at U. S. races. One in the gentle hands of Carroll Shelby ran away with the under 1500 cc race at Cumberland, Maryland. On the west coast Bill Pollack proved that he could deal out rations of fits to all but the most experienced Porsche Spyder drivers at Santa Maria, California.

Called *Veloce*—the fast—this little bomb has the advantage of being able to run in production car races. It's significant, however, that the technical committee of the Southern California Region of the SCCA was a little difficult to convince about its eligibility as a production model when Pollack raced it at Santa Maria. The fact that the only two real Veloce Spyders imported are still apparently under factory control and not really available for testing leaves a couple of interesting questions unanswered.

First, it seems they have difficulty in finishing a race with any binders left to speak of. Second is the question raised by the laconic statement of an English mechanic on their handling after witnessing the '56 Mille Miglia. Simple and to the point he said, "They 'ad 27 of 'em 'anging about in the trees between Brescia and Rome."

Conjecture aside, this model is doing an excellent selling job because of its agility, phenomenal speed and acceleration. You can be sure that every small specials builder in the U. S. has got his eye on the little Giulietta two-cam mill since its potential has been so well demonstrated by the Veloce.

In the same category as the question "Which came first, the chicken or the egg," you can put the question that comes to mind when you review the history of Alfa. Are they the world's best at hot rodding production models and turning them into racing cars or the world's best at detuning racing cars and turning them into production models? The relationship between the Giulietta and the Veloce does nothing to answer this question. Under superficial examination—standing still—you'd have a hard time telling them apart, but you'd have a tougher time turning a Giulietta into a Veloce.

The body and chassis are practically the same, the principal difference being that the Veloce uses aluminum for hood, doors and deck lid. The chassis-body integral construction precludes the more extensive use of light alloys. The chassis is of the platform type which consists of a pressed steel floor that is reinforced along each edge. Forward of the firewall there is a system of channeled sections that carry the engine and gearbox unit and the front suspension layout. This system follows in general the shape of a sloping shallow box with an upper and lower member on each side. The lower members are tied together by a simple cross member and forward of this is a channel section diaphragm that ties all four members together. To this is welded the body shell, consisting of the front fenders, cowl, firewall bulkhead, rear turtle deck and rear fenders.

Integral construction offers a great deal of rigidity in torsion, but may suffer in beam stiffness. Remembering that a beam's resistance to bending varies as to the cube of its depth, the lack of a steel top to take compression loads and the deep cutaways for the doors has effectively reduced the depth of the beam. However, this body is well reinforced around the opening in an effort to compensate for any loss of strength. The disadvantage of this type of structure in a racing car is that even a minor collision could cause damage extremely difficult to repair.

(Continued on page 36)

Driver can watch his instruments through wheel rings without difficulty. Left-hand gauge cluster includes water temp., fuel and oil temp; center gauge is tach calibrated to 8000 rpm and incorporates oil pressure gauge; dial at right is speedometer marked in mph.

This is the Giulietta piston which exerts a compression of 8:1. Pistons in the Veloce do not have fourth ring around the skirt, but otherwise are the same. Veloce slug exerts about 9:1 compression.

Veloce takes hard turn at better than 70 mph. Although car feels flat through curve, it does lean.

ALFA ROMEO SPRINT VELOCE

SPECIFICATIONS

POWER UNIT:

Type	In-line four.
Valve arrangement	Double overhead camshaft.
Bore & Stroke (Engl. & Met.)	74 by 75 mm
Displacement (Engl. & Met.)	78.6 cu. ins; 1290 cc.
Compression Ratio	8.5 to 1.
Carburetion by	Two Weber double choke side draught—Tipo 40DCO3.
Max. bhp @ rpm	90 at 6500; factory figures (might be conservative).
Idle Speed	2000 rpm (car never runs under this speed).

DRIVE TRAIN:

Transmission ratios I	3.59
II	2.10
III	1.36
IV	1.00
Final drive ratio (test car)	4.1
Other available final drive ratio	4.9 4.5 5.1

CHASSIS:

Wheelbase	7 ft. 10 ins.
Front Tread	1286 mm (50.5 ins.).
Rear Tread	1270 mm (49.9 ins.).
Suspension, front	Independent by wishbones and coil springs.
Suspension, rear	Solid axle; coil springs located by "A" frame offset to left.
Shock absorbers	Telescopic.
Steering type	Worm and finger ZF box.
Steering wheel turns L to L	2½
Turning diameter	215.6 ins.
Brake type	Four wheel hydraulic—front drums heavily ribbed—front: two leading shoes.
Brake lining area	264 sq. ins.
Wheel studs, circle diam.	(four studs).
Tire size	5.50x15 Pirelli-Cinturato.

GENERAL:

Length	156.6 ins.
Width	60.2 ins.
Height	51.8 ins.
Weight, test car	1962 pounds (curb).
Fuel capacity—U.S. gallons	21

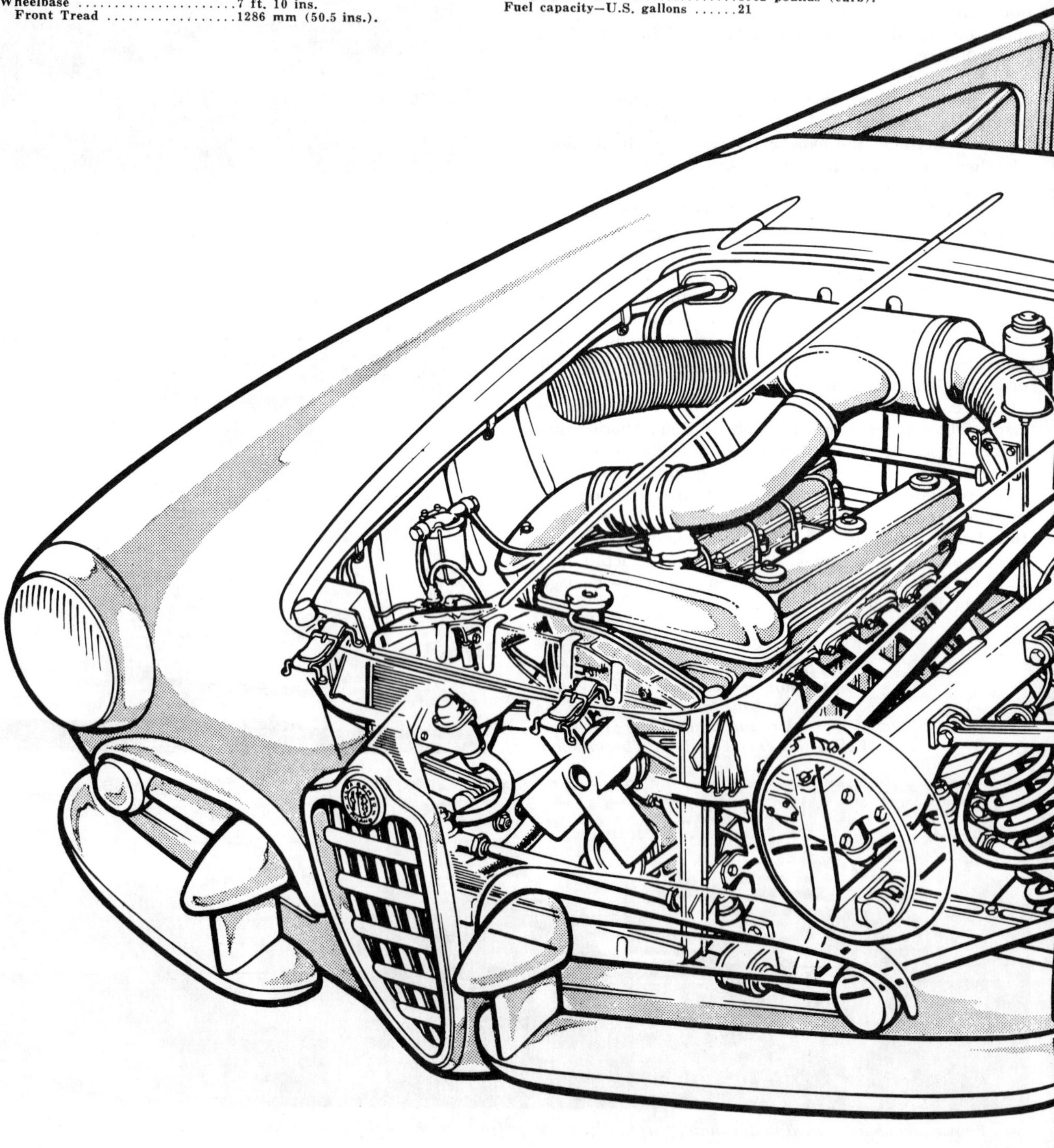

ALFA ROMEO SPRINT VELOCE

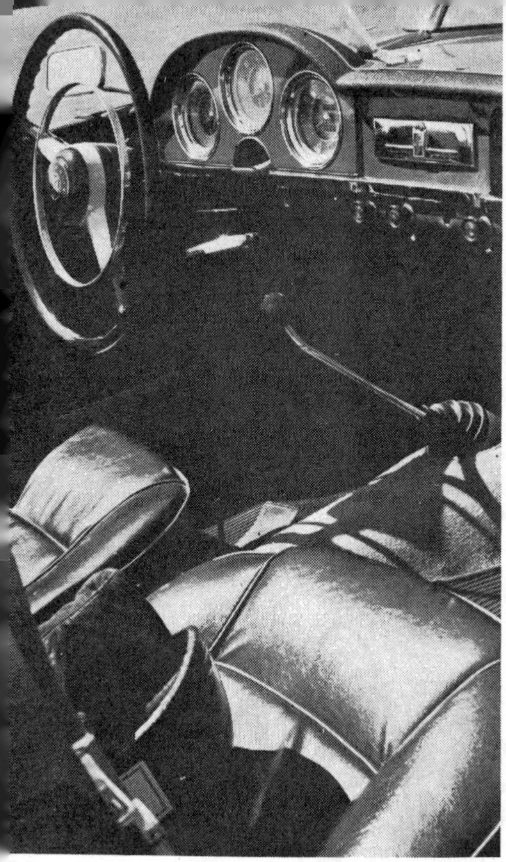

Cockpit is luxuriously appointed in leather. Note deep seats, and husky and graceful shaped gear lever for easy shifting.

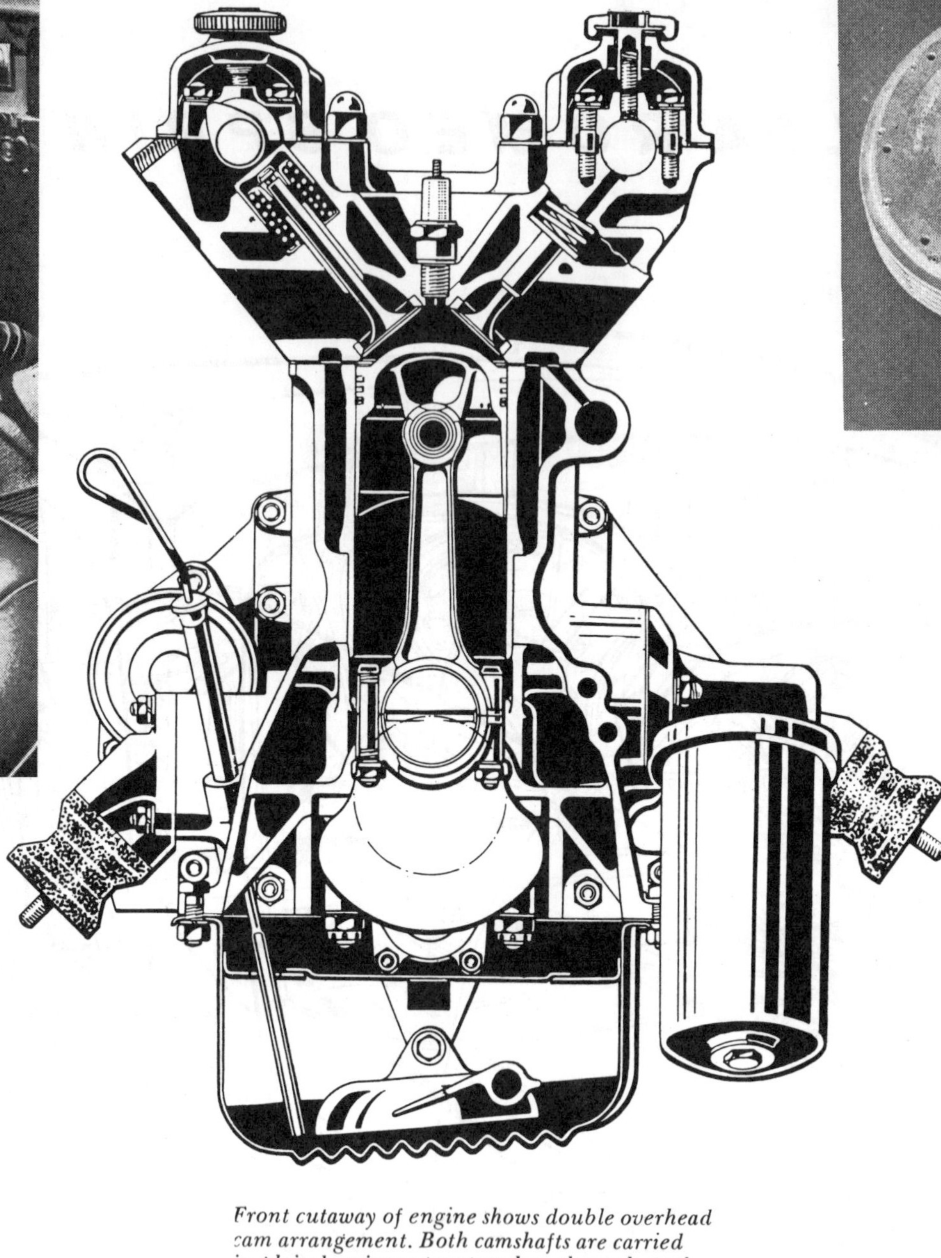

Front cutaway of engine shows double overhead cam arrangement. Both camshafts are carried in plain bearings-operate valves through steel cups fitted over the valve springs.

Front suspension is independent by unequal-length wishbones and coil springs. The coil springs are steeply inclined and surround telescopic type shocks. A steel cable controls rebound and rubber bumpers establish the limits of full bump. This front end layout is unusual in that instead of a kingpin carrier tying the upper and lower "A" frames together with a stub axle attached to this, Alfa practice is to incorporate this in one unit that is bushed and turns at the points where it is attached to the "A" frames.

The steering box is of the familiar ZF worm and roller type. However, the steering linkage is quite different from the usual type. Attached at right angles to the vertical output shaft of the steering box is what amounts to a double-ended pitman arm. To the rear of this arm is attached the short track rod for the left, or driver's side, front wheel. Attached to the front end of this arm is a drag link that crosses the chassis slightly above and to the front of the axle center line.

The drag link actuates an exact duplicate of the steering box pitman arm and track rod for the right side wheel. This places all the steering linkage ahead of the front axle center line. A heavy stabilizer bar links the two lower "A" frames. The front end layout impresses you immediately as being clean and well executed. It isn't hard to see that Alfa has been turning out quality merchandise for a long time.

The live rear axle is suspended by coil springs that enclose telescopic shocks. The spring disposition is vertical and they are mounted in heavy steel cups welded to the rear section of the body shell. Radius rods on each side locate the final drive unit under acceleration and braking loads. These radius rods are attached to brackets that extend downward from the rear axle housing, forming in effect a half Watt's link. From the housing they pass forward on a horizontal plane to brackets attached to the chassis platform.

Front, rear brake drums. Rear is 10 inches inside, has ribbed exterior. Front drum, right, is finned.

Large diameter flexible hoses carry air to cockpit for ventilation and to air-filter before it's carried to balance box.

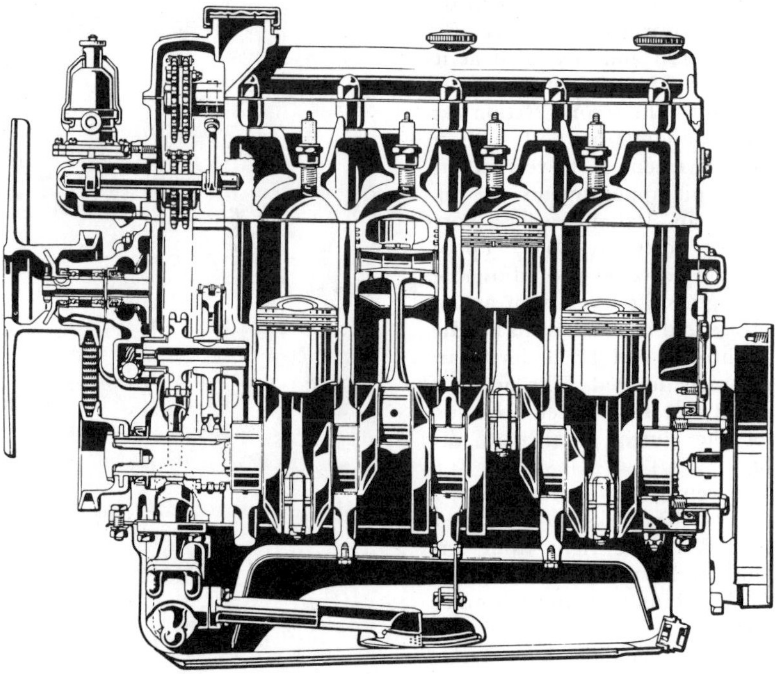

Side cutaway view shows internal works. The cast aluminum block is fitted with iron wet liners. Note crankcase and sump line is well below the center line of the five-main-bearing crankshaft.

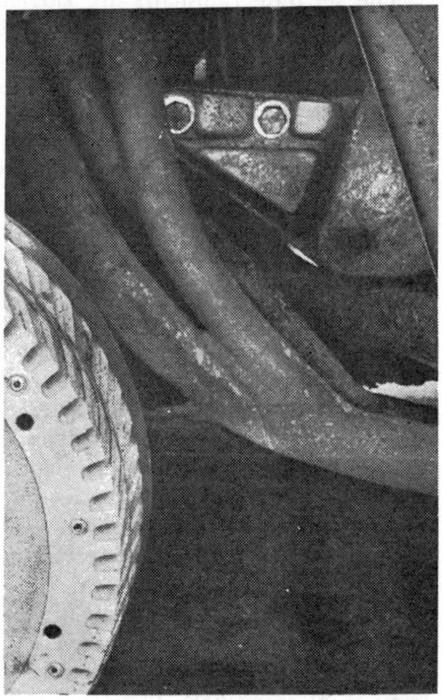

Split headers join 2 common collector boxes and terminate in one pipe at rear.

The reaction to braking and acceleration loads inherent in this sort of layout is absorbed by a triangular tube frame attached by a ball joint at its apex to the driver's side of the differential housing and extending forward. Its effective length is about half that of the radius rods. This must be a busy little unit since it serves as torque reactor, stabilizer bar and panhard rod. The heavily ribbed differential housing is of cast aluminum and incorporates a small sump. The steel axle housings are bolted to this casting. Full bump and rebound is limited by straps and rubber bumpers.

The brakes that boast an impressive amount of lining area (264 square inches) are hydraulically operated. In the front, two leading shoes expand in a 10.5-inch drum that is of bimetal construction. The aluminum muff that is shrunk around the ferrous lining is liberally vaned to dissipate heat. The rear drums are .5 inches smaller in diameter than in the front. Also of bimetallic construction, they are ribbed peripherally. The usual practice for sports car brakes is followed by using one loading and one trailing shoe in the rear. The hand brake operates only on the rear wheels by a system of cables and levers.

The engine gearbox unit has three mounting points. It is not mounted vertically in the chassis but lays over slightly to the right. In addition to other modifications, this engine has been carefully assembled of selected components and is noticeably smoother than the production Giulietta. Extensive use of alloys makes this four cylinder twin camshaft engine extremely light.

The cast aluminum block is fitted with iron wet liners and is extremely deep. The joining line of crankcase and sump is well below the center line of the five main bearing crankshaft. The cam chain cover carries the cross shaft that drives the distributor and the gear type oil pump. Two stages of chain are used in the cam drive, thereby reducing

Veloce

the speed by 50% for the longer upper chain. The upper chain is easily adjustable by means of an idler sprocket. The aluminum head has ferrous valve seat inserts with the valves themselves disposed at an included angle of 80°. The 14 mm spark plug is located in the center of the hemispherical combustion chamber equi-distant between the two valves. Porting is typically Alfa in that it is as "straight through" as you can get it.

The separate camshafts are carried on plain bearings and operate the valves through the now almost universal process of hardened-steel cups that fit over the valve springs. Adjustments for tappet clearances are made by use of small steel buttons between cup and valve end. Double-nested helical springs are used with conventional retainers and Chevrolet type keepers.

The alloy Mondial pistons are domed supplying a ratio of nine to one. Reliefs are cut in the dome for the valves and two compression rings and one scraper, or oil control ring, is used. The short connecting rod carries a wrist-pin with a tapered I.D. that is retained by circlips. The big end of the rod splits at right angles to the cylinder bore with two bolts retaining the cap.

The electrical system is 12 volt Morelli throughout with the generator driven by pulley and belt from the front of the crankshaft. The distributor has a centrifigal advance mechanism that accounts for 30 of the 40 degrees advance. The usual vacuum brake on the advance mechanism has been omitted.

INTAKE-EXHAUST

Carburetion layouts on Italian racing machinery have a way of making you wonder where the money comes from. This one is no exception. Two dual-throat Weber sidedrafts almost dwarf the engine. Called Tipo 205, the throat diameter is 40 mm (1.6 inches). These bolt directly to the head fulfilling the one cylinder—one carburetor theory. Ram tubes are dispensed with by the use of a massive cast alloy balance box that is fed cool air through a large diameter flexible tube after passing through an air cleaner. Apparently there is a choice of either picking the air up under pressure through the ventilator aperture in the left front fender, or by re-routing the tubing, from the interior of the cockpit.

The exhaust system features split headers with two common collector chambers and terminates in a single pipe in the rear. Pipes one and four join gracefully about 20 inches from the port and enter the first collector chamber just under the cockpit. Pipes two and three follow the same pattern. The second collector chamber is near the rear of the chassis with the single tail pipe ending at the body line. In case anyone is interested, the Alfanese for collector chamber is "lung."

The clutch is a conventional single dry plate. The gearbox has four forward speeds—all syncromesh. The gearbox housing is cast alloy. The method of transmitting the drive from gearbox to rear axle is of special interest in that the driveshaft is in two separate sections. The first section attaches to the back of the gearbox by means of a rubber universal and is carried at its other end by a ball bearing mounted to the chassis. The second section carries universal joints on both ends to allow for movement of the rear axle.

The cockpit appointments are luxurious for an Italian racing sports car. Bucket seats are softly upholstered, leather covered and adjustable. The floor gearshift lever is on an impressive size and convenient to handle. Clutch, throttle and brake levers are conventionally located. The addition of a raised rest for the left foot just clear of the clutch pedal is a nice thought.

Cockpit ventilation is provided by a flexible tube that picks up air from a vent located in the right front fender. The wind screen is of the small pre-1956 Le Mans type and is formed of plastic.

Instruments include a large tachometer and speedometer in addition to the usual temperature, oil pressure and electrical gauges. The tachometer is not red-lined, but it would seem reasonable that 6500 rpm would be maximum revs as claimed. However, one owner casually mentioned seeing 7600 in top at an indicated 130 mph—something the mechanics in Milano, who carefully assembled this engine, would be glad to hear.

The longer you consider it the more obvious it becomes that this car, for all its tininess is a true Alfa and will make its mark on the automotive world as surely as did the 2.9.

It's inevitable that production 1290 cc chassis and engines will fall into the hands of the U. S. specials builders and with or without the lead supplied by the factory in the Veloce, Alfa Specials are going to be tough contenders for the silverware.

— Russ Kelly

CONTINUED FROM PAGE 31
accelerates from 30 in 2nd and 3rd gears to 50 mph in 6 seconds, and then goes on up in 3rd and 4th to 80 mph in 22 more seconds. The unfortunate part is that you *do* have to justify this performance; on a performance/displacement chart it looks great, while on a performance/dollar chart it looks quite weak. It depends on how much your enthusiasm can be dampened by being blown off from a signal by a Plymouth (tho it would take a '56 to do it).

It took quite a bit of practice to get the times posted here. The best ones were finally reached by jazzing the throttle between 3-4000, popping the clutch, hoping you'd get a good bite without laying too long a delaying strip of rubber, then shifting in each gear at 6000 rpm. Sometimes Bob D'Olivo and I went thru the ¼-mile in 3rd gear and sometimes in 4th; the shift point into 4th was an indicated 68 mph. Somewhat better times could be gotten with more practice and surely with the optional floor gearshift, which should be much more positive. The column shift is definitely satisfactory, but doesn't lend itself to snapshifting.

The acceleration runs didn't seem to faze the engine, clutch, or brakes one bit. The engine stayed nice and cool, the single-plate dry clutch stayed its positive self, and the hydraulic, finned-drum brakes never once gave any indication of fading after repeated hard stops.

As to comfort, part of the story is told by the padding and shape of the individual seats. After a 300-mile round trip, neither my passenger nor I had any aches or pains. (My only complaint is that true bucket seats would better contain you in fast cornering). A good feature is their separate adjustment to provide more than enough legroom. Headroom there's less of. When you 1st bend and twist your way into the seat you're suddenly aware that your head is practically touching the head-lining. You then sort-of slouch down and never again (until you get out and back in) do you notice the low roofline.

Small fry (singular, up to 8 years old; plural, up to 6) can be fairly comfortable on the rear platform with its optional jump seat. Or, if you're a traveling man, it's better suited to the fitted luggage, which is also optional. This gives you an opportunity to carry more than the night case and briefcase you'd otherwise have to be content with storing in the trunk.

The rest of the comfort story is told in the ride, which is fine on a smooth road, but begins to chop over swells. Adding to the annoyance factor are the thumps and whumps you hear going over tarstrips, railroad tracks, and potholes. Fortunately, all of these noises are well insulated from the steering gear.

One of the more "ordinary" things about the Giulietta is its fantastically good vision, provided by the curved one-piece windshield, sloping hoodline, small quarter panels, and rear window with its cleavage. Even in the rain vision is good, for the electric wipers sweep almost the entire windshield, overlapping each other in the center.

Definitely not one of the "ordinary" items of the Giulietta is the minute engine, consisting of an aluminum block with wet sleeves, an aluminum head with hemispherical combustion chambers, overhead valves that are directly opened and closed by twin camshafts chain driven from the statically and dynamically balanced crankshaft that runs on 5 bearings. Fuel is fed to the engine from the 11⅝-gallon tank thru a mechanical pump and a twin-barrel Solex downdraft carburetor. If you're economy conscious you'll be happy to know that the engine doesn't gulp fuel. In highway driving you can expect close to what we got, particularly if you drive the Giulietta hard. Tank mileage in town netted an average 22.3 mpg, while highway averages went to 29.3 mpg.

—W.A.W.

ALFA ROMEO GIULIETTA

THE basic four-door Giulietta is the costliest light car in this roundup of the field. Even though it's Alfa's most inexpensive touring machine, it can't be classed as an economy car on grounds of first cost. It can be operated for nickels and dimes, but actually you don't buy the Giulietta for its fuel economy.

You have to be a special sort of enthusiast to pay a big-car price for this machine. You have to prize its competition-quality chassis, its superb brakes, its precise ZF steering, and its beautiful, smooth and efficient dual overhead camshaft engine. If you're a student of racing engines and chassis, chances are that nothing else in the light-car field will appeal to you as much.

The mass-produced Giulietta four-door is a scaled-down version of the fleet, formidable, safe and costlier 1900 T.I. Alfa. The unit-construction principle is the same in both cars, and the engines are basically alike, although the 1300's carburetion puts more emphasis on fuel economy. The Giulietta has huge, deeply-ribbed aluminum drum brakes, and coil spring suspension all around. Its rear axle is set up like a racing car's, with rugged radius arms and A-frame on the final drive housing to maintain perfect suspension geometry under *all* operating conditions.

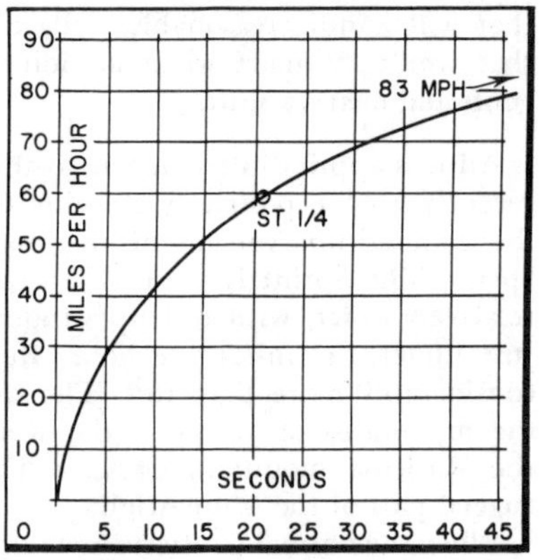

Although this "economy" Giulietta develops peak power at 5200 rpm – the hotter Sprint models cost more – its engine is still loafing at this speed and gives the impression that it's barely stressed. For acceleration and top speed the 1300 stands among the leaders in this survey. But there's much more to it than mere straightaway "go." In roadholding, steering and brakes, the Giulietta stands aloof, hardly to be compared with any other machine in its class.

The car's body is sober and conservative, somewhere between the plain and the distinguished. Luggage room is excellent, front-leg room good, rear-seat leg room minimal. The interior of the body is surprisingly austere and suggests that the four-door is made primarily for the home market. There are now large parts depots for the car on the east and west coasts, and the dealer organization is growing in proportion to the increasing supply of the little thoroughbreds from Milan.

SCI

ROAD TEST: the amazing Alfa Veloce Coupe

Resemblance of the Sprint Veloce to the Alfa Romeo Giulietta is limited to exterior appearance. Terrific weight savings have been achieved through extensive use of aluminum in hood, trunk and door panels, and bumpers

ITALIAN manufacturers know that as soon as they build a car that it's going to be raced in the Mille Miglia; because of this, they are forced to build an automobile that will handle reasonably well, and to design an engine that won't fly apart when an imitator of Tazio Nuvolari goofs and misses a shift.

All this applies in particular to the small Fiats and, more recently, to the positively amazing Giulietta Sprint Veloce, the competition version of the normal 1300 cc Giulietta Sprint. The Sprint is a fine, high speed, touring-sports car, relatively quiet, with roll up windows, and a top speed of just under 100 miles per hour. But the Sprint Veloce is considerably more than this. The car was literally born in the mountains of Italy—on the Radicofani, the Futa and the Raticosa mountain passes which make up the most rugged part of the Mille Miglia.

When the normal Giulietta first came out, it had its baptism in the 1955 Mille Miglia, but was badly beaten by several well driven 1300 Porsches. Alfa was already designing a competition version for the next year. Thus was born the Sprint Veloce, a 1300 cc sports car which is a real Italian racing car with dual Weber double choke 40DC03 carburetors, a centrally mounted four speed gearbox, and a hot-rodded 1300 cc mill which conservatively produces 90 bhp at 6000 rpm. The car's maximum speed was increased to an honest 110 mph which resulted in a car that swept the 1300 cc Gran Tourismo class in victory after victory. One indication of the car's capabilities is the fact that it immediately broke all previous 1300 Porsche records on the Nürburgring; the Giulietta Sprint Veloce is, in short, the most sensational competition car to come out of Italy in 1956.

Veloces are scarce and it was only through the generosity of Bob Tappan, owner of this car, that we were able to make the test. It was immediately taken over one of the roughest stretches of mountain road in all of Italy—the bit of highway between Florence and Bologna which twists up and over mountain passes, the Futa and the Raticosa. This is the kind of road which shows up the merits and demerits of any car that is driven hard over it. When you first get into the Veloce, you find the seats are extremely soft, so soft in fact that if you're short, you find yourself looking through

Air intake hides second Weber carburetor. Plugs are accessible for easy changing. Fuel filter and brake master cylinder are handy for servicing.

Under hard cornering conditions, normal understeer characteristic can be neutralized by playing with gas pedal and steering wheel. Car is literally steered by provoking rear-end breakaway.

Gear shift lever is centrally located, as in more civilized version. Bucket seats are on the soft side, but give good back and lateral support to driver. Steering is only 2½ turns lock to lock.

the upper half of the steering wheel, but a little padding in the right places takes care of this. Lateral support is good and the rugged, centrally mounted gear shift lever fits into the palm of your hand—exactly the way a competition shift would. In fact, the more you look at this baby Alfa, the more you realize that it was expressly designed for competition. Plastic windows, lightweight aluminum doors, the bucket seats, and under the hood the dual Webers and the extractor-tuned exhaust manifold, are visible signs of the Veloce's role in life. But it's not until you get the car out onto the road that you fully appreciate what Alfa has created. The tach is red-lined at 6700 rpm and with this limit even in normal driving, the needle hardly ever sees six. If you're used to driving a car where you have to keep more than one eye on the rev counter, the Veloce will be a revelation. At 5000-6000, all 1290 cc's sing in harmonious contentment and the driver gets the impression of almost unlimited rpm, let alone unlimited acceleration. Power comes in fully at 4500 rpm on the Veloce and in second gear, the short space of time that it takes to go from 4500 to 6500 rpm is quick enough to pull out of any demanding situation. On the road, the car gives the driver a feeling of complete security. The quick, extremely light steering, the extraordinary braking ability and the car's compact size all make for fantastic maneuverability and allows the driver to put the Veloce through slots where even 300 SL's fear to tread. This is what a real Gran Turismo machine should be able to do and this Alfa never lets the driver down.

The Veloce has a steering characteristic that is difficult to find among high performance machinery. Under normal driving conditions, the car has a decided understeer. However, under hard cornering conditions, a transition from understeer to neutral steer can easily be effected by playing with the gas pedal and the steering wheel. This, of course, results in an extremely flexible automobile in which the driver is at home in a minute. The steering box is of ZF manufacture which accounts for its lightness. What is interesting is that it takes the same amount of effort to hold the Veloce on full lock as it does when on half lock. Steering return is not vicious and has this constant feel to it which

Entry and exit are considered very good for a sports car. Use of plexiglass windows eliminates window mechanism in doors, results in added elbow room inside car. Although rear quarter windows open, ventilation is not tops. Sprint Veloce body is built by Bertone.

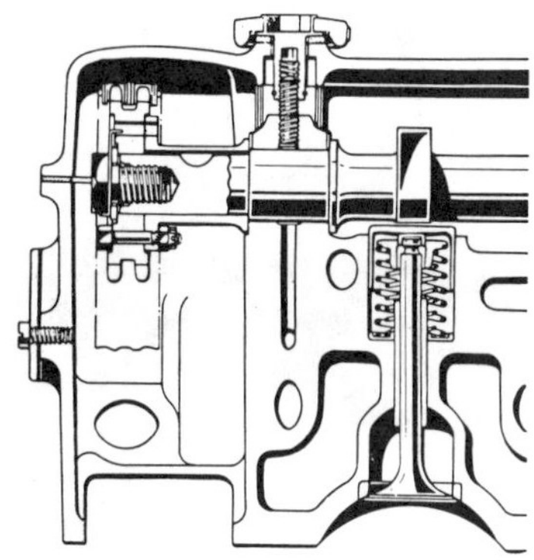

Details of overhead cam system. Chain drive sprocket is at left. Double valve springs are counter-wound. Valve adjustment is through use of shim caps over valve stem, under lifter cup.

Trunk compartment carries spare, jack and lug wrench, plus filler cap for 21-gallon fuel tank. Luggage space is at a premium compared to barn-like interiors in current Detroit cars.

makes for really pleasant driving. Two and a half turns from lock to lock is just about right, and with no play in the wheel nor shock from the road, the Veloce is undoubtedly one of the easiest cars to maneuver that we've ever experienced. But that's getting ahead of the logical procession of things, so let's take a close look at how the car is put together.

Overall quality of workmanship on the Veloce is not as good as on competitive German cars nor is it as good as the production normal Sprints. The plexiglass window slides have a tendency to stick shut and are not the most convenient things to open and shut. One big advantage to the sliding windows is that the area in the doors where the roll ups on the Sprint fit, is completely cut out and allows for an extra amount of elbow and storage room. This, and the huge amount of space behind the seats make up for the small trunk and practically useless glove compartment. The dash is laid out well, but the instruments, especially the tach, are difficult to read. The hand brake is probably the poorest item on the car—being stiff and awkward to release. But defects like this can be excused when one realizes that the hand brake and the glove compartment just aren't meant to be used on the Veloce since the car is a competition machine pure and simple.

For a sports car, entry and exit is very good. The seats, despite the fact that they're too soft, offer good lateral and vertical support and the offset driving position is hardly noticeable until pointed out by an observant passenger. The car is equipped with a heater and defroster unit that could not be put through a test in the hot Italian summer but would probably prove to be adequate enough for a harsh New England winter. Wind noise is quite noticeable over 90 miles an hour. At that speed, a loud whistle emanates from around the windshield pillars. Ventilation also leaves a bit to be desired on the Veloce coupe despite the presence of fresh air ducts under the dash and plexiglass rear quarter windows that crack open.

Engine accessibility is really terrific. The plugs can be changed in a flash and the dip stick does not hover close to the exhaust manifold as is standard practice on many cars. Fuel filter, master cylinder, and carburetor intake air filter are close at hand as soon as the aluminum, forward hinged, hood is lifted. The trunk contains the gas filler for the 21 gallon tank which is special on all Veloces, as well as the

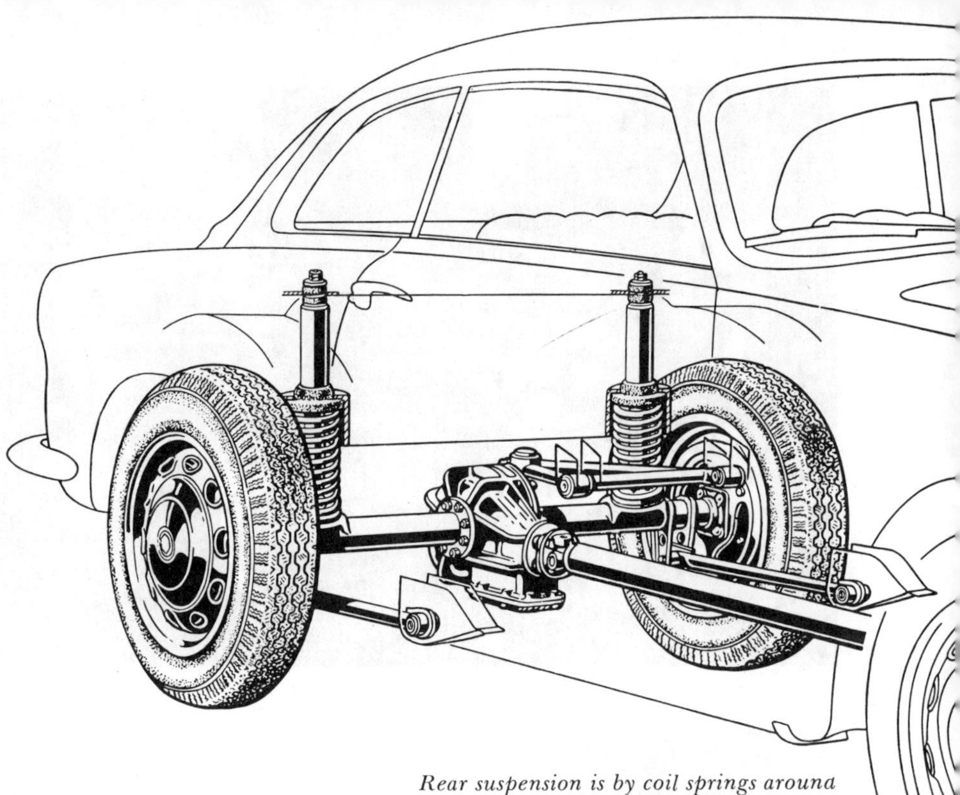

Rear suspension is by coil springs arouna tubular shock absorbers. Triangle bar arrangement to the right of the differential acts as a lateral stabilizer.

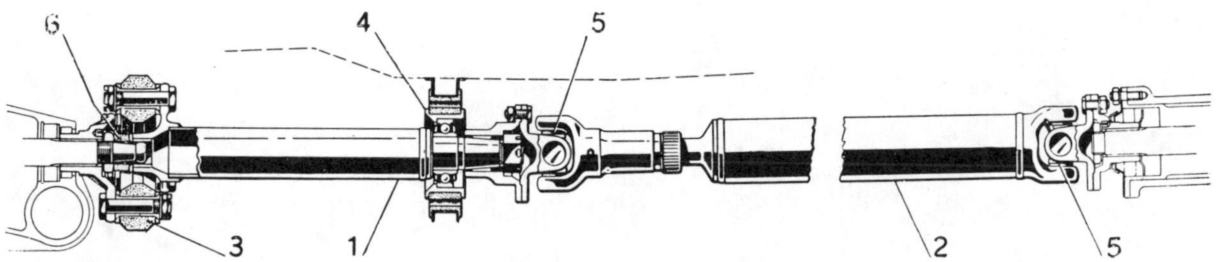

Two section drive shaft is used in the Sprint Veloce. 1—front section; 2—rear section; 3—rubber bushing point at transmission; 4—intermediate support; 5—universal joints; 6—grease seal at end of transmission main shaft.

battery, spare tire, and tools. The bumpers are of aluminum—another difference between the Veloce and the normal Sprint. Alfa has let no item go unseen to cut the weight factor.

But let's take the Veloce out onto the road and see how she goes. The engine is almost instantaneous in its response to the key starter. A rough idle makes you note that the tach does not read under 2000 rpm—the double overhead cam engine is meant to be kept at 5000. Getting away from rest and jabbing the gas pedal you feel the surge of the throttle response. With the windows open, and when passing through a tunnel or down a street ringed by tall buildings, you can hear the singular exhaust noise that is pecular to the Veloces; it's sort of a high pitched rasp that only the Veloce owns and when you get a bunch of them together on a circuit, you get buzzing not unlike a bunch of angry hornets.

Maximum speeds in gears are as follows: 1st, 37 mph; 2nd, 55; 3rd, 80 plus; 4th, a good 111 mph. In fierce acceleration the tach is taken right up to the red line at 6700, and if really pressed, on up to 7000 rpm. The gear shift is sheer pleasure, being very solid and rugged to the touch and very definite. There is never any question as to which gear is doing the pulling. Rapid down shifts can be made with or without double clutching and first is completely synchronized giving the driver excellent control in hard and fast mountain driving. The clutch seems stiff at first but this is due to the relatively high low gear.

When accelerating, you are overwhelmed by the fact that this is only 1290 tiny cc's, every one of which is really doing more than its maximum bit of work. The acceleration of the car is fantastic. Zero to sixty mph in 10.5 seconds; zero, to one hundred in 26.5 seconds. The standing quarter was covered in 18.1 seconds—very fine going for a small displacement machine. Cruising speed is limited only by the driver's adversity to noise since over 90 mph the wind noise begins to become more than noticeable. Tracking on the straight was found to be anything but good above 90. At that speed the car developed a slight wander tendency. Safe all day cruising could be held at 85 with complete security and ease.

In hard cornering, the Veloce, as does the normal Sprint, exhibits marked roll habit, but this roll is noticeable only to outsiders—passenger and driver being completely oblivious to it. The aforementioned characteristic of understeer on the Veloce is good in that while holding the car on neutral steer

CONTINUED ON PAGE 82

GIULIETTA SPIDER

GIULIETTA T. I. BERLINA

GIULIETTA SPRINT COUPE

ENGLISH Specifications:	DEUTSCH Techn. Daten:	FRANÇAIS Spécification:	ITALIANO Dati tecnici:	ESPAÑOL Especificaciones:	PORTUGUES Dados Técnicos:
Name: Giulietta Spider	Name: Giulietta Spider	Nom: Giulietta Spider	Nome: Giulietta Spider	Nombre: Giulietta Spider	Nome: Giulietta Spider
Type: Convertible	Type: Cabriolet	Type: Cabriolet	Tipo: Spider	Tipo: Conversivel	Tipo: Conversivel
Doors: 2	Türen: 2	Portes: 2	Porte: 2	Puertas: 2	Portas: 2
Seats: 2	Sitzplätze: 2	Places: 2	Posti: 2	Asientos: 2	Lugares: 2
Motor: 4 cyl.	Motor: 4 Zyl.	Moteur: 4 cyl.	Motore: 4 cil.	Motor: 4 cil.	Motor: 4 cil.
Bore: 2.9156 ins.	Bohrung: 74 mm	Alésage: 74 mm	Alesaggio: 74 mm	Diámetro: 74 mm	Diâmetro: 74 mm
Piston stroke: 2.955 ins.	Hub: 75 mm	Course: 75 mm	Corsa: 75 mm	Carrera: 75 mm	Curso: 75 mm
Cub. cap.: 78.68 cu. ins.	Zyl.-Inh.: 1290 ccm	Cylindrée: 1290 ccm	Cilindrata: 1290 ccm	Cil. Cub.: 1290 ccm	Cilindrada: 1290 ccm
Compression: 8 : 1	Kompression: 8 : 1	Compression: 8 : 1	Rapp. di compr.: 8 : 1	Compresión: 8 : 1	Compressão: 8 : 1
Brake H. P.: 65	Brems-PS: 65	Ch. au frein: 65	Pot. al freno: 65	Pot. cab. al freno: 65	Pot. ao. freio: 65
R. P. M.: 6000	U/min: 6000	T/min.: 6000	Giri/min.: 6000	Rev./min.: 6000	Rotações/mín.: 6000
Wheelbase: 86.7 ins.	Radstand: 2200 mm	Empattement: 2200 mm	Passo: 2200 mm	Dist. entre ejes: 2200 mm	Dist. entre eixos: 2200 mm
Track: 50.7 ins.	Spurweite: 1286 mm	Voie av. et ar.: 1286 mm	Carreggiata: 1286 mm	Vía delantera/trasera: 1286 mm	Via: 1286 mm
Turning circle: 32.8 ft.	Wendekreis: 10 m	Diam. de braquage: 10 m	Diametro di sterzata: 10 m	Diámetro del círculo de viraje: 10 m	Diâmetro mín. de viragem: 10 m
Tyre size: 155 x 15	Bereifung: 155 x 15	Pneus: 155 x 15	Pneumatici: 155 x 15	Neumáticos: 155 x 15	Pneus: 155 x 15
Radiator capacity: 14 Imp. pts. 8.5 US qts.	Kühlerinhalt: 8 l	Cap. d'eau: 8 l	Acqua radiatore: 8 l	Cap. del radiador: 8 l	Cap. do radiador: 8 l
Engine oil capacity: 8.8 Imp. pts. 5.3 US qts.	Ölinhalt: 5 l	Cap. d'huile: 5 l	Olio motore: 5 l	Cap. del carter de motor de aceite: 5 l	Cap. de óleo: 5 l
Gastank capacity: 11.7 Imp.gal. 14.1 US gal.	Tankinhalt: 53 l	Cap. d'essence: 53 l	Serbatoio benzina: 53 l	Cap. combustible: 53 l	Cap. de gasolina: 53 l
Max. speed: 96.25 m.p.h.	Höchstgeschw.: 155 km/h	Vit. max.: 155 km/h	Velocità mass.: 155 km/h	Velocidad max.: 155 km/h	Velocidade máx.: 155 km/h
Gas cons.: 31.3 mpg Imp. 26.1 mpg US	Benzinverbrauch: 9 l	Consommation: 9 l	Consumo benzina: 9 l	Cons. de carburante: 9 l	Consumo de gasolina: 9 l
Price ex-works: Lit. 1,900,000	Preis ab Werk: Lit. 1.900.000	Prix départ usine: Lit. 1.900.000	Prezzo loco fabbrica: Lit. 1.900.000	Precio desde fabrica: Lit. 1.900.000	Preço desde fabrica: Lit. 1.900.000

ENGLISH	DEUTSCH	FRANÇAIS	ITALIANO	ESPAÑOL	PORTUGUES
Name: Giulietta t. i.	Name: Giulietta t. i.	Nom: Giulietta t. i.	Nome: Giulietta t. i.	Nombre: Giulietta t. i.	Nome: Giulietta t. i.
Type: Saloon	Type: Limousine	Type: Conduite intérieure	Tipo: Berlina	Tipo: Sedan	Tipo: Sedan
Doors: 4	Türen: 4	Portes: 4	Porte: 4	Puertas: 4	Portas: 4
Seats: 4	Sitzplätze. 4	Places: 4	Posti: 4	Asientos: 4	Lugares: 4
Motor: 4 cyl.	Motor: 4 Zyl.	Moteur: 4 cyl.	Motore: 4 cil.	Motor: 4 cil.	Motor: 4 cil.
Bore: 2.9156 ins.	Bohrung: 74 mm	Alésage: 74 mm	Alesaggio: 74 mm	Diámetro: 74 mm	Diâmetro: 74 mm
Piston stroke: 2.955 ins.	Hub: 75 mm	Course: 75 mm	Corsa: 75 mm	Carrera: 75 mm	Curso: 75 mm
Cub. cap.: 78.68 cu. ins.	Zyl.-Inh.: 1290 ccm	Cylindrée: 1290 ccm	Cilindrata: 1290 ccm	Cil. Cub.: 1290 ccm	Cilindrada: 1290 ccm
Compression: 8 : 1	Kompression: 8 : 1	Compression: 8 : 1	Rapp. di compr.: 8 : 1	Compresión: 8 : 1	Compressão: 8 : 1
Brake H. P.: 65	Brems-PS: 65	Ch. au frein: 65	Pot. al freno: 65	Pot. cab. al freno: 65	Pot. ao. freio: 65
R. P. M.: 5500	U/min: 5500	T/min.: 5500	Giri/min.: 5500	Rev./min.: 5500	Rotações/mín.: 5500
Wheelbase: 93.8 ins.	Radstand: 2380 mm	Empattement: 2380 mm	Passo: 2380 mm	Dist. entre ejes: 2380 mm	Dist. entre eixos: 2380 mm
Track: 50.7 ins.	Spurweite: 1286 mm	Voie av. et ar.: 1286 mm	Carreggiata: 1286 mm	Vía delantera/trasera: 1286 mm	Via: 1286 mm
Turning circle: 32.8 ft.	Wendekreis: 10 m	Diam. de braquage: 10 m	Diametro di sterzata: 10 m	Diámetro del círculo de viraje: 10 m	Diâmetro mín. de viragem: 10 m
Tyre size: 155 x 15	Bereifung: 155 x 15	Pneus: 155 x 15	Pneumatici: 155 x 15	Neumáticos: 155 x 15	Pneus: 155 x 15
Radiator capacity: 14 Imp. pts. 8.5 US qts.	Kühlerinhalt: 8 l	Cap. d'eau: 8 l	Acqua radiatore: 8 l	Cap. del radiador: 8 l	Cap. do radiador: 8 l
Engine oil capacity: 8.8 Imp. pts. 5.3 US qts.	Ölinhalt: 5 l	Cap. d'huile: 5 l	Olio motore: 5 l	Cap. del carter de motor de aceite: 5 l	Cap. de óleo: 5 l
Gastank capacity: 8.8 Imp. gal. 10.7 US gal.	Tankinhalt: 40 l	Cap. d'essence: 40 l	Serbatoio benzina: 40 l	Cap. combustible: 40 l	Cap. de gasolina: 40 l
Max. speed: 96.25 m. p. h.	Höchstgeschw.: 155 km/h	Vit. max.: 155 km/h	Velocità mass.: 155 km/h	Velocidad max.: 155 km/h	Velocidade máx.: 155 km/h
Gas cons.: 33.2 mpg Imp. 27.7 mpg US	Benzinverbrauch: 8,5 l	Consommation: 8,5 l	Consumo benzina: 8,5 l	Cons. de carburante: 8,5 l	Consumo de gasolina: 8,5 l
Price ex-works: Lit. 1,520,000	Preis ab Werk: Lit. 1.520.000	Prix départ usine: Lit. 1.520.000	Prezzo loco fabbrica: Lit. 1.520.000	Precio desde fabrica: Lit. 1.520.000	Preço desde fabrica: Lit. 1.520.000

ENGLISH	DEUTSCH	FRANÇAIS	ITALIANO	ESPAÑOL	PORTUGUES
Name: Giulietta Sprint	Name: Giulietta Sprint	Nom: Giulietta Sprint	Nome: Giulietta Sprint	Nombre: Giulietta Sprint	Nome: Giulietta Sprint
Type: Coupe	Type: Coupé	Type: Coupé	Tipo: Coupé	Tipo: Coupé	Tipo: Coupé
Doors: 2	Türen: 2	Portes: 2	Porte: 2	Puertas: 2	Portas: 2
Seats: 2	Sitzplätze: 2	Places: 2	Posti: 2	Asientos: 2	Lugares: 2
Motor: 4 cyl.	Motor: 4 Zyl.	Moteur: 4 cyl.	Motore: 4 cil.	Motor: 4 cil.	Motor: 4 cil.
Bore: 2.9156 ins.	Bohrung: 74 mm	Alésage: 74 mm	Alesaggio: 74 mm	Diámetro: 74 mm	Diâmetro: 74 mm
Piston stroke: 2.955 ins.	Hub: 75 mm	Course: 75 mm	Corsa: 75 mm	Carrera: 75 mm	Curso: 75 mm
Cub. cap.: 78.68 cu. ins.	Zyl.-Inh.: 1290 ccm	Cylindrée: 1290 ccm	Cilindrata: 1290 ccm	Cil. Cub.: 1290 ccm	Cilindrada: 1290 ccm
Compression: 8 : 1	Kompression: 8 : 1	Compression: 8 : 1	Rapp. di compr.: 8 : 1	Compresión: 8 : 1	Compressão: 8 : 1
Brake H. P.: 70	Brems-PS: 70	Ch. au frein: 70	Pot. al freno: 70	Pot. cab. al freno: 70	Pot. ao. freio: 70
R. P. M.: 6000	U/min: 6000	T/min.: 6000	Giri/min.: 6000	Rev./min.: 6000	Rotações/mín.: 6000
Wheelbase: 93.8 ins.	Radstand: 2380 mm	Empattement: 2380 mm	Passo: 2380 mm	Dist. entre ejes: 2380 mm	Dist. entre eixos: 2380 mm
Track: 50.7 ins.	Spurweite: 1286 mm	Voie av. et ar.: 1286 mm	Carreggiata: 1286 mm	Vía delantera/trasera: 1286 mm	Via: 1286 mm
Turning circle: 32.8 ft.	Wendekreis: 10 m	Diam. de braquage: 10 m	Diametro di sterzata: 10 m	Diámetro del círculo de viraje: 10 m	Diâmetro mín. de viragem: 10 m
Tyre size: 155 x 15	Bereifung: 155 x 15	Pneus: 155 x 15	Pneumatici: 155 x 15	Neumáticos: 155 x 15	Pneus: 155 x 15
Radiator capacity: 14 Imp. pts. 8.5 US qts.	Kühlerinhalt: 8 l	Cap. d'eau: 8 l	Acqua radiatore: 8 l	Cap. del radiador: 8 l	Cap. do radiador: 8 l
Engine oil capacity: 8.8 Imp. pts. 5.3 US qts.	Ölinhalt: 5 l	Cap. d'huile: 5 l	Olio motore: 5 l	Cap. del carter de motor de aceite: 5 l	Cap. de óleo: 5 l
Gastank capacity: 11.7 Imp.gal. 14.1 US gal.	Tankinhalt: 53 l	Cap. d'essence: 53 l	Serbatoio benzina: 53 l	Cap. combustible: 53 l	Cap. de gasolina: 53 l
Max. speed.: 96.25 m.p.h.	Höchstgeschw.: 160 km/h	Vit. max.: 160 km/h	Velocità mass.: 160 km/h	Velocidad max.: 160 km/h	Velocidade máx.: 160 km/h
Gas cons.: 31.3 mpg Imp. 26.1 mpg US	Benzinverbrauch: 9 l	Consommation: 9 l	Consumo benzina: 9 l	Cons. de carburante: 9 l	Consumo de gasolina: 9 l
Price ex-works: Lit. 1,925,000	Preis ab Werk: Lit. 1.925.000	Prix départ usine: Lit. 1.925.000	Prezzo loco fabbrica: Lit. 1.925.000	Precio desde fabrica: Lit. 1.925.000	Preço desde fabrica: Lit. 1.925.000

Such perfect proportions would be notable in a car of any size. The top contributes to claustrophobia.

ROAD TEST ALFA VELOCE

IT HAS BEEN nearly two years since we gave a test report on the tiny car with the big name, the Alfa Romeo Giulietta Sprint Spyder. That report was a favorable one, and in the intervening months the car has become quite popular, considering its price and somewhat limited market. However, some months ago, the Turin firm announced a new super-sprint version, to which they have given the name "Sprint Veloce."

The 1300-SV model, with engine modifications, develops 90 bhp, as compared to 65 in the Sprint model and 50 in the regular sedan. A few of the early Veloce cars were built as lighter models, with a weight saving of over 200 pounds as compared to the standard 65-bhp Sprint. However, currently, the 90-bhp Veloce engine is supplied only in the standard-weight body-chassis. Our test car had the Spyder-type body (actually a true convertible coupe), and, as often happens during a production run, it weighed slightly more than our first Spyder test car, with identical-type body.

Also, our test car was supplied to us by a well known West Coast driver, Fred Woodward of Continental Motors in Whittier, Calif. The odometer read 2100 miles; the car had been raced three times. Each time it secured first in class, including one third place overall in the under 1500-cc production class at Riverside. The only apparent effect of this activity on the car was a synchromesh which did not synchro in the 2nd gear and a brake pedal that was slightly closer to the floor than normal. The only non-standard feature of the car was a 4.55 axle in place of the 4.10 unit listed as standard for the Veloce.

The axle change accounts for 11% of the speedometer error, so the makers should not be blamed for the extreme optimism which appears in the tabulation. Also, as shown in the tabulated data, the mean maximum speed recorded (104.6 mph) is not a great deal faster than that of a normal 1300-S, which will reach 100 honest mph under favorable circumstances. However, the axle ratio was changed to give better acceleration on short circuits, and during the best run of 107.1 mph the tachometer was touching 7000 rpm, 300 rpm beyond the recommended limit and 1000 rpm beyond the peak power speed. Incidentally, the speedometer indicated 123 to 125 mph during the best high-speed run.

Our instructions from the owner-driver were to ignore the 6700-rpm red line. "It will go, and has gone, to 8000 rpm, but I normally use 7500," he said. We used 7200 rpm as a limit for the acceleration tests, which unfortunately were somewhat retarded because of the slow shift into 2nd gear. However, as this willing little engine gets into its stride, the acceleration improves substantially over the 65-bhp model. The Veloce version develops more torque (at a higher speed) and the standing start times

The giant knurled gearshift knob is a delight to use.

PHOTOGRAPHY: POOLE

Small packages can go in with the top and the spare.

The trunk is larger than one would expect it to be.

a car has no business being so desirable

are substantially improved to 60, 70, 80 and 90 mph. This additional punch is well illustrated by a comparison of Tapley pull readings in pounds per ton, bearing in mind that the 4th-gear figure is approximately the same on both models because the Veloce engine torque-peaks at a higher speed, where air drag is also appreciably higher.

gear	65 bhp	90 bhp
4th	230	235
3rd	310	330
2nd	430	470
1st	560	off scale

Without a doubt, the engine of this Alfa is its outstanding feature. The Veloce version simply seems to have no drawbacks. It is smooth and peppy, and appears to idle and pull at low speeds as well as the 65-bhp unit. The exhaust system on our test car was definitely raucous, with a rap at over 3000 rpm which was great fun but would have to be toned down considerably for legal city use. And, this is one of the very few engines which can be run up to 7000 rpm and more without wincing—it just doesn't seem to mind at all.

Perhaps the reason for this engine's charm is its silky smoothness, for without a doubt the Alfa 1300 is one of the smoothest 4's ever built. There is just a trace of typical in-line 4-cylinder vibration during deceleration at about 3000 rpm that would not even be noticed unless the driver were extremely critical. Likewise, overhead cam clatter is virtually non-existent, but a hi-fi ear can detect some mechanical noise at idle.

First gear gives a maximum speed of 35 mph when pressed and despite a slight initial flat spot, the Tapley meter goes out of sight with a reading beyond the scale provided. The 2nd-gear syncromesh would work satisfactorily on downshifts with only a blip of the throttle, but upshifts were impossible to achieve quickly without a crunch. We crashed through but lost a few tenths of a second on each trial. Third gear, with its 86-mph maximum, is useful both in town and when balked on the highway. In 4th gear the minimum speed is about 20 mph, but no one in his right mind would try to accelerate hard from this speed without shifting down to at least 3rd, if not 2nd. In fact this is the real charm of the machine, the freedom to treat it like a little lady, or alternatively, to beat the living tar out of it without really being brutal, and with little or no fear of expensive noises.

ROAD & TRACK ROAD TEST NO. 161

ALFA ROMEO 1300 SV SPYDER

SPECIFICATIONS

List price	$3686
Wheelbase, in.	86.6
Tread, f and r	50.0
Tire size, mm.	155-15
Curb weight, lb	2050
Distribution, %	55/45
Test weight	2320
Engine	4 cyl, dohc
Bore & stroke	2.91 x 2.95
Displacement, cu in.	78.7
cu cm.	1290
Compression ratio	9.50
Horsepower	90
peaking speed	6000
equivalent mph	93.0
Torque, lb-ft	86.8
peaking speed	4500
equivalent mph	69.7
Gear ratios, overall	
4th	4.55
3rd	6.16
2nd	8.92
1st	15.1

CALCULATED DATA

Lb/hp (test wt.)	25.8
Cu ft /ton mile	76.0
Engine revs/mile	3870
Piston travel, ft/mile	1905
Mph @ 2500 ft/min.	78.9

PERFORMANCE, Mph

Top speed, avg.	104.6
best run	107.1
3rd (7500)	86
2nd (7500)	59
1st (7500)	35
see chart for shift points	
Mileage range	24/32 mpg

ACCELERATION, Sec

0-30 mph	3.8
0-40 mph	6.8
0-50 mph	9.7
0-60 mph	14.1
0-70 mph	18.5
0-80 mph	23.7
0-90 mph	30.6
standing start ¼ mile	17.6

TAPLEY DATA, Lb/ton

4th	235 @ 63 mph
3rd	330 @ 55 mph
2nd	470 @ 45 mph
1st	off scale
Total drag at 60 mph, 84 lb	

SPEEDOMETER ERROR

Indicated	Actual
30 mph	25.0
40 mph	33.5
50 mph	42.0
60 mph	50.8
70 mph	59.1
80 mph	67.4
90 mph	76.2
100 mph	85.1
123 mph	107.1

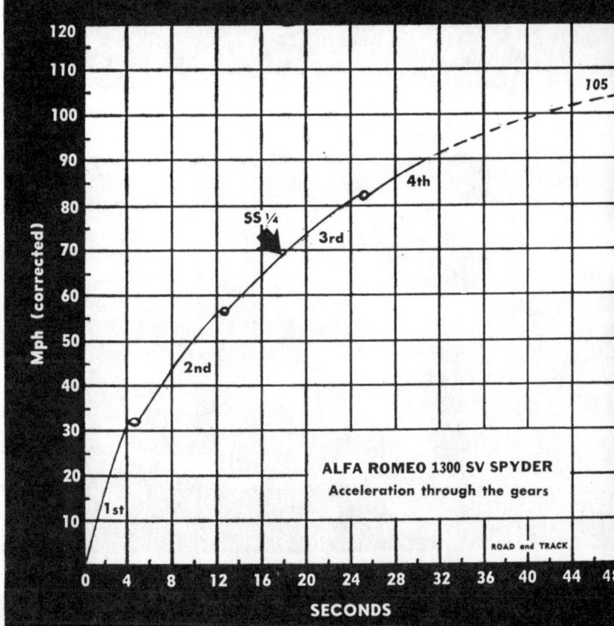

ALFA

Front and rear bumpers, though beautiful, would give little protection in a fracas with a bigger car. Inclined slightly from the vertical, the small double-overhead-cam engine lets the hood slope sharply. Meticulous finish is evident.

On the other side of the ledger, the deceleration department is fully capable of dealing with the performance capabilities. As hinted earlier, the brakes were untouched since new, and produced no strange odors from the finned bi-metal brake drums, though the shoes did need adjusting.

As for comfort, the Giulietta is typical sports car in its ride, and though not at all harsh it is at its best at high cruising speeds. Even over rough, undulating highways it sits on the road surface as if it weighed 2½ tons, which proves the lie in the popular American misconception that "a heavy car sticks to the road." And, of course, it hardly seems necessary to add that the Veloce is one of the best handling sports cars you can find, at any price. With its forward weight bias and low roll center i.f.s. it understeers, but this characteristic as well as the roll angle are both moderate, at least until corners are approached at the absolute ragged edge. The steering is very pleasant with just over 3 turns lock to lock, and a wonderfully convenient turning circle of only 30 feet.

The interior seats two only and cushions which appear to be very close to the floor are actually quite comfortable with the proviso that headroom is somewhat restricted. The seating position feels a little strange at first, but after a few minutes everything feels just right, making this an easy, natural car to drive. There is a heater, which we did not try, and winding windows which we did find convenient on a winter day in California. The cloth top folds simply and easily with no fussing or fuming, though not in a matter of seconds or with one hand—or via a pushbutton.

As for economy we obtained one overall check on 209 miles of varied driving at 24.1 miles per gallon. Despite the improved performance a fuel rate of 30 mpg is achievable under favorable conditions at a steady 50 mph, and a similar figure should be possible at 60/65 mph with the standard Veloce 4.10 axle. Incidentally, the standard axle ratio would also increase the acceleration times in the lower speed ranges, but the honest top speed should improve to possibly 112 mph.

Italian cars are typically different in many ways from the products of other countries, and the Alfa in any model is well put together, beautifully finished, and sheer pleasure to drive. We returned the bright red Veloce with real regret.

IMPORTANT NEWS ABOUT

ALFA ROMEO!

FACTORY PRICE REDUCTIONS...

Save up to $892 now on your new ALFA ROMEO—price cuts made possible by increased production to meet the rising U. S. demand.

GIULIETTA SPIDER	GIULIETTA COUPE	GIULIETTA VELOCE SPIDER	GIULIETTA VELOCE COUPE
Previously $3533	Previously $4195	Previously $4578	Previously $5084
NOW $3298	NOW $3784	NOW $3686	NOW $4194

ALL PRICES P.O.E.

TWO GIULIETTA SERIES COUPES:
Sprint Coupe 1300 cc, 100 mph.
Super Sprint Coupe (VELOCE), 90 hp, 115 mph.

TWO GIULIETTA SERIES SPIDERS:
Sprint Spider, 1300 cc, 96 mph.
Super Sprint Spider (VELOCE), 90 hp, 115 mph.

See your local dealer, or call U. S. Distributor

HOFFMAN

NEW YORK: 433 Park Avenue, PLaza 9-5120
BEVERLY HILLS, California, 9130 Wilshire Blvd., CRestview 4-8231

The Motor Road Test No. 12/58 (Continental)

Make: Alfa Romeo **Type:** Giulietta T.I.

Makers: Alfa Romeo S./A.: via M.U. Traiano, Milan, Italy.

Concessionaires: Thomson and Taylor Ltd., Portsmouth Road, Cobham, Surrey.

Test Data

CONDITIONS: *Weather: Wind under 10 m.p.h., dry. (Temperature: 41–48° F., Barometer: 29.6 in. Hg.). Surface: Tarmac. Fuel: 98–100 octane petrol.*

INSTRUMENTS

Speedometer at 30 m.p.h.	18% fast
Speedometer at 60 m.p.h.	19% fast
Speedometer at 90 m.p.h.	17% fast
Distance recorder	5.4% fast

WEIGHT

Kerb weight (unladen, but with oil, coolant and fuel for approx. 50 miles)	18¼ cwt.
Front/rear distribution of kerb weight	52/48
Weight laden as tested	21¼ cwt.

MAXIMUM SPEEDS

Flying Quarter Mile

Mean of four opposite runs	98.3 m.p.h.
Best one-way time equals	100.8 m.p.h.

"Maximile Speed" (Timed quarter mile after one mile accelerating from rest).

Mean of four opposite runs	94.8 m.p.h.
Best one-way time equals	97 m.p.h.

Speed in gears (at 6,000 r.p.m.)

Max. speed in 3rd	75 m.p.h.
Max. speed in 2nd	52 m.p.h.
Max. speed in 1st	31 m.p.h.

FUEL CONSUMPTION

39 m.p.g. at constant 30 m.p.h. on level.
39 m.p.g. at constant 40 m.p.h. on level.
36½ m.p.g. at constant 50 m.p.h. on level.
32 m.p.g. at constant 60 m.p.h. on level.
30 m.p.g. at constant 70 m.p.h. on level.
24½ m.p.g. at constant 80 m.p.h. on level.
22 m.p.g. at constant 90 m.p.h. on level.

Overall Fuel Consumption for 85 miles, 18 gallons, equals 27 m.p.g. (10.4 litres/100 km.)

Touring Fuel Consumption (m.p.g. at steady speed midway between 30 m.p.h. and maximum, less 5% allowance for acceleration) .. 29½ m.p.g.
Fuel tank capacity (maker's figure) .. 9 gallons

STEERING

Turning circle between kerbs:

Left	35 feet
Right	35½ feet
Turns of steering wheel from lock to lock	3¼

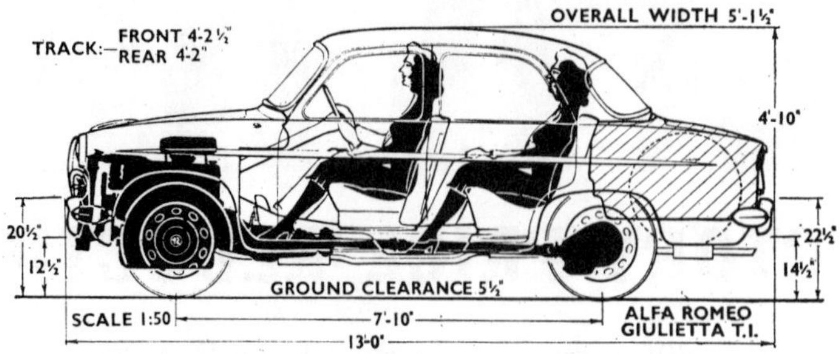

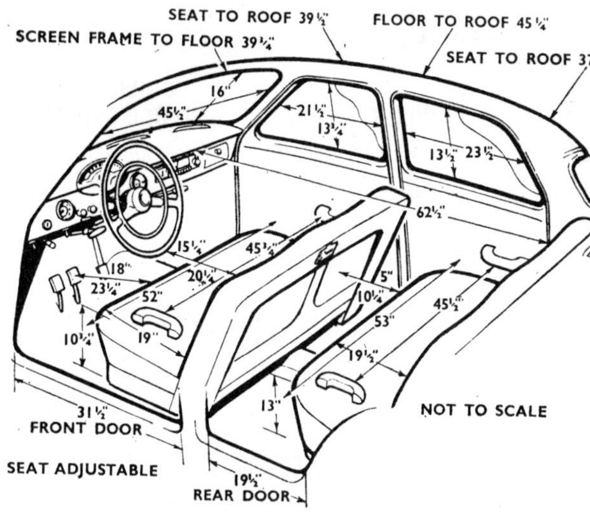

ACCELERATION TIMES from standstill

0–30 m.p.h.	4.6 sec.
0–40 m.p.h.	7.5 sec.
0–50 m.p.h.	11.5 sec.
0–60 m.p.h.	18.0 sec.
0–70 m.p.h.	22.9 sec.
0–80 m.p.h.	37.4 sec.
0–90 m.p.h.	51.4 sec.
Standing quarter mile	19.9 sec.

ACCELERATION TIMES on upper ratios

	Top	3rd
20–40 m.p.h.	13.2 sec.	9.0 sec.
30–50 m.p.h.	11.9 sec.	8.4 sec.
40–60 m.p.h.	12.9 sec.	9.4 sec.
50–70 m.p.h.	15.2 sec.	9.4 sec.
60–80 m.p.h.	19.5 sec.	—
70–90 m.p.h.	24.8 sec.	—

BRAKES from 30 m.p.h.

1.00g retardation (equivalent to 30 ft. stopping distance) with 110 lb. pedal pressure.
0.77g retardation (equivalent to 39 ft. stopping distance) with 75 lb. pedal pressure.
0.50g retardation (equivalent to 60 ft. stopping distance) with 50 lb. pedal pressure.
0.16g retardation (equivalent to 190 ft. stopping distance) with 25 lb. pedal pressure.

HILL CLIMBING at sustained steady speeds

Max. gradient on top gear	1 in 10.5	(Tapley 215 lb./ton)
Max. gradient on 3rd gear	1 in 7	(Tapley 315 lb./ton)
Max. gradient on 2nd gear	1 in 5	(Tapley 470 lb./ton)

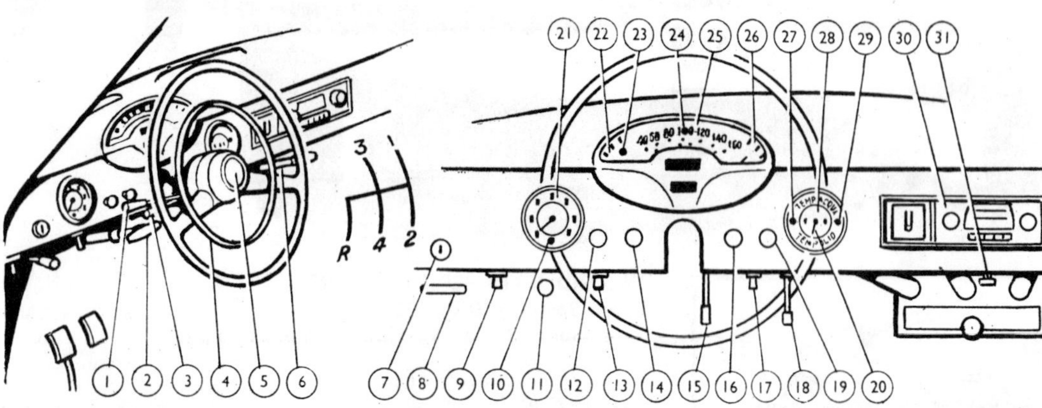

1, Headlamp dip-switch lever. 2, Handbrake. 3, Direction indicator control. 4, Horn ring. 5, Headlamp flasher button. 6, Gear lever. 7, Ignition and starter switch. 8, Bonnet catch release. 9, Windscreen wipers control. 10, Dynamo charge warning light. 11, Windscreen washer control. 12, Lights switch. 13, Panel light switch. 14, Choke control. 15, Hand throttle. 16, Fresh air control. 17, Heater control. 18, Trip re-setting knob. 19, Heater/ventilator fan switch. 20, Oil temperature gauge. 21, Tachometer. 22, Fuel contents gauge. 23, Low fuel level warning light. 24, Direction indicator warning light. 25, Speedometer and distance recorder. 26, Oil pressure gauge. 27, Main beam warning light. 28, Water thermometer. 29, Heater fan indicator light. 30, 31, Radio controls.

The Alfa Romeo Giulietta T.I.

An Economical Family Four-seater with Outstanding Performance and Safety Factors

ANY car in the under-1½-litre class which can reach 50 m.p.h. from rest in less than 15 seconds, 70 m.p.h. from rest in under 30 seconds, which will combine a true road speed of 85 m.p.h. with a fuel consumption of, say, 27 m.p.g., can be considered in the top class so far as acceleration, speed and fuel economy are concerned.

When therefore it is claimed that a 1.3-litre car has reached 50 m.p.h. in less than 12 seconds, 70 m.p.h. in under 23 seconds, and has combined 27 m.p.g. with but a fraction under 100 m.p.h. (a magic figure that can actually be exceeded in slightly favourable conditions), the reader's reaction may well be that of incredulity. He may think that distances and instruments were in error, or if reassured on this point, that some serious sacrifices have been made such as the use of exceedingly low-drag, light-weight body construction at the expense of longevity and comfort, or by fitting an engine so highly tuned that the peak of performance is only a passing phase whereas inflexibility, noise, and vibration are permanent features.

In the case of the Alfa Romeo Giulietta T.I. recently tested by *The Motor*, it is scarcely necessary to state that all the instruments were carefully calibrated. As usual two stopwatches were used simultaneously to record the times and as a first step it was discovered that the speedometer had a constant speed error of between 18 and 19% up to 60 m.p.h., afterwards falling to 14% at 90 m.p.h., at which speed therefore the needle was indicating substantially over 100 m.p.h. Despite the remarkable exaggerations the slowest time recorded over a known km. was the equal of 95 m.p.h. and the fastest very nearly 101 m.p.h., giving a mean of no less than 98.3 m.p.h. Of even greater consistency, the times for the quarter mile following a standing-start mile varied by less than 5% and in the case of the exceedingly rapid standing quarter-mile tests the stopwatch variation was less than 1%.

Accepting, then, the validity of the figures we must seek to explain them. This is indeed a hard task, for although the power claimed by the makers represents nearly 50 b.h.p./litre the actual 63 h.p. available to propel the car are no more than is offered by several competitive types. Moreover, the acceleration times are not derived from a particularly favourable power to weight ratio, as the Giulietta goes on the scales at 18¼ cwt. and as tested weighed a little over 21 cwt. Dimensionally, also, an overall width of 5 ft. and a height of 4 ft. 7 in. are normal, permitting an interior width of 52 in. over the front and 55 in. over the rear seats, whilst the wheelbase of 7 ft. 10 in. permits the back passengers 9 in. of leg room when the driver has 19 in. between the pedals and the leading edge of his seat.

Italian cars have a world reputation for sleek lines with the expectation of low wind drag, but it must in fairness be said that this applies only to the specialized two-seater, two-door bodies which are the particular forte of the Italian coachbuilder; the normal production four-door saloon bodies are trimly shaped and show evident attention to the problems of wind resistance in detail, but the Giulietta in no way sacrifices comfort or convenience in the cause of aerodynamics.

Externally it is neat rather than beautiful and internally the fabric seats and metal finished facia do not offer immediate

In Brief

Price (including heater and screen washer as tested) £1,460 plus purchase tax £731 7s. equals £2,191 7s.

Capacity	1,290 c.c.
Unladen kerb weight ...	18¼ cwt.
Acceleration:	
20-40 m.p.h. in top gear ...	13.2 sec.
0-50 m.p.h. through gears	11.5 sec.
Maximum direct top gear gradient	1 in 10.5
Maximum speed	98.3 m.p.h.
"Maximile" speed	94.8 m.p.h.
Touring fuel consumption ...	29.5 m.p.g.

Gearing: 17.0 m.p.h. in top gear at 1,000 r.p.m., 34.5 m.p.h. at 1,000 ft./min. piston speed.

PLAINLY PRACTICAL.—The sensibly arranged facia panel, the large Delaney-Gallay heater unit, also the small-diameter, elegantly designed steering wheel of the car are shown above and to the right can be seen that the rear seats offer adequate leg- and ample elbow-room for two persons.

appeal to the eye, although the seats give good support including a degree of side location despite the bench-type layout. The arcuate speedometer is hard to read as well as being inaccurate, but the provision of a tachometer together with thermometers, both oil and water, is much to be commended. The instruments are all clearly visible and a feature deserving praise is the centre button in the steering wheel which flashes the headlamps to make the visual signals to other cars which, on the Continent, are fast displacing the audible warning of approach which remains under the control of the conventional horn ring.

In front of the passenger-carrying compartment of the car the main organs of the engine are readily accessible and behind there is an adequate luggage locker with a lightweight support for the locker lid.

In respect of furniture, fittings and general amenities, the Giulietta does not rise above the average for this class of car and certainly it offers no more than may be expected in view of the list price.

This comment may also be applied to smoothness, silence, and top-gear ability. The car cannot climb a 1-in-10 gradient without recourse to third speed, and here it is appropriate to note that the steering column gear change is by no means the best of its kind and that the standing start acceleration figures would be even better than they are if not marred by a slow shift from the second to the third gear position.

As is often the case with Continental cars the indirect ratios are widely spaced, but within the normally used range of 10-60 m.p.h. thereon the engine was extremely smooth. On top gear, however, vibrations were noticed above 60 m.p.h. and these may have emanated from the rather long propeller shaft, despite the fact that this has a centre steady bearing. Noise increased in the upper ranges of speed, to merit serious criticism as one felt that, on this ground alone, road speeds above 80 m.p.h. would not long be sustained and thus that the long-distance average speeds of the car would fall below expectations based upon general performance and road ability.

It is impossible to attribute this high noise level to any particular component or factor. The exhaust note is not unduly obtrusive and the engine is reasonably quiet mechanically, although when cold it suffers from the rather old-fashioned disease of piston slap. With an 8 : 1 compression ratio it is necessary to use premium fuel but if this is done pinking is absent without recourse to the hand control on the ignition timing which is provided adjacent to the steering column. Here also may be found a hand throttle with which the engine can be set on a fast idle for warming up or run light at high r.p.m. during tuning operations. These should be required rarely, if ever, for although the engine used on the T.I. develops 63 h.p. compared to 50 b.h.p. of the standard model, other versions of the same unit deliver no less than 90 b.h.p., so there is an evident margin of stamina and reliability.

If some aspects of the car are no better than they should be, and some deserving of criticism, when we turn to the realms of safety, control and roadworthiness we have again to speak in superlatives.

The braking figures show that 100% retardation is provided by the Girling system, given 110 lb. pedal pressure; what the figure does not show is that a combination of large diameter but stiff light-alloy drums, and a friction-lining area of 125 sq. in./laden ton, gives a virtual immunity from fade even under extreme conditions such as a fast descent of over

BEFORE AND BEHIND.—The neatly designed twin o.h.c. 63 b.h engine offers excellent accessibility to the carburetter, fuel pur sparking plugs, fuse box and valve gear. In the tail the lugga capacity is rather greater than suggested by external appearan

SIMPLICITY.—Classic columns of an Italian square set off the simple but balanced lines of this very high-performance family car.

3,000 ft. with no assistance from the engine. If braking power be recognized as exceptional, cornering and control can well be thought unrivalled. The small diameter steering wheel seems in exactly the right place, a reaction obtaining with persons of widely varying dimensions and physique. The pedals are adequately spread yet permit the easy manipulation, if an inappropriate word may be permitted, of the brake and accelerator pedal by the heel and toe of one foot. A few yards of travel display the delicacy of the steering, and the sensation of a completely positive series of links between the driver's hand and the road wheels, but there is a natural mental reaction that such supremely easy response, savouring indeed of power steering, must be bought at the cost of serious over-steer; and knowledge that castor angle is virtually nil heightens a suspicion that the car may become progressively unstable on the open road.

In fact the car has excellent straight-running qualities at speed and, with increasing sideways force, demonstrates substantial under-steer on corners. One has therefore the remarkable combination of high sensitivity at low speeds, on limited lock, with complete stability at high speeds or with large angles of lock. In this fashion the driver, centaur like, becomes part of the car which he handles with such added skill and confidence that he rises a class, or perhaps two, above his normal form.

The fact that the rear axle is positively located in relation to the frame doubtless plays a part in this technical victory and the general feeling of safety is enhanced by the extremely stiff structure of the vehicle which permits rough roads to be negotiated without trace of shake. On the other hand, the relatively large mass of the live rear axle necessitates a stiff, high-frequency, suspension system, and back-seat passengers find reading difficult, and writing impossible.

But if the Alfa Romeo Giulietta T.I. cannot claim to be the most comfortable car in its class it is, in the opinion of experienced testers, the most roadworthy production car in the world. As further matters of opinion there are a few smarter and better-looking cars in this class, and a considerable number with more stylish and more luxurious internal appointments. In statistical fact there are a limited number offering better fuel consumption.

There are none which can in speed or acceleration approach the brilliant performance of the Giulietta T.I. which can thereby claim the best combination of utility, running economy, safety and speed in the world today.

Specification

Engine

Cylinders	4
Bore	74 mm.
Stroke	75 mm.
Cubic capacity	1,290 c.c.
Piston area	26.65 sq. in.
Valves	Double o.h. camshaft
Compression ratio	8/1
Carburetter	Solex AIPAIG
Fuel pump	Fispa mechanical
Ignition timing control	Vacuum
Oil filter	Fram
Max. power (gross)	63 b.h.p.
at	6,000 r.p.m.
Piston speed at max. b.h.p.	2,950 ft./min.

Transmission

Clutch	Single plate
Top gear (s/m)	4.55
3rd gear (s/m)	6.16
2nd gear (s/m)	8.92
1st gear (s/m)	14.25
Reverse	15.32
Propeller shaft	Open two piece
Final drive	Spiral bevel
Top gear m.p.h. at 1,000 r.p.m.	17
Top gear m.p.h. at 1,000 ft./min. piston speed	34.5

Chassis

Brakes	Girling
Friction lining area	132 sq. ins.
Suspension:	
Front	Wishbone and coil i.f.s.
Rear	Coil springs and radius arms (rigid axle)
Shock absorbers	Girling telescopic
Steering gear	ZF
Tyres	Pirelli 155—15

Coachwork and Equipment

Starting handle	None
Battery mounting	In rear boot
Jack	Post type
Jacking points	On side of body

Standard tool kit: Radiator drain tap spanner, jack, plug spanner and bar, carburetter crowfoot spanner, wheel brace, axle and gearbox filler, air intake wrench, socket wrench (carburetter), plug spanner, tool roll, wheel disc lever.

Exterior lights: 2 headlamps, 2 side lamps, 2 rear stop, tail and reversing lamps, 1 number plate lamp.

Number of electrical fuses	6

Direction indicators: Flashing type, incorporated in side lamps, rear lamps and wing side lamps (6).

Windscreen wipers	Electric, self cancelling, two blade
Windscreen washers	Tudor
Sun vizors	2 cloth covered

Instruments: Speedometer with decimal trip, tachometer, petrol gauge, oil pressure gauge, oil and water temperature gauges.

Warning lights: Ignition, direction indicators, heater, petrol level, lamps on.

Locks:	
With ignition key	Door
With other keys	Rear boot, door and glove box
Glove lockers	1 on passenger's side
Map pockets	In front doors
Parcel shelves	Behind rear squabs
Ashtrays	1 dashboard, 1 rear of front seat
Cigar lighters	None
Interior lights	Roof
Interior heater	Yes
Extras available	Radio
Upholstery material	Cloth and leather cloth
Floor covering	Carpet
Exterior colours standardized	6
Alternative body styles	None

Maintenance

Sump:	$8\frac{3}{4}$ pints S.A.E. 40 summer, 30 winter
Gearbox	$2\frac{3}{4}$ pints, S.A.E. Dentax 90
Rear axle	$2\frac{1}{2}$ pints, S.A.E. Spirax 90 EP
Steering gear lubricant	Shell Spirax 90 EP
Cooling system capacity	13 pints (2 drain tap)
Chassis lubrication:	By grease gun every 1,250 miles to 23 points
Ignition timing:	Fixed 8° Max. advance at 5,000 to 5,200—44°
Contact-breaker gap	0.014-0.015 in.
Sparking plug type	Lodge HLN
Sparking plug gap	0.024-0.027 in.

Valve timing: Inlet opens 22° b.t.d.c., inlet closes 65° a.b.d.c., exhaust opens 55° b.b.d.c., exhaust closes 12° at.d.c.

Tappet clearances (cold) Inlet	.017 in.
Exhaust	.019 in.
Castor angle	0° 10′
Steering swivel pin inclination	8° 35′
Tyre pressures:	
Front	Pirelli $19\frac{1}{2}$, Michelin 18.2 lb.
Rear	Pirelli 21, Michelin $19\frac{1}{2}$ lb.
Brake fluid	B.P.

Battery type and capacity: Tudor (Italian) 12-volt type 6 SP 3F, 38 amp. hr.

SCI ROAD TEST: ALFA ROMEO 1300 TI

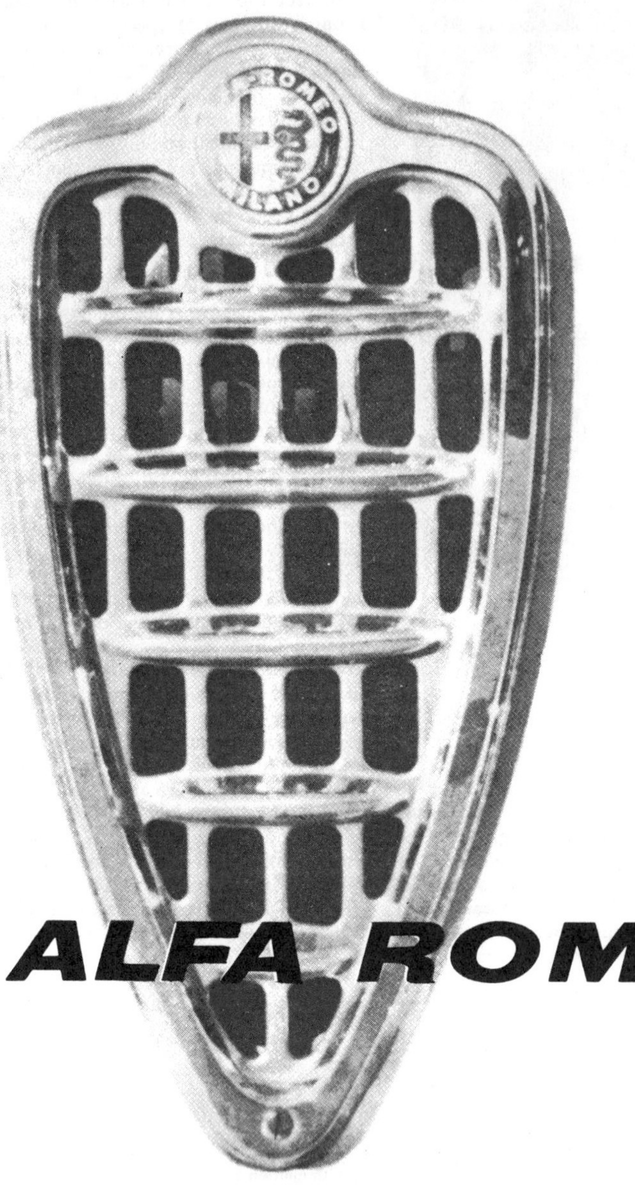

INTRODUCED at the Frankfurt Motor Show in 1957, the Alfa Romeo "TI" Berlina is a four-door 1300 Giulietta fitted with the 65 hp Sprint engine. With the introduction of the "Tourismo Internationale", Alfa has added a fast machine designed specifically for international touring car competitions in 1958.

Remember the last Mexican Road Race in 1955, when a team of the famous Alfa 1900 "TI" machines swept the board in their class as well as showing their heels to numerous higher powered cars in the mountains? Well, Alfa Romeo is at it again, going after the 1300 "family car" laurels with this souped-up, sawed-off little bomb that is a tremendous surprise from behind the wheel.

Mario Bernasconi, Alfa press relations chief, put a TI test car at our disposal for a week to do a tour of Italian factories. We picked it up at the Lugano (Switzerland) distributor, and with over 5000 miles already on the clock, it was obviously a well broken-in factory test car in an excellent state of tune.

The machine is basically a Giulietta from the ground up. Chassis, suspension and running gear are the same as the Sprint, except for the four-door sedan body welded to the platform frame. Distinguishing marks are hard to find. A plate on the dash says that it is a TI but there's nothing on the exterior. A half-pint rev counter is set into the left side of the dash an easy glance away by the driver. The rest of the instruments are simple and well grouped, especially easy to read at night when illuminated. With the 100 mph speedo are fuel gauge, oil gauge, and a fuel warning light. In a separate cluster at the right are the oil and water temperature gauges. Toggle switches actuate wipers, heater blower and panel light while conventional pull controls operate choke and interior air distribution, which operated well in the cold Italian winter.

The most pleasant surprise was the discovery of a hand throttle just under the steering column. This proved to be a

Really barreling this family sedan into a turn on the Modena Autodromo, *the lock on the wheels indicates understeer. Increasing pressures in the tires improved handling.*

Although the TI uses the same chassis and running gear as the Sprint coupe and Spyder roadster, it is a very roomy 4-door sedan.

Plenty of room for passengers and their luggage. Trunk opens upwards, exposes Pirelli-fitted spare, tools and space.

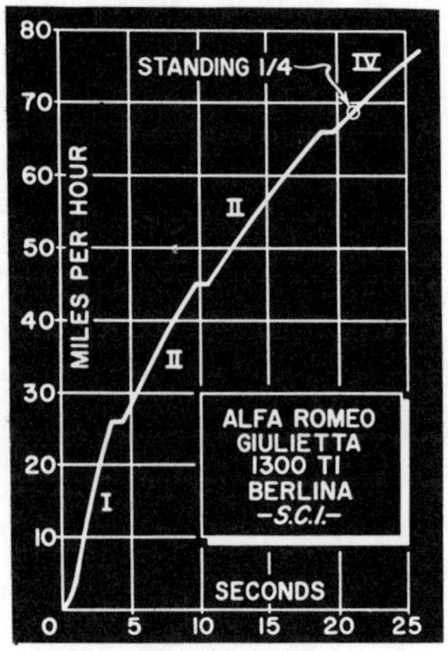

big help when warming up the car after a freezing night. Instead of having to stay in the car and spend a minute playing footsie with the accelerator pedal, you could step up the idle and go back inside for a second cup of coffee. It seems to us that any sports-touring car should have a hand throttle, and Alfa is to be congratulated for providing this small but very convenient control. Turn indicators and a hand-operated dip switch complete the instrumentation.

Driving position was comfortable and pedal placement so ideal that it was almost a pleasure to heel and toe. Alfas are famous for this, even the pre-war machines had closely spaced pedals, but in those days the accelerator was between the clutch and brake. Thought had gone into seat design of the TI also, for there were slight naturally-fitting side braces in the seat squab offering a lot of lateral stability.

Physically, the TI lacks the beauty of line that the Sprint possesses: it has a stubby appearance that detracts from its Italian birthright. But like most Italian products, appearance does not necessarily indicate what is below the surface. The TI proved to be a lot of fun, from the spot behind the wheel.

It is surprising how much space Alfa has been able to provide inside their four-door Giulietta, for it is still a small car. It has the same wheelbase as the Sprint series, but there is ample room for four passengers and their luggage. The trunk is surprisingly large. Storage compartments inside the car are small: the glove compartment and the map pockets in the doors are too narrow to be of much use.

The hood latch was hard to find, but was finally located on the firewall directly over the pedals. An orthodox safety catch up front released the hood, revealing the compact 65 hp Sprint engine sitting well down in the engine room. There is easy access to plugs, oil dip stick and the like, but the battery is in the trunk away from engine heat. Instead of the single-throat Solex that comes on the conventional

(Continued on page 85)

ALFA ROMEO GIULIETTA 1300TI BERLINA

PERFORMANCE

TOP SPEED:

Two-way average	90 mph

ACCELERATION:

From zero to	Seconds
30 mph	5.4
40 mph	8.1
50 mph	12.4
60 mph	16.3
70 mph	21.7
Standing ¼ mile	21.1
Speed at end of quarter	69 mph

SPEED RANGES IN GEARS:

I	0-26 mph
II	6-45 mph
III	10-66 mph
IV	22-top

FUEL CONSUMPTION:

Hard driving	23 mpg
Average driving (under 60 mph)	27-28 mpg

BRAKING EFFICIENCY:

(10 successive emergency stops from 60 mph, just short of locking wheels)

Stop	Per cent
1	59
2	59
3	57
4	59
5	57
6	59
7	50
8	50 right rear locking
9	50 right rear locking
10	59

POWER UNIT:

Type	In line 4
Valve Arrangement	80° Dohc
Bore & Stroke	2.85 x 2.89 in (74 x 75 mm)
Stroke/Bore Ratio	1.01/1
Displacement	78.7 sq in (1290 cc)
Compression Ratio	8/1
Carburetion by	Solex 35 APAI-G
Max. Power	65 bhp @ 5500 rpm
Max. Torque	79.5 lb ft @ 3500 rpm
Idle Speed	1000 rpm

DRIVE TRAIN:

Transmission ratios	
I	3.31
II	1.96
III	1.35
IV	1.00
Final drive ratio (test car)	4.55
Other available final drive ratio	4.10
Axle torque taken by	radius arms and A bracket

CHASSIS:

Wheelbase	93.7 in
Front Tread	50.6 in
Rear Tread	50.0 in
Suspension, front	Coils and wishbones, anti roll bar
Suspension, rear	Coil springs, radius arms, A-bracket
Shock absorbers	Telescopic
Steering type	Worm and roller
Steering wheel turns L to L	3.5
Turning diameter	36 ft
Brake lining area	139.5 sq ins
Tire size	155 x 15 (equiv to 5.60 x 15)

GENERAL:

Length	157 in
Width	61 in
Height	55 in
Weight, test car	2050 lbs
Weight distribution, F/R	49/51
Weight distribution, F/R, with driver	48/52
Fuel capacity	14 U. S. gallons

RATING FACTORS:

Bhp per cu in	0.83
Bhp per sq in piston area	2.41
Torque (lb-ft per cu in)	1.01
Pounds per bhp — test car	31.6
Piston speed @ 60 mph	1830 fpm
Piston speed @ max bhp	2640 fpm
Brake lining area per ton (test car)	136 sq in

TI engine is fitted with one double-choke downdraft Solex.

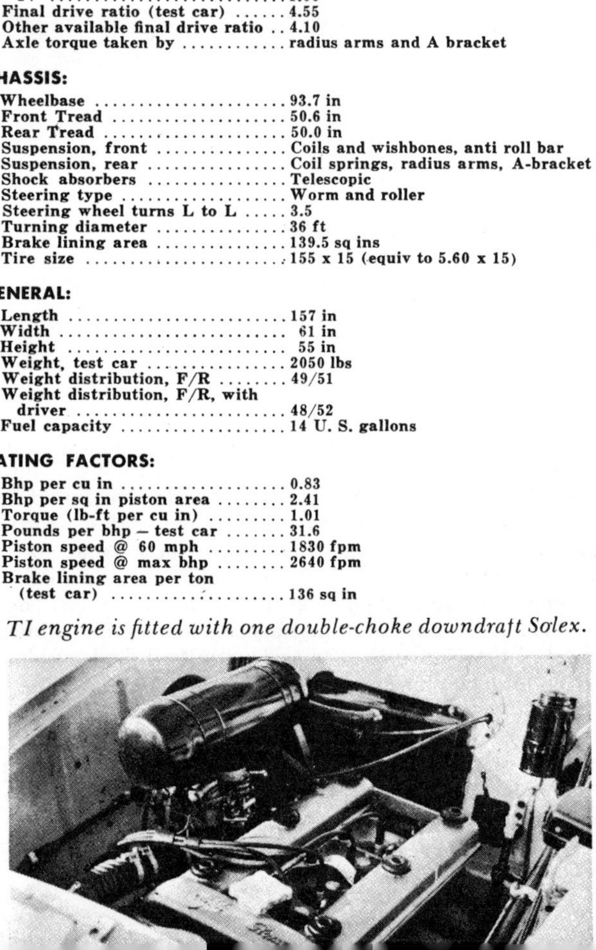

Alfa addicts place car's grillework on the same artistic level as a painting by Di Vinci.

Upright grille links modern Alfa to past examples of the marque. For racing hood is secured with straps through four top fittings.

Giulietta sings around bend in the upper fifties with commendable stability. Body assumed angle early in curve—kept same degree of lean all the way through corner.

The little machine with the long name proves to be one used car equally at home in the parking lot or on the starting grid.

IF you subscribe to the belief that an automobile must reflect the nature of the people who designed and built it, it is not difficult to understand why the Giulietta is both an exciting and excitable motor car.

A short anecdote will reveal the Alfa Company's approach to things automotive. When the 1930 P 3 Grand Prix car was trampled into the tar by the rampaging Mercedes and Auto Union racers, the Alfa firm found itself overstocked with highly tuned, fantastically expensive GP engines. A less sporting outfit might simply have retired them to the company museum. Not so Alfa. For their engine they built a new auto—a sports car chassis which would stand up to the GP's power—and then sold the new units to Italians who had both money and courage.

This all-out sporting heritage is immediately evident when you slip behind the wheel of the Giullietta. The seat with its dished cushion and curved back holds the driver firmly, allowing him to use the steering wheel for control rather than for support.

The man-sized knob topping the centrally located gear shift lever falls readily to hand, while the tach and speedo dials are mounted up high where they can be read without the driver's consciously having to take his eyes from the road. There can be no doubt that the driving position of the Alfa is meant for people who like to get in a car and go from point A to point B in the shortest possible time.

by Harry Fanelli

USED CAR

ALFA ROMEO GIULIETTA SPRINT SPYDER

It took a while to get the starter to react to the key-start mechanism—possibly a bad connection, or faulty switch—but once the juice got to its destination the little double overhead cam mill crackled into life without a cough or sputter.

With a cold engine, the revs had to be held around 1500 to keep from stalling. As the mill warmed up, it would sit and idle around 950 with very little noise coming from the two mufflers and resonator.

When he gave me the rundown on this particular 1957 Alfa, the former owner stressed the fact that it was fitted with Veloce cams and special pistons giving it a 9:1 compression ratio. The Veloce cams, tailored for racing, are wilder than those on the stock Giulietta and probably account for the high idling rate when cold.

Without this bit of background information my first impression of the Alfa on the road would have been mildly disappointing. Around town—at modest revs—it seemed that we were being out dragged by any VW that caught us at a stop light.

The steering, brakes and general feel of the car on the road were, however, wonderful and convinced me that it was a desirable piece of property. While closing the deal with the owner I mentioned that the car didn't seem to have much torque. Rather wistfully, I watched him pocket my check while he explained that he always ignored the 6700-rpm red line. His statement recurred to me when I took the car to a deserted road shortly after it became mine. I decided to find out if the car's elusive torque really lurked in the never-never land of 7000 rpm.

Apparently, I had been dealing with an honest Alfa owner. When I forgot about the 6700 rpm limit, first showed 30 mph; second, 50; third, 75; and fourth, a shade over 100. The engine just did not seem to have any point at which it became fussy; the harder you pressed with your right foot, the higher the revs without strain or pain. Reflecting afterwards on the engine's ability to operate unflustered at a rev range meant for racing cars, it occured to me that this is typical of most Italian vehicles—from buses to minute sedans.

Manufacturers in Italy have to face the fact that their products will be raced—if not formally in a recognized event, then most certainly in casual dust-ups between two happy Latins who happen to be going in the same direction on a bit of road. Woe unto the builder of a *machina* that pokes a piston through its hood and lets a novice Nuvolari down.

The more I drive the Giulietta the more I become enamoured of it. It is difficult to decide which is the best feature of the car—its engine or brakes. Both are quite a few notches above similar components on most medium-priced sports cars. The three-inch-wide brake drums are bimetallic and finned—both features helping to dissipate frictional heat. The braking surface is inordinately large for a car weighing only 2320 lbs. The brakes bring the car to a safe, unswerving stop from any speed the Alfa is capable of, and keep on doing it long after your foot gets tired from punching the pedal.

A silky-smooth, rev-happy engine and tremendous brakes would be of little use if they weren't mounted on a road-worthy chassis, and that is just what the little Alfa has. A very un-vintage like approach was taken by the Alfa engineers when they decided to adopt a monocoque construction for the Giulietta series. That the Alfa's punt-like chassis-body was a wise choice of the design staff is proved by the way the car handles over choppy roads. There is no cowl shake or squeaks or rattles; and the entire shell is free from twisting or weaving. This immensely strong structure provides a stable base for both front and rear suspension.

The Alfa's ability to cling to the road stems in part from the ability of the wheels to work undisturbed by chassis flexing. Coil springs, front and rear, and a solid rear axle tied down by twin traction rods at the ends and a husky A-frame—with two legs mounted on the chassis pan and the point attached to one side of the differential—comprise the other reasons for the Giulietta's limpet-like road holding.

When cornered hard the Alfa has a marked roll—which is evident only to outside observers. You just don't seem to notice when you are behind the wheel or in the passenger seat.

The ZF steering box gives two and one half turns lock to lock, transmits no shock from the road, and has absolutely no play. The steering is unusual in that it requires very little effort, whether for small corrections or full-lock, full-bore bends.

Giuliettas all exhibit marked understeer. My particular version of the marque with its extra power—74 hp as opposed to 65 in stock trim—can be steered with the throttle. Neat foot work on the loud pedal can convert this semi-race Alfa to a neutral steering machine. All of this only adds to the fun-power—besides making the car fantastically easy to control in any type of corner.

As with any man-made machine the Alfa has faults—although I think most of them are minor. Under sudden acceleration on rough surfaces there is some rear axle bounce, although to be truthful, this doesn't seem to faze the road holding one bit. The hand brake wouldn't hold my son's coaster wagon on a slight downgrade; but then why does an *automobili da coursa* need a handbrake? On our test car the muffler rapped the belly pan when going over dips in the road, and the right tail-light lens is broken. After six months or so of ownership two other faults have cropped up: the paint requires constant attention to keep it a bright Italian racing-red, and the battery supplied by the factory died without putting up much of a fight.

To offset these bad features the car has veloce cams, 9:1 pistons, cleaned up ports, and electric pump to prevent fuel starvation at high revs. One

other asset to put in the credit column—and it is an important quality—is the car's flexibility. The better half of the Fanelli family can putter the Alfa through town without double clutching or alerting the law with a racing exhaust system. On the weekends I can play Stirling Moss with the car without having to spend hours changing bits and pieces to make it race worthy.

To sum up: anyone with a yen for a Ferrari, and the price of an Alfa shouldn't hesitate. You can have a car that has—like most Italian cars at any price—all the frustrating little faults, but one which also provides pure, race bred road performance at a fraction of the tag on a Ferrari. •

Alfa top is unusual among sports car lids in that it is leak proof, wrinkle free, and not unattractive. Additionally—it doesn't drum at high speeds.

Sprint scrambles for traction as it is thrown into right angle turn. Loose surface didn't seem to faze car's suspension in the least.

The joys of owning a sports car are fully demonstrated here as Alfa slices its way through a herd of Detroit behemoths.

Black crackle finish on cam covers typifies workmanship lavished on Alfa's engine. Little power plant is one of the smoothest fours ever built.

SPECIFICATIONS CHART

Alfa Romeo Giulietta Sprint (Used)

PERFORMANCE:

Acceleration thru the gears

0-30	4.0 secs.
0-40	6.9 secs.
0-50	9.8 secs.
0-60	16.0 secs.
Top speed	103 mph

FACTORY SPECIFICATIONS:

No. of cyls.	4
Arrangement	in-line.
Valve system	double overhead cam.
Head	Light alloy with hemispherical combustion chambers.
Bore and stroke	2.91 x 2.95 in.
Displacement	78.7 cu. in. (1290).
Comp. ratio	9:1
Max. bhp @ rpm	74 @ 6000.
Max. torque @ rpm	80 lbs/ft @ 4500
Carburetion	Webber side draft
Electrical system	12 volt

CHASSIS:

Unitized body and chassis

Suspension front	Independent with coil springs.
Suspension rear	Solid rear axle positioned by track rods and coil springs.
Steering	Worm and Roller
Brake lining area	132 in.

DIMENSIONS:

Wheelbase	86.6 in.
Tread front	50 in.
Tread rear	50 in.
Overall length	151 in.
Width	62 in.
Height	49 in.
Ground clearance	7½ in.
Turning circle	27 ft.
Steering wheel turns lock to lock	3
Tire size	155 x 15
Gas tank capacity	12 gal.

REFERENCE FACTORS:

BHP per cu. in.	.94
Weight	2320 lbs.
lbs. per bhp	31.4
Piston speed @ peak rpm	
Weight distribution	55 percent on front wheels.
Gas consumption (hard driving)	24.1 mpg.
Price	$2500

SCI ROAD TEST:

super spider

Photos by Don Typond

"OF COURSE it's a road machine. Alfas have a long racing history, true, but the factory also has to sell cars, and to do this we have to give customers a decent ride and enough power at the bottom end so they can handle traffic", said Karl Grassow, Service Director of Hoffman Motors. Later, Lou Comito (Huntington Imported Cars), owner and usual driver of our test car, added, "Karl is right. The car is a production car and is designed as a road car. This particular one was set up for *go;* but with a few simple detuning devices we use it in close traffic, too. And you wouldn't want a better machine to take to the corner drug store or around Lime Rock!" This about sums up our impressions of the Super Spider.

The term Super Spider is the terminology preferred by the factory. Actually, the SS is the newest, and by far the hottest, in the Veloce series. But let's take a look at Alfa offerings one at a time.

The production Giulietta Sprint engine is fitted with a single Solex two-stage carburetor, which with an 8.0 compression ratio produces 70 hp @ 6000, as it appears in the least expensive Spider. The other increment is the Veloce or Super engine with dual Webers and 9.1 CR producing 103 hp @ 6000 rpm, SAE (identical with the previously publicised 90 Italian). Our Alfa began to red line at 6700, which according to the factory is the absolute safe maximum; however Lou Comito said that for unsustained operation we could go as high as 7200 rpm.

Coming down the chute at Lime Rock and accelerating into the straight, it occurred to us that 1300 ccs unblown shouldn't go this fast, and certainly shouldn't accelerate so savagely. The Alfa Super Spider is by far the fastest 1300 the writer has driven, one of the sweetest handling, and the car in which he felt the most confident at high speed on the twisty Lime Rock bends. After a few careful laps, one pounds up to the corners and literally throws the car at them. The car will forgive you if you overextend; the "line" is not nearly so important as it is with less tolerant cars. To the onlooker, the car may appear dreadfully out of shape because of its heavy lean, but the driver is undisturbed by it and always feels that the car is completely the slave to his slightest whim, which it is.

Once, roaring down the straight at one-hundred-plus, the writer "chickened-out", taking the escape road. The next time around, the approach was not much better, but a stab at the brake and a hard turn of the wheel threw the tail out to put the car into a perfect attitude for the sweeping right hander. Where it tightens up at the end, a dive into the apex gives the impression of extreme velocity, but an educated toe at work on the accelerator and hands that move quickly over but short distances on the wheel will seem to get one around any turn every time. As we said before, the lean, though appreciable, is imperceptable to the driver.

Lean, or body roll, has a bad reputation that it doesn't deserve. It is the result of phenomena that have certain very

Diving into apex at speed, Super Spider leans heavily but doesn't worry the driver. Tail eases out gradually.

Graceful, elegant Farina body, the ideal compromise, blending road comfort, pride with racing functionality.

Engine is tilted a few degrees to make room for dual Webers. Collection box is fed through tubing over engine. Instrumented interior has quality tone about it.

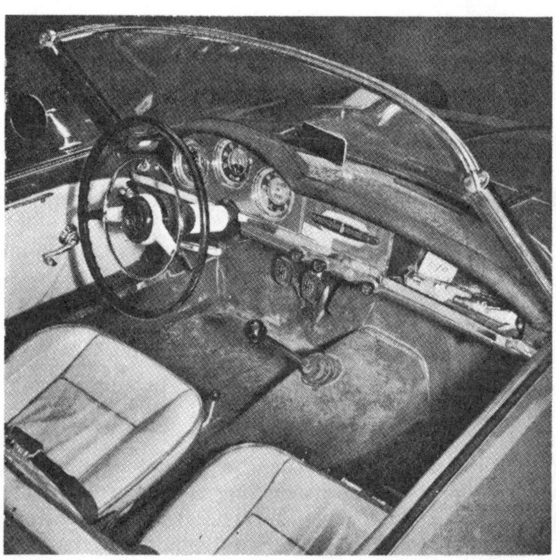

important advantages – among which is a gradualness in change of attitude. The Alfa rolls under really hard cornering, loading up the outside tires heavily. At the same time, the tail slides out gently, breaking the rear away so gradually that there is ample time to apply throttle and make steering corrections. In other words, you don't have to be a great driver to "press on" through a corner.

Now, let's take a look at the suspension that contributes to this excellent roadability. The front *ifs* is comprised of a lower wishbone with a single ball-jointed upper control arm. The bottom of the soft coil spring is based on the A-frame, the spring running diagonally upward and inward into a frame member. A tubular-type double-acting shock absorber is anchored inside the coils. Upward wheel travel is limited by the conventional rubber bumper; downward travel by a flexible cable anchored between the frame and the lower wishbone. Each front wheel is fitted with a monstrous aluminum-bonded diagonally-finned brake drum that almost completely fills the recess inside the wheel, which is vented to allow the passage of cooling air.

The rigid rear axle is anchored by two coil springs with concentric tubular double-acting shocks inside. Under extreme loading, the axle will bottom out against rubber bumpers, and two flexible metal straps limit the rebound. The axle is located by a radius rod on each side that attaches to a bracket on the underside of the axle, and anchors to the chassis at a forward point. For sideways location, there is a welded-tube triangular-shaped bracket located above the axle immediately to the left of the differential. This fastens to the frame at two points; the third point terminates at a ball joint attached to the axle housing. At either end of the axle, and nestling inside each vented wheel, is a large drum, slightly smaller than the front and vertically finned for cooling.

As may be expected from a car designed for competition, the brakes are impeccable. The writer made at least a dozen fast laps at Lime Rock, really boring into the corners before standing on the brake, often at 100 mph speeds, and there was never any indication that the brakes were being overworked. Fade was nil, and there was always the same constant high-decelleration with very little pressure on the pedal. The car is made to stop at whatever speed the engine will drive it.

Thirteen hundred cubic centimeters is not the displacement of a big engine. By American standards, it's about equal to the volume of *two cylinders* of a modern vee-eight; even by European standards the engine is on the small side

Most of the loading transfers to the outside tires, insuring lots of bite.

Left to right: Alfin drum, upper control arm, spring-shock, lower A-frame with flex rebound arrester, exhaust headers.

Rear live axle is suspended by soft coil springs, located by sturdy radius rods. Pirelli "Cinturato" tires are standard.

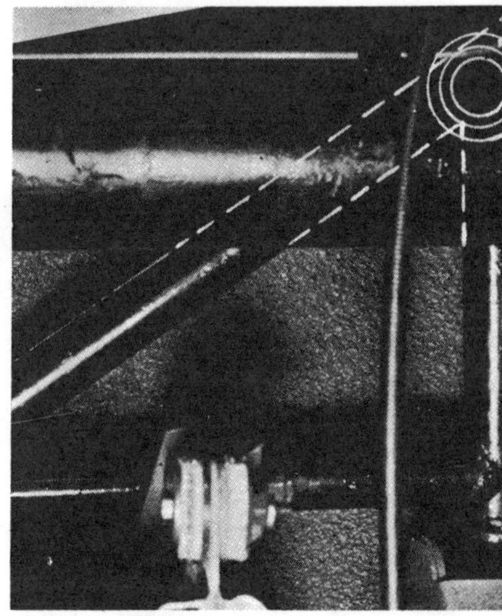

Side location of rear axle is controlled by triangular bracket that attaches to frame, ball jointed at axle (drawn).

Alfa Romeo Super Spider Continued

of average. How, then, can it develop 103 hp and move a full-sized sports car through violent acceleration to better than 100 mph speeds? Well, let's take a look at the engine.

The engine is an in-line four that is just about square, with an edge of a single millimeter going toward the stroke. Compression ratio is 9.1 to 1; but perhaps the 1.28 hp/cu. in. output is possible because of the double overhead chain-driven camshafts that give the engine what seems like unlimited rpm. Revs in any gear are limited only by the amount that the driver is willing to overrev. Stroke is approximately three inches; therefore the piston speed in ft/min numerically approximates one half the rpm.

In the carburetion department, two dual-throat Webers sit on the right side of the engine, which is slightly angled to the left of the vertical to make room for the Webers. These carbs are attached by a small manifold that acts, in effect, as a spacer, and the carbs follow the tilt of the engine.

The jets in our test car were a bit too big; the engine would flood and choke up under 3000 revs, and on any sharp left hand turn. To accelerate cleanly out of the only left hander at Lime Rock, it was better to use 2nd gear so that the engine would clear as quickly as possible. Lou has since found the correct combination.

Fresh cool air is ducted from a scoop on the left through flex tubing into a large filter mounted atop the rear of the engine. Another flex tube feeds from the filter into a collection box closed over the air openings of the carbs, supplying cold air.

Exhaust is ducted from the bottom side of the tilted engine through headers that form into a four-into-two, two-into-one that leads into a small expansion chamber. Normally, another large chamber is located amidship, and there is a still smaller one immediately before the end of the pipe. It is a very efficient way to eliminate high back pressure at high revs without shattering the ear drums of the people in the cockpit. However, Lou's car is "SCCA-stock" which means straight pipes all the way. With the racket-buster off, the exhaust took on a beautiful hammering note up around 6000.

The driver sits in a spacious bucket seat with the shifting lever at the spot where his right hand would fall naturally

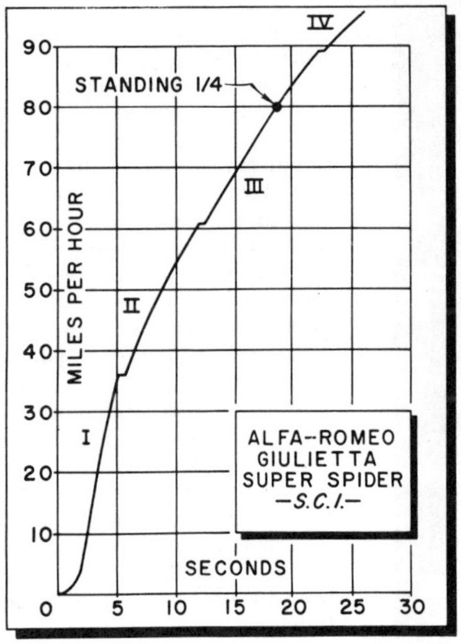

Huge diagonally-finned Alfin front drum almost completely fills inside wheel. Wheel dished, liberally vented, for air flow.

(The transmission is synchronized in all four forward gears). The instruments are hooded to prevent reflection on the windshield, and are contained within three circular faces. To the left are the small instruments — amps, heat, generator and fuel; center gauge is the tachometer; to the right the speedometer-odometer combination. Seats are fully adjustable fore and aft, and the windows are of the roll-up variety. Even though the car goes, it doesn't do so at the expense of road comfort.

The car is not without fault; no machine is. The one thing that impressed us is that this particular model is downright unhappy creeping through traffic unless it is detuned. The cam, carburetors, and clutch much prefer to be given their head. But to say that we liked the Super Spider is not enough; it is a truly marvelous machine. It has enough power on tap to sustain a drift: handling is superlative; predictable and forgiving. The interior has a quality tone about it, and the body, designed and built by Farina, has not sacrificed graceful elegance to the expediency of competition styling. It is unquestionably worth the moderate ($3685 POE N. Y.) asking price.

Len Griffing

ALFA ROMEO GIULIETTA "SUPER SPIDER"

Price	$3686
U.S. Importer	Hoffman 443 Park Ave., New York 22, N.Y. 9130 Wilshire Blvd., Beverly Hills, Calif.

PERFORMANCE

ACCELERATION:

From zero to	Seconds
30 mph	4.4
40 mph	6.5
50 mph	8.8
60 mph	11.7
70 mph	15.4
80 mph	18.8
90 mph	23.3
Standing ¼ mile	18.8
Speed at end of quarter	80 mph

SPEED RANGES IN GEARS: (2000-6800 rpm)

I	11-36 mph
II	18-61 mph
III	26-89 mph
IV	35-top

SPEEDOMETER CORRECTION:

Indicated Speed	Timed Speed
30	25
40	35
50	43
60	52
70	63
80	73

SPECIFICATIONS

POWER UNIT:

Giulietta	In-line four cylinder, water-cooled
Valve Operation	Double overhead cams, vee-inclined valves
Bore & Stroke	2.91 x 2.95 in (74 x 75 mm)
Stroke/Bore Ratio	1.01/1
Displacement	79 cu in (1290cc)
Compression Ratio	9.1/1
Carburetion by	Two Weber 40DCO3 twin choke side-drafts
Max Power	103 bhp (SAE) @ 6000 rpm (90 net bhp)
Idle Speed	1200 rpm

DRIVE TRAIN:

Transmission ratios	
I	3.31
II	1.96
III	1.35
IV	1.00
Final drive ratio	4.10 (4.55 optional)
Axle torque taken by	Radius rods upper A-frame
Platform frame	Welded to body panels
Wheelbase	86½ in
Tread, front and rear	50 in
Front suspension	Coils, lower wishbone, single upper arm, anti-roll bar
Rear suspension	Rigid axle, coil springs, lower radius rods, upper A-bracket
Shock absorbers	Girling telescopic
Steering type	ZF worm end roller
Steering wheel turns L to L	3
Turning diameter, curb to curb	33 ft
Brakes	Finned, bi-metallic drums, 2LS front
Brake lining area	139.5 sq in
Tire size (metric)	155 x 380 (equiv. to 5.60 x 15)

GENERAL:

Length	134 in
Width	61 in
Height	51 in
Fuel capacity	14 U.S. gallons

RATING FACTORS:

Specific Power Output (SAE)	1.34 bhp/cu in
Piston speed @ 60 mph	1765 ft/min
Speed @ 1000 rpm in top gear	17.7 mph

alfa romeo giulietta spider

9 DUAL-PURPOSE SPORTS CARS FOR 1959

IT WAS William Shakespeare who had Romeo say of Giulietta that "A rose by any other name would smell as sweet." Indeed, these sweet little cars from Milano are known by a myriad of names; Spyder, Sprint, Berlinetta, Spyder Veloce, Sprint Veloce and now Super Spyder. Perhaps it would clarify things a bit to straighten out this semantic spaghetti.

super spider

Upper car is normal-engined Spider, lower one is more powerful Spider Veloce (now called Super Spider). Power available drastically alters handling and roll.

The first three cover the basic body styles available; roadster, coupe and sedan. The next two were originally reserved for lightened, aluminum-panelled versions of the first two, complete with Weber-carbureted editions of the husky twin-cam mill. We ran tests on each of them in SCI, in December '56 and February '57 respectively. Subsequently the Alfa factory quit exporting the lightened jobs to the States (seems they weren't considered stock) but continued to use the same names for the all steel jobs when they were fitted with the high output engine. This served to confuse just about everybody. The latest step may help sort things out, though it's a mite late. The Spyder Veloce (steel) is now sold as the Super Spyder and, as mentioned in our October road test, it experienced a paper power increase from 90 to 103 hp by changing from CUNA to SAE rating. Clever, these Italians.

Clever indeed, for the two Giuliettas loaned to us by the brothers O'Brien of Paterson, N. J. are remarkable cars indeed. Both are fairly early examples, Tom's red Spyder Veloce being slightly senior to Jim's white Spyder. Both therefore enjoy slightly stiffer suspension than is now offered. Both too, have been brought up to "SCCA stock" specifications which interestingly enough entails more changes to the Spyder than to the SV. The factory obligingly issued a list of suggested steps to enliven the Spyder's cooking engine and not surprisingly, Jim has followed every one of them.

Accordingly, he milled the head .040 inch, matched the ports to correct misalignment—enlarging to the size of the gasket hole is in this case not a drastic step at all—and polished the ports to a "glass-like finish." He replaced the 22 and 24 mm venturis with 25 and 27 mm ones (as on the 190SL, the twin-chokes are opened progressively and therefore have unequal venturis), raised the float level and fitted larger main jets. The secondary throttle is opened by manifold vacuum working against a weight on a little lever; this weight was reduced by half.

The shims under the valve springs were brought up to Veloce specs to allow higher revs. For structural reasons, no machining should be done on the flywheel to lighten it as the clutch pressure plate attaches at the rearmost surface of its outer rim.

One thing that was unchanged is the ring and pinion set. Both cars have 41 tooth ring gears, but the normal Giulietta has a nine tooth pinion giving a 4.55 final drive ratio. The Veloce, or Super Spyder, comes with a ten tooth pinion. For the short twisty circuits so prevalent in this country, the 4.55 really seems to be the best for either car.

But Tom's red Veloce still has the 4.10, and it was interesting to note the difference in power plants and gear ratios such that we were reaching the same revs in each car at different landmarks on the Lime Rock circuit. Since some of these were just after sharp corners, this bears some looking into. Can it be that two cars with identical suspensions corner at different speeds?

Yes, it can. In actual fact, the red car was running "sticky" retreads on Pirelli racing carcasses. Its shock absorbers had been recalibrated to make them damp a little more fiercely, but this does not even begin to indicate the vast difference in these two cars when cornering them near the limit.

Both of them sit somewhat high and therefore roll considerably. Both need lots of steering lock, and at full bore the front end begins to slide out. Let's face it, these Alfas understeer in a big way but coming off the gas pedal controls it easily and safely.

The principle difference between these sister cars is this: despite the float level in the normal Giulietta having been raised just as high as can be, its engine just plain runs out of gas on hard corners. When this happens, it is just as if the throttle had been suddenly and completely closed; the nose drops slightly and dives for the inside of the corner. As I said, it happens suddenly. You find yourself frantically unwinding the wheel while hopefully jabbing at the throttle to get the accelerator pump to squirt a little gas into the venturis.

On the other hand, with the Veloce there is no interruption in the power flow. In fact, the extra power enables the rear end to be worked out despite the stickier treads. Though the car still rolls extravagantly, the whole process of cornering seems to be spruced up. The tires scream still, the engine roars even more loudly, and cornering technique is now a happy blend of balancing throttle position with steering wheel movements.

While economy may dictate the purchase of a Spyder, those who want to really get something out of racing this handsome breed are well advised to save up for the Super Spyder. For those who are not interested in such strenuous competition—G production seems to be *the* competitive class these days—the extra $388 for a power pack seems a bit heavy. But those who are going to race will find it hard to be satisfied with an engine that cuts out in the corners. It's cheaper to buy what you want in the first place than to modify later. The two Weber carbs are $212 each, the manifold is $81. Parts for converting *completely* to Veloce specs run over $1600 while an assembled Veloce engine is only $1400. Apparently everything but the crankshaft is slightly different!

Either way, these 1300 cc machines are delightful cars to drive on the road and are easily the most attractive looking small sports car available. But a word of caution: when tuned for racing they loose their ladylike docility. They *can* be driven on the road; but when there's traffic it's closer to a chore than a pleasure.

Upon reflection it does seem odd that the rules of most sports car clubs prohibit all but the most minor changes to the suspension. Less stringent rules here could make the cars safer on *both* highway and race track. Today's rules allow unlimited alterations to jets and venturis which make some "stock" sports cars undrivable at low and medium speeds. A strange situation for organizations dedicated to dual-purpose cars and improving the breed.

sfw

Veloce engine is tilted slightly more to make room for two dual choke Weber carbs.

Double-barreled single Solex tends to "lean" out in long, hard turns.

Super Spider (Veloce) will always have lead over its kid sister.

Photos by Don Typond

IN ONE WORD—BEAUTIFUL! The sleek, simple lines of the little Alfa make it undoubtedly one of the loveliest cars ever made.

JOHN BOLSTER TESTS

The Alfa Romeo Giulietta

The Sprint Veloce version of one of the world's most desirable cars

THERE are certain cars which are acknowledged throughout the motor sporting world as being thoroughbreds, and among these the Alfa Romeo Giulietta stands very high. Indeed, I would go so far as to suggest that, if a poll of the readers of AUTOSPORT were undertaken, the Sprint Veloce version would be voted just about the most desirable car that money can buy. This much-coveted model is the subject of the present road test.

As the concessionaires, Messrs. Thomson and Taylor of Cobham, Surrey, did not have a demonstrator of the latest Veloce type, I borrowed an immaculate example, which had only recently been run in, from a private owner, with their full approval. I have to thank Richard Shepherd-Barron for the loan of his delightful car, and for providing me with a memorable experience.

The Sprint Veloce is a two-seater coupé by Bertone of superb appearance, and in "Alfa-red" it really looks a picture. It is a practical car, with wide doors and ample luggage space, while a very large fuel tank allows for long non-stop journeys. The beautiful finish of every detail is a joy to behold. The machine is small and compact, but it is extremely sturdily constructed for continuous high-speed use on Continental roads. It is therefore not exceptionally light.

The engine is a magnificent piece of work. The sturdy counterbalanced crankshaft runs in five main bearings, an almost unheard-of refinement in an engine of only 1.3-litre capacity. The bore and stroke are virtually "square", and the three-ringed domed pistons run in wet liners. Naturally, the light-alloy head has the "four-per-pot" arrangement of holding-down studs, which is standard on all the best 2-o.h.c. engines.

The camshafts operate the inclined valves through inverted pistons. A two-stage duplex roller chain drive is employed, with a jockey sprocket on the upper run. The steel-backed main and big-end bearings are of lead-bronze, with a lead-indium overlay. There is a large oil filter, and the light alloy sump has what is, in effect, an oil radiator beneath it. This is a separate ribbed compartment, containing a labyrinth through which all the oil must pass. It is detachable for cleaning.

The Veloce engine is rated at over 90 b.h.p. at 6,000 r.p.m., but it can run considerably faster than this. It differs from the 70 b.h.p. Sprint engine in having a higher compression ratio and extremely radical camshafts, giving no less than 64 deg. of overlap. It also has two Weber twin-choke horizontal carburetters, which are fed from a cold-air box.

In the past, the gearbox has been the least satisfactory feature of the Alfa Romeo. Now, an entirely new box, with Porsche-type synchromesh on all four speeds, has been standardized. The propeller shaft is in two sections, the front end having rubber universal joints for damping. The rear axle is very light, being built-up with steel side tubes and a light alloy centre. It is located positively by a pair of sturdy trailing arms beneath, and above, just to the left of the light alloy centre section, there is a tubular triangular member with very widely spaced pivots. Obviously, immense trouble has been taken in locating the axle so that cornering and torque reaction forces may be well resisted.

Helical springs and telescopic dampers are the suspension medium all round. In front, the wishbones have a very wide base to give excellent stability to the car. The present steering column could not readily be transferred to the right, without interposing several universal joints in the shaft. Indeed, the Veloce engine has to be canted 8 deg. to the left, for the big Weber carburetters fill the whole of the right side of the bonnet. The brake drums, with turbo-fins, are literally enormous for the size of the car. The two pairs of exhaust pipes do not join until they are about to enter the first of the two silencers.

This Alfa Romeo is thus a small sporting car of immense technical interest. When I took my seat, I found that entry is easy, and that the all-round visibility is excellent, aided by the deletion of triangular ventilating panels from the front windows. The rear windows may be opened slightly outwards to give a draught-free extractor effect. The driving position is very "Dr. Farina".

The engine takes a few moments to fire regularly on a cold morning, and twin batteries had been fitted to the test car for this reason. At all other times it is an instant starter, and idles regularly almost at once. At moderate speeds, the smoothness and silence are literally beyond belief, which I put down largely to the five-bearing crankshaft. As the revs. mount, a hard and purposeful note is heard, though the unit is

ALTHOUGH a good deal of chromium is used at the front the result is pleasing and not vulgar.

POWER UNIT: The twin-cam, 1,290 *c.c. engine develops* 94.8 *b.h.p. at* 6,000 *r.p.m., which is a lot of horses from so small an engine!*

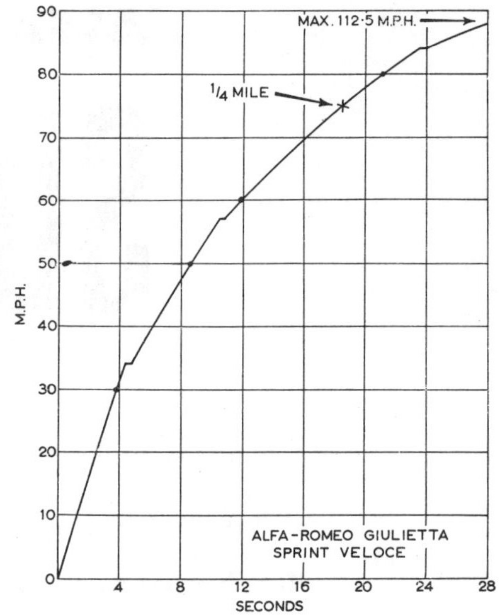

Acceleration Graph

never noisy, and remains smooth throughout the range. The red mark on the rev.-counter dial begins at 6,750 r.p.m.

Naturally, a race-bred unit of this type does its best work above 4,000 r.p.m. It behaves impeccably in traffic, however, if one uses a little consideration with the accelerator pedal. The performance figures, as shown on the data panel and the graph, are really outstanding for a 1.3-litre car. However, I treated the clutch gently and observed the rev.-limit scrupulously as I was driving a private owner's pride and joy. With a little more brutality, my figures could be beaten.

The maximum speed of 112.5 m.p.h. is tremendous, and the little car certainly felt as though it was travelling. As is usual, the speedometer became rather gay at the top end, indicating as much as 125 m.p.h. The engine was running well within its limit and seemed entirely happy to maintain this considerable velocity. Around 90 m.p.h. is a very pleasant cruising speed.

The Alfa handles particularly well, especially on wet roads. The cornering power is very high, and one gets a splendid feeling of being fully in command of the little machine. There is a certain amount of roll, but this is scarcely noticed from inside the car. The suspension, steering and roadholding are all of a very high standard. The rear suspension may appear complicated, but it achieves results that could never be approached by a conventional semi-elliptic layout. Larger cars can equal the performance figures of the Veloce, but they cannot be flung around with the abandon which one may employ with this machine.

It is necessary to emphasize that this is a meticulously constructed luxury car, and not an ultra-light machine which will need rebuilding after one season. The controls work beautifully, and in this connection one must give full marks

(*Continued on page* **82**)

SPECIFICATION AND PERFORMANCE DATA

Car Tested: Alfa Romeo Giulietta Sprint Veloce 2-seater coupé, price £2,698 7*s*. including P.T.

Engine: Four cylinders 74 mm. x 75 mm. (1,290 c.c.). Twin overhead camshafts. 8.5 to 1 compression ratio. 94.8 b.h.p. at 6,000 r.p.m. Two twin-choke Weber carburetters. Coil and distributor ignition.

Transmission: Single dry plate clutch. Four-speed gearbox with central lever and synchromesh on all gears. Open two-piece propeller shaft. Hypoid rear axle.

Chassis: Platform chassis in unit with steel body structure. Independent front suspension by wishbones, helical springs, and telescopic dampers, with torsional anti-roll bar. Worm and roller steering with three-piece track rod. Rear axle on lower trailing arms and single upper triangulated member, with helical springs and telescopic dampers. Girling hydraulic brakes, 2 L.S. in front, in large light alloy turbo-finned drums. Bolt-on pierced disc wheels fitted 155 x 15 Pirelli tyres.

Equipment: 12-volt lighting and starting. Speedometer. Rev.-counter. Oil pressure and temperature, water temperature, and fuel gauges. Self-parking windscreen wipers and washers. Self-cancelling flashing indicators. Heater.

Dimensions: Wheelbase 7 ft. 6½ ins. Track 4 ft. 2 ins. Overall length 13 ft. 0½ in. Width 5 ft. 0½ in. Weight 18¾ cwt.

Performance: Maximum speed 112.5 m.p.h. Speeds in gears: 3rd 84 m.p.h., 2nd 57 m.p.h., 1st 34 m.p.h. Standing quarter-mile, 18.4 secs. Acceleration: 0-30 m.p.h. 3.8 secs., 0-50 m.p.h. 8.6 secs., 0-60 m.p.h. 11.8 secs., 0-80 m.p.h. 21.2 secs.

Fuel Consumption: 25 m.p.g. (approx.).

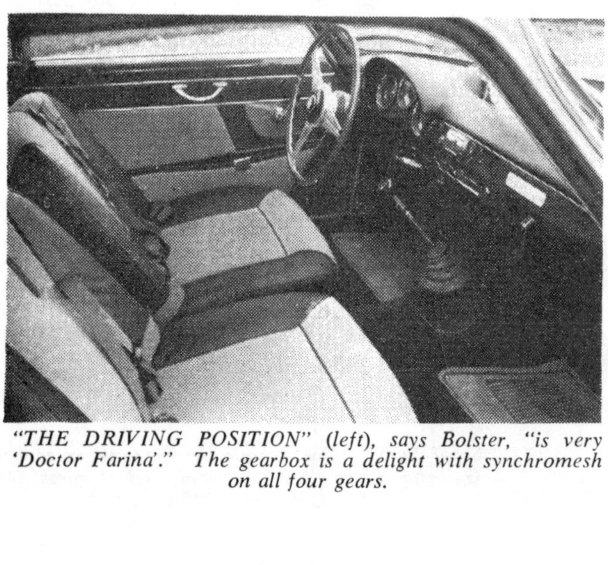

"THE DRIVING POSITION" (left), says Bolster, "is very 'Doctor Farina'." The gearbox is a delight with synchromesh on all four gears.

THE BOOT is not very large, and the spare wheel takes up a lot of space, but there is plenty of room behind the seats.

ALFA GIULIETTA S.S.:

SUPERCHARGER,

A hundred brake and 120 per from a tiny blob of a thing you could almost put in your pocket. Man, how we envy that Continental Correspondent of ours . . .

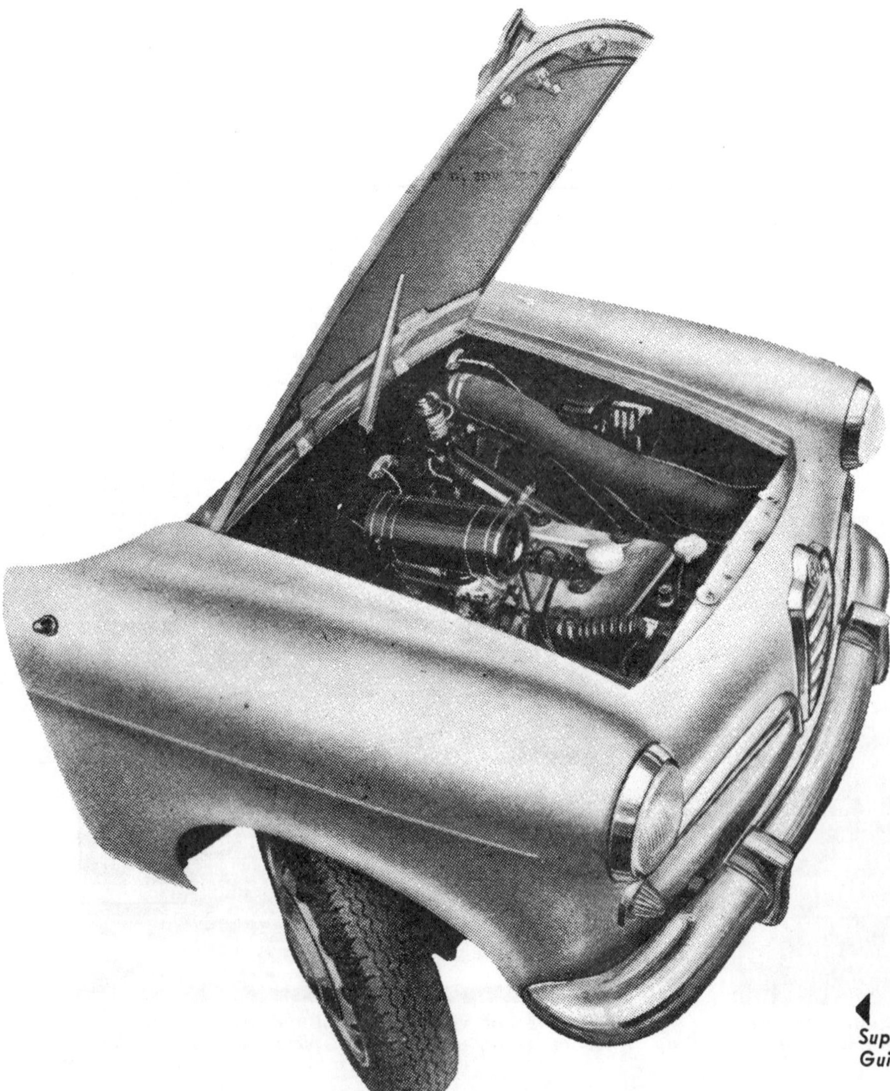

Super Sprint motor is basically the same as normal Guilietta d.o.h.c. unit (pictured) yet it gives 120 m.p.h. from 1,290 c.c.

WITH coupes, roadsters and detachable hard tops on two lengths of wheelbase and various power ratings, the Giulietta range already catered for a wide selection of sports car buyers. Yet at the Turin Show in 1957 a new prototype appeared which brought enthusiasts jostling like bees round a honey pot.

It was a long, low, streamlined coupe of rare beauty, built by Bertone to designs by Scaglione — who knows more than most stylists about the science of wind-cheating. Wheelbase lay between that of the standard coupe and the short Pinin Farina roadster, but the long tapered nose and finely drawn tail gave it an overall length 14 in. more than that of the coupe.

Few details were available, but power was announced as 100 b.h.p. at 6,000 r.p.m. on a 9 to 1 compression. There was a new five speed gearbox and the price was 2.6 million lire, an advance of 15 per cent. over the normal Sprint Veloce coupe.

At this stage the new car was clearly intended as a competition model, with aluminium panels, plastic windows, lightweight interior trim and minimum frontal area with overall height a little over 47 in. But it was difficult to see how anyone of normal size could wear a crash helmet in it; my head was firmly wedged against the roof with no hat at all in that model and drivers seemed destined to develop acute spinal curvature.

For a year only occasional snippets of information about modifications filtered through, but production has now begun and I had a short test run on one of the first batch of 20 cars. Quite a lot of changes have been made. Length has been reduced by cutting overhang at front and rear and head room has been gained by changing the contour of the roof. Unless you saw the proto-

ANYONE?

from GORDON WILKINS

type and production models side by side, you wouldn't notice it. More obvious is the addition of a miniature replica of the traditional Alfa grille in the middle of the wide oval air intake, giving positive identification.

The whole character of the car has changed, too. What was originally conceived as a lightweight racing model with aluminium panels and plastic windows is now a steel-panelled, comfortably upholstered Gran Turismo coupe with wind-up glass windows and an engine very close to that of the latest Sprint Veloce, which has gained extra power in the past year.

Where lies the advantage over the Sprint Veloce coupe? In three points at least: lower frontal area, lower drag from better streamlining and a superb five-speed gearbox with Porsche servo-ring synchromesh engaging a set of ratios which seem to serve up exactly the right sections of the engine performance curve.

The roof is only elbow high and you stoop low to enter, but once inside I had 1½ in. headroom. The seats are subtly shaped, with soft rolls at sides and edges, fitting snugly but without restraint like a well tailored suit. The small wheel is fairly well away from the driver, to suit the extended-arm driving style and in front of him are three main dials grouped under a matt black cowl. Pillars are exceptionally slim and the view forward over the drooping nose is excellent. Rearward vision is good but cannot include the extremities of the downward sloping tail. The hand drops naturally on to the short central gear shift

Revised lines for its first production run. Alfa's front end differs from that of pilot car (which was our cover for July last year, remember?) and headroom is greater, observes a thankful Wilkins.

ALFA GIULIETTA S.S.: SUPERCHARGER, ANYONE?

on a prominent central tunnel and the gear ratios are ranged in the correct order for instinctive operation; first and second forward and back nearest to the driver; third and fourth forward and back in the middle; fifth and reverse forward and back on the far side. So many needless problems are created for drivers by five-speed gearboxes which depart from this basic shift pattern. The pedals are just right, with a long spoon-shaped gas pedal right where it is wanted for a quick jab with the heel when braking and cramming the car quickly into a tight corner.

Steering is light and accurate, but with a little over three turns from lock to lock it seems low geared at first on sharp mountain corners. Driving out of town with Alfa test driver Bonini at the wheel I noticed he took sharp turns hand over hand and then let the wheel spin back through his hands under the strong caster action while accelerating away. His own countrymen are far from unanimous on this procedure and I found that by swinging the wheel a full turn with co-ordinated use of the gas pedal the car could be placed exactly on the sharpest bends without changing the position of the hands on the wheel.

Balance and stability are excellent. The car seems to drift or slide without vice, and there is no sideways shake or apparent roll.

The inside rear wheel may squeal for a moment on a hairpin bend but wheel grip is extraordinarily good for a rigid rear axle; the radius arms and diagonal control bar keep the rear wheels in firm contact with the road, as you notice when braking hard on a bumpy stretch.

The brakes, with spiral-finned front drums, seemed well up to

Superb streamlining and good visibility — a rare combination. You might call the Super Sprint a development on the Porsche theme; but don't ask us who stuck that odd whitewall on the front.

the job as far as a short test could show.

They presumably had anti-fade linings of fairly low friction co-efficient, but stopped the car quickly without excessive pedal pressure.

The clutch handled full throttle snap gear changes happily and the box, with the powerful Porsche synchromesh, was excellent.

Purists of the old school may still scorn such aids to easy driving but a modern sports car like the Giulietta has such fantastic acrobatic abilities on winding, hilly roads that the driver has plenty to keep both hands and feet occupied without unnecessary movements in double-declutching for each gear shift. With 100 horsepower available in a nimble little coupe which they say weighs only 1,430 lb., things happen quite quickly and the driver has to concentrate.

Not that this new Giulietta is at all intractable. Noise level is not unusually high for a sports coupe and the car will roll along quietly through town traffic at 15 m.p.h. in fourth gear. Fifth is treated rather like an overdrive and kept for the open straight stretches. There is a modest amount of luggage space which could make it a practical vehicle for trips short of the full holiday grand tour.

It is not so long since the typical Italian sports car was an aggressive, harsh red machine which flashed past to the scream of gears and the snarl of the exhaust, often accompanied by the whining variations provided by the blower. Now they are quieter — for Italy, too, has imposed noise limits — and more flexible, but they have not lost anything of their fine accurate steering, correct control placement, and extraordinary stability and inbuilt safety margins.

For the owner who wants to race his Giulietta, Bertone is planning to build replicas with aluminium panels which should save about 100 lb. of weight.

But racing and rally men already have a potent alternative, which has proved itself brilliantly during the 1958 season; the lightweight streamlined Zagato Giulietta coupe with engine modified by Conrero to give 113 b.h.p. It has already been timed at 128 m.p.h. against about 120 claimed for the Sprint Special — but it does not have the five-speed gearbox. #

Inside, the perfect setting for many an ecstatic hour of high-speed motoring. Gearbox has five speeds. That's the screen washer over by driver's left knee.

CUSTOMISING,

...ITALIAN STYLE

When the Italian coachbuilders take it into their head to "pretty-up" a production car, just about anything can result — and usually does. Here's GORDON WILKINS' report on what they did to the Giulietta Alfa Romeo.

THE Alfa Giulietta sedan illustrates the surprising gulf which often exists between the style of Italian production models and the superb creations of the specialist coachbuilders. It is no sleek, low-built beauty; it is high and chunky, high not only beside Bertone's Giulietta Sprint coupe but alongside a humble family sedan like the Ford Anglia. It also has some surprising practical faults like the tiny lid which gives only limited access to the luggage trunk.

Many leading Italian specialists now offer customised versions of the Giulietta for those who want appearance which is more in line with its fine performance and superb road holding. Here is how the experts tackle a difficult job. Most of them provide a bigger trunk lid and cut away the crude air intakes stamped in the sheet metal on each side of the radiator grille, filling the holes with fine mesh grilles in a cnromium frame. Allemano and Bertone go further, reshaping the openings and adding fog lamps to produce an appearance like that of the Alfa Romeo 2000. Allemano and Fissore raise the centre grille and build behind it a tapered fairing on the hood to break up the flat expanse.

The high flat side of the standard model is decorated with a single straight moulding sloping towards the rear as though added in a fit of absentmindedness by a designer whose ruler had slipped. The specialists submerge this in a large area painted in a second colour with a bright plated frame, tapered to give a level top line. The treatment helps to reduce the apparent height and increase the effect of length. Rear wings are modified and sometimes slightly peaked to give the effect of embryo tail fins. The interior is often specially upholstered in two colours. Tail lamps are changed and external access is sometimes provided for the fuel filler. Vignale changes the rear end of the roof, producing the fashionable overhang above the rear window and slightly increasing the headroom. He and Zagato add plastic draught deflectors to the windows, while Moretti offers swivelling ventipanes. Lombardi adds tall chromium overriders which form the end face of the rear wings and fairs them into new more prominent tail lamps. Most practitioners add flashing indicators on the sides and two—Fissore and Scioneri—build out the front wings to form hoods over the headlamps.

Cost of the treatment varies between £125 and £250, according to the work involved, but if you want something really different Colli will take a stock sedan and transform it into a small, high speed station waggon. #

Colli got busy with a stock Giulietta saloon and turned it into a handsome little station waggon, complete with luggage grid on the roof and a few other mods.

In standard form, the Alfa Romeo Giulietta T.I. is not a particularly handsome car.... Italian stylists periodically shudder a little and do something about it.

Neatly restyling of the whole front end of this Giulietta was the work of Allemano. See how he has moulded the bonnet to make the radiator grille more pronounced.

Francis Lombardi altered the back end of this one by wrapping around the bumpers and chroming extremities, besides a paint job.

Moretti, who also makes little sports cars, did over this Giulietta mildly by two-toning it and adding a strip of colour down the sides.

ROAD TEST ALFA SUPER SPIDER

A few adjustments, and the Super really is

IT HAS BEEN our custom not to repeat the road test of any make or model automobile unless the manufacturer effects substantial improvements over a previous model. Our road test procedure has been refined, over the past 10 years, to a point where we can be reasonably sure that our results are the best obtainable with any given car. It goes without saying that if we are given a substandard example of an automobile for our tests, we return the car to have the cause of its deficiencies corrected. Before any car is given a test, we know approximately how well it will perform. Experience with previous models of the same make and theoretically predicted acceleration times give us a criterion by which we can determine if our test machine is up to a norm for the type. There are also cases in which an engine misfire or a slipping clutch leaves little doubt regarding a test car's shortcomings.

Unfortunately not all such below-normal examples can be detected by either theory or observation, and sometimes we reject a car on pure hunch. Even so, no matter how many safeguards we set up, there is still the chance

The Giulietta engine, Weber carburetors and all.

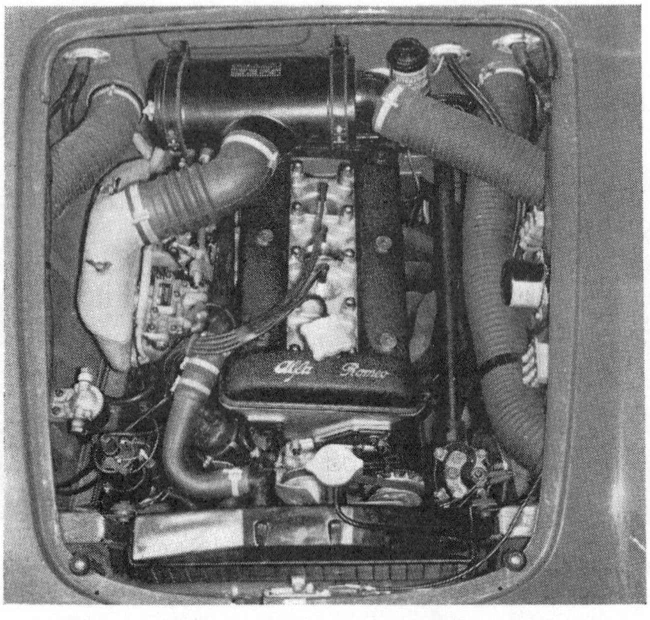

PHOTOS BY POOLE

Good heavens! This sports car has a real trunk.

that a slightly below-par test machine (or one slightly above, for that matter) could slip through. A perfect car, from our point of view, would be one that would most nearly represent the best cars in the hands of private owners. Although this is usually what we get, we seldom publish the results of any test without getting a flood of letters taking us to task for reporting such poor results. We are sincerely sorry for any inadvertent blows to our readers' pride of ownership, but we do not create performance figures. We just report what our instruments have to say.

The Alfa Romeo Veloce Spider test (March 1958) was a classic example of our unwittingly arousing owner anger. The results of the test, which pleased us no end, were evidently not good enough for the Alfa contingent, because what seemed like every Giulietta fancier in the world promptly came down on our heads. But the evidence pointed the other way. The first Veloce test was on a car fitted with non-standard rear-axle gearing (4.55:1 in place of the usual 4.10), but otherwise quite stock. At that time the car seemed to be very fast for a 2000-plus-lb, 1290-cc automobile, an impression further substantiated by the fact that the car was one used for racing and had just recently gained its 3rd straight class win. This seemed to us to be all of the evidence needed to support the assumption that it was a reasonably fast example of the type.

Evidence or not, the mail poured in and recently, when our Technical Editor acquired a new Veloce (now redesignated by Alfa Romeo as a Super), it was decided that such an opportunity to settle the whole business should not be passed by. The new car is almost identical to the previous test machine, the exceptions being in the transmission (an all new Porsche design), the rear-axle ratio (a standard 4.10), and the power rating, although at 103 SAE bhp it is quite identical to the 90 bhp (CUNA method) it had before.

Pre-test preparations were a bit more elaborate than are usual. The first job was to send our Technical Editor back to the classroom—at Hoffman Motors service school (Hoffman is the U.S. distributor for Alfa Romeo). The purpose was to learn the proper method of tuning the Super engine for best performance. Tuning is a term used to cover a variety of operations, from a simple adjust-

Look out, Reginald! He's going to spring!

ROAD & TRACK ROAD TEST 220

ALFA ROMEO SUPER SPIDER

SPECIFICATIONS

List price	$3887
Curb weight	2040
Test weight	2350
distribution, %	55/45
Dimensions, length	152
width	62
height	52
Wheelbase	86.6
Tread, f and r	50.6/50.0
Tire size	155-15
Brake lining area	132
Steering, turns	2.8
turning circle, ft.	32
Engine type	4 cyl, dohc
Bore & stroke	2.91 x 2.95
Displacement, cu in.	78.7
cc	1290
Compression ratio	9.5
Bhp @ rpm	103 @ 6000
equivalent mph	111.1
Torque, lb-ft.	86.8 @ 4500
equivalent mph	83.3

GEAR RATIOS

O/d (n.a.), overall	
4th (1.00)	4.10
3rd (1.33)	5.45
2nd (1.96)	8.04
1st (3.31)	13.6

CALCULATED DATA

Lb/hp (test wt)	22.8
Cu ft/ton mile	62.7
Mph/1000 rpm (4th)	18.6
Engine revs/mile	3240
Piston travel, ft/mile	1593
Rpm @ 2500 ft/min	5090
equivalent mph	94.3
R&T wear index	51.6

PERFORMANCE

Top speed (4th), mph	n.a.
best timed run	112.5
3rd (7000)	98
2nd (7500)	71
1st (7500)	42

FUEL CONSUMPTION

Normal range, mpg	22/28

ACCELERATION

0-30 mph, sec	2.9
0-40 mph	4.9
0-50 mph	7.3
0-60 mph	11.0
0-70 mph	14.9
0-80 mph	19.0
0-90 mph	23.6
0-100 mph	29.9
Standing ¼ mile	17.1
speed at end, mph	75

TAPLEY DATA

4th, lb/ton @ mph	190 @ 70
3rd	275 @ 64
2nd	415 @ 52
1st	580 @ 36
Total drag at 60 mph, lb	82

SPEEDOMETER ERROR

30 mph	actual 26.4
40 mph	35.8
50 mph	45.2
60 mph	54.6
70 mph	63.9
80 mph	73.2
90 mph	82.7
100 mph	92.0

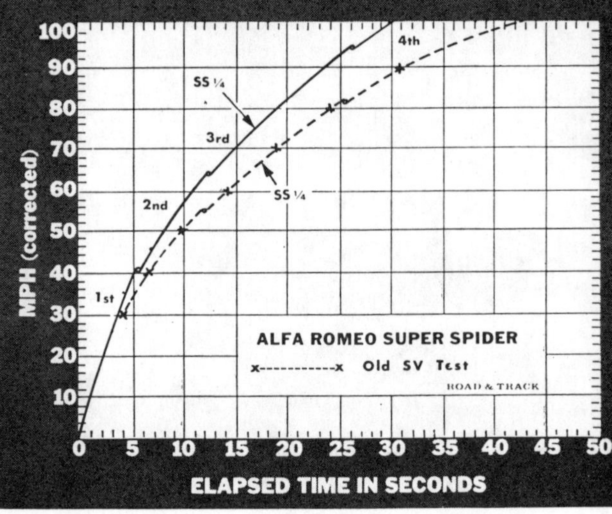

The luxurious and unusually roomy interior.

ment of carburetion and timing to the wildest modifications imaginable. We should like to stress the fact that the tuning taught us was strictly adjustment: not even the carburetor jets were changed. The value of good stock settings is very often overlooked by people who are trying to get the most from their cars. We hope that the performance increases noted on the acceleration graph will encourage some of our readers to give more careful attention to their manufacturers' recommendations.

Immediately after the new test car was given its tune-up, and with 4800 miles on the odometer, it was taken out to show what it could do. As it turned out, this was much more than we had anticipated because it turned out to be a *much* faster machine than the first Veloce. There seems to be no explanation for the enormous increase in speed. The Alfa Romeo factory claims no more power for this new car, and with the numerically lower 4.10 rear-axle ratio to handicap the car on its acceleration runs, we didn't expect to better previous figures.

All of the runs were made without difficulty. The dohc engine was exactly as we remembered it: marvelously smooth over most of the rpm range, with just a trace of roughness on deceleration. Flexibility is good, but over a higher range than most people are accustomed to (3000 rpm is pretty much a minimum figure but the engine is willing and smooth up to 6000 or even 7000 rpm). After one stops thinking in terms of 1500 to 3500 and makes the change to 3000 and 6500 rpm, everything becomes quite docile and ready to go.

The new gearbox virtually eliminates the 2nd-speed gear crash that plagued earlier Guiliettas. The ratios are unchanged, which pleases us because the gear spacing was perfect already. Shifting is still a bit short of being foolproof; a sharp, hacking shift from first to 2nd gear will produce a crunch, but just the barest hesitation in neutral will overcome this.

The chassis also has not been changed, but then why should it be? It is one of the best to be found anywhere, although it embodies nothing radical in its live rear axle at the rear and unequal-length A arms at the front. Here is a perfect example of how the modern sports car can be softly suspended, yet handle better than its cart-sprung forebears. Brakes deserve a special mention here. The drum-type brake is probably on its way out, but a good drum is better than a bad disc, at least for the present, and the huge aluminum drums fitted on the Alfa are the best to be found.

All things taken into consideration—even the high purchase price of the Alfa Romeo Super Spider—it is as fine a production sports car as may be purchased anywhere today. Without going into its many features, which are quite adequately covered in our previous road test and appear both in the March 1958 issue of R&T and in our 1959 Road Test Annual, we would like to say that it is an outstanding design. Not many years ago, only the outright competition machines in the under-1500-cc category would perform so well.

Two factors are responsible for this particular car's showing such a wonderful turn of speed: the very sound design by Alfa Romeo, and the excellence of the Hoffman service school, where they furnished us with the data needed to get the most from the car. Perhaps the speeds shown may not seem very high to those who own Corvettes, but for a 1300-cc machine with rather heavy touring bodywork, they are utterly fantastic.

A CAR ROAD TEST

Alfa Romeo Giulietta Sprint

THE name of Alfa Romeo began to be really well-known in 1923, when the company first showed serious interest in motor racing. But the marque is considerably older than that. Its history dates back to 1906 when Alessandro Darracq, who owned a garage in Portello, Italy, planned to build cars. He, however, encountered many difficulties, and, in 1909, his garage was taken over by the A.L.F.A. company—the initials standing for Anonima Lombarda Fabbrica Automobili.

In 1911 Nicola Romeo, an engineer, acquired the firm and called it Anonima Ing. Nicola Romeo. Then, in 1918 the company was reorganised and received the title under which it subsequently became world famous, winning laurels in road and track races in virtually every civilised country.

We think that it is appropriate, in a road test report such as this, to recall—however briefly—the long history of Alfa Romeo, for the car with which we are here concerned is very clearly the result of great experience in the art of designing and manufacturing high-performance vehicles of first-class quality.

The Giulietta Sprint is a connoisseur's car. The very thought of such a model being possessed by someone incapable of *loving* a car sends shudders down the spine of any enthusiastic, knowledgeable and sensitive motorist.

Some cars—and very admirable they are, too, in their own particular way—may be thought of as modern magic carpets. Others, even fewer in number, offer their fortunate owners something more akin to the wings on the feet of Mercury than to the flying carpets of Persia : their drivers feel that they are a part of them. Such a car is the Alfa Romeo Giulietta Sprint; any driver who is capable of appreciating a vehicle of its type and quality soon finds himself not merely admiring it because of what it does but loving it because of what it enables him to do.

Cold figures do not—cannot—express the joy of driving that this "one-ness" of man and machine brings, nor can they even begin to convey the utter safety of such a vehicle in the hands of an experienced driver. Here, however, are some figures which illustrate what the Sprint does; we shall return later to the subject of how it behaves.

The manufacturer recommends that, in the interest of mechanical reliability, certain speeds be not exceeded in the various gears. These speeds are : 103 m.p.h. in top, 77 m.p.h. in 3rd, 53 m.p.h. in 2nd, and 31 m.p.h.

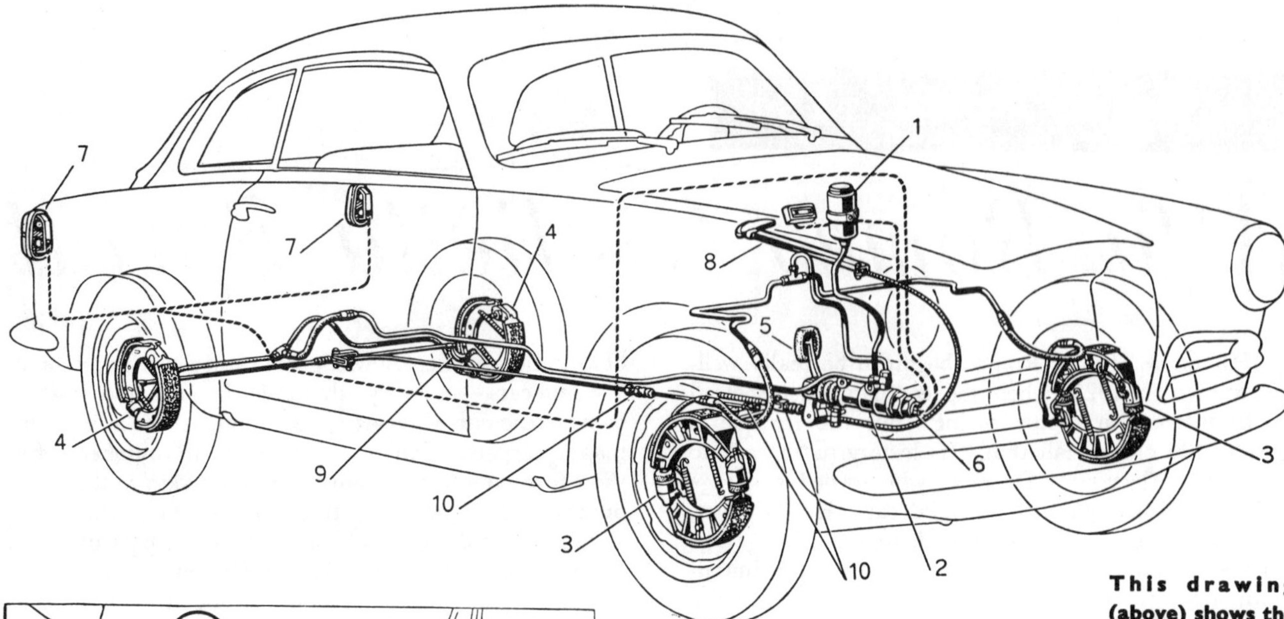

This drawing (above) shows the braking system. Feed tank (1), hydraulic master cylinder (2), front wheel hydraulic cylinders (3), rear wheel hydraulic cylinders (4), brake pedal (5), stop-lights pressure switches (6), stop-lights (7), hand-brake lever (8), hand control levers for rear brakes (9), hand-brake adjustment device (10).

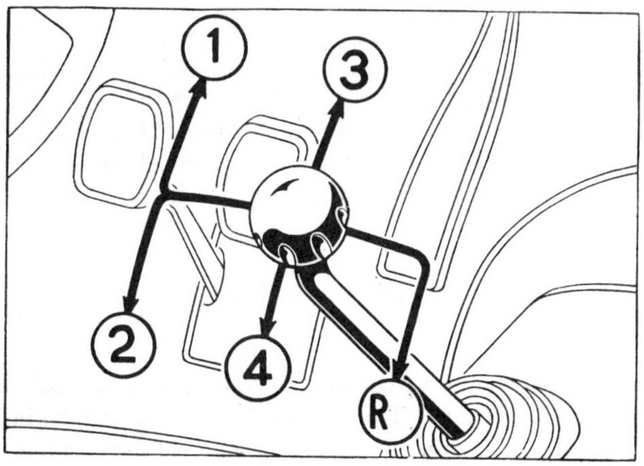

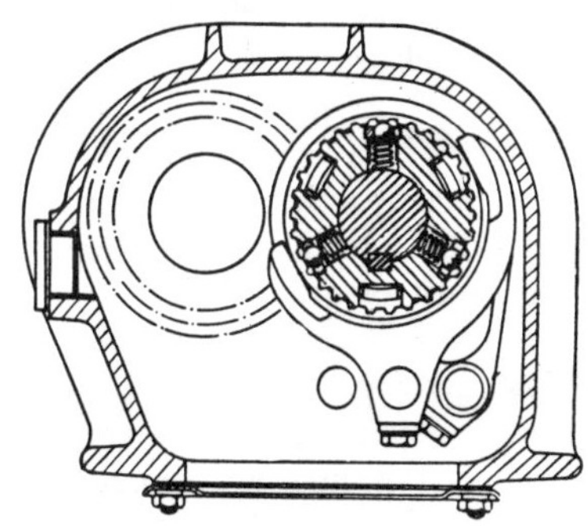

Gear lever positions are shown here — also a cross-section through the synchronising arrangement for 3rd and top gears. (All four forward speeds are equipped with synchromesh).

The rear suspension employs coil springs, the axle being positively located by radius arms.

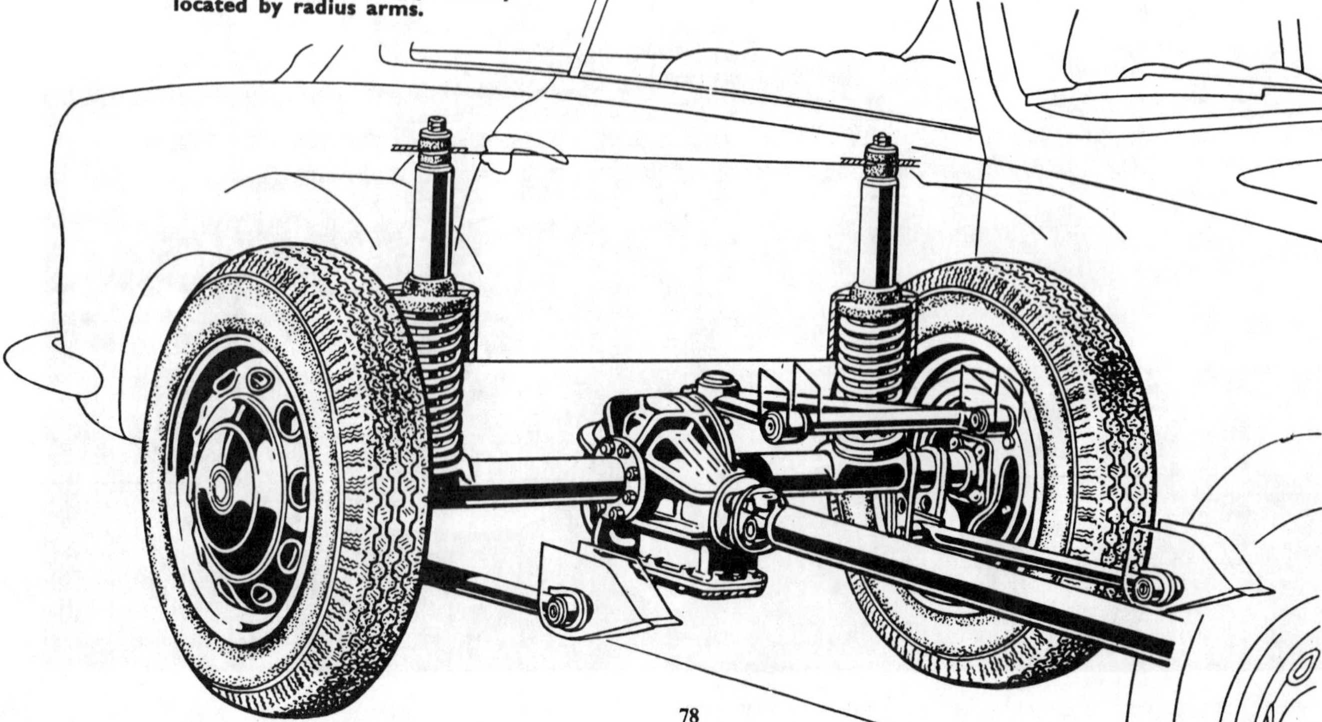

in 1st. These road speeds, which correspond with about 6,650 crankshaft r.p.m., are all easily attained. The beautiful smoothness of the engine at all speeds from the idle to the recommended maximum r.p.m. is due to such features as: fully-balanced crankshaft running in five main bearings, twin overhead camshafts (directly operating the valves), light-alloy cylinder block with inserted cast-iron "wet" cylinder liners—and, of course, to the accuracy with which the components are manufactured from high-grade materials and to the care with which they are assembled. The thin-wall crankshaft and connecting-rod bearings are of steel lined with lead-indium; the alloy pistons have two compression rings, the uppermost of which is chromium-plated, and one oil scraper ring.

The pressed-steel oil sump is finned for cooling. The front brakes are of two-leading-shoe type and their drums are equipped with herring-bone cooling fins. As owners of motor cycles and other items with air-cooled engines are well aware, finning is an excellent means of offering a maximum area to the air and thereby obtaining cooler operation.

The value of the Sprint's finned sump is evident when the car is driven very fast mile after mile: we found that in these conditions the oil temperature (shown on a gauge on the facia panel) remained steady at a trifle under 190-degrees Fahrenheit—almost precisely the same temperature as that of the radiator water.

As for the brakes, they are magnificent. The harshest use fails to diminish their power—and this is such that even when the Sprint is fitted with Michelin "X" tyres (as was the model that was submitted for test) it is possible to decelerate so swiftly that the steel-mesh-backed treads howl on the road's surface.

Proportions of front-rear braking effort seem to be ideally planned, and there is utter evenness of braking at both the front and the rear. Brake-pedal pressure is reasonably light, and the deceleration that is obtained is entirely progressive: after a few miles spent at the controls, the driver instinctively applies in all conditions precisely the correct amount of pressure to obtain the desired degree of deceleration.

A fast car with a small frontal area and a good aerodynamic shape has great need of good brakes. The Sprint's brakes are so good that, in an emergency demanding the rapid loss of speed, the scene in the rear-view mirror is much more likely to cause the driver anxiety than that seen through the windscreen. . . .

The value of such powerful braking would, to a very large extent, be lost if the car's roadholding and controllability were not of parallel quality. The Sprint—as we have already sought to convey—is without peer in its class regarding the manner in which it handles.

With the exception of the hand-brake (which, in our opinion, would be better placed between the two front seats), all the controls—minor as well as major—are excellently positioned and operate in altogether delightful fashion.

Light at all speeds above a walking pace, the steering is high-geared (three turns of the steering wheel move

A delightful view greets the enthusiast when he raises the bonnet top, the underside of which is padded with sound-deadening material.

Quick-action clamps secure the spare wheel and the petrol-filler cover. The battery is also located in the boot.

the front wheels from one extreme lock to the other) and so precise that an able driver rightly feels that he can "place the car to an inch". Fast cornering is a delight, and only extremely forceful drivers will discover that when the cornering-limit (which is very high) is closely approached the tyres howl in warning and protest whilst the car still remains obedient to a light, sure touch at its controls. There is very little body roll; as the passenger is likely to react far more than the car to the effects of *G*, there is a grab-handle positioned above the off-side front window (the Sprint has left-hand drive).

The suspension is deliberately planned to "favour" roadholding rather than riding comfort, but—with the exception of slow travel over badly-worn tarred surfaces, which may produce fairly marked vertical movement at the rear of the vehicle—the occupants of the excellently upholstered front seats enjoy a ride that is exceptionally comfortable for a sports car with a wheelbase of 7 ft. 9¾ ins. Surprising, too, in view of its other virtues, is the suspension system's ability to swallow-up the effects of quite large bumps and depressions when motoring on untarred roads.

Our acceleration figures, as has been usual in this series of road tests, were taken without driving the car any harder than we judge that an enthusiastic owner would be willing to treat his own vehicle. Nevertheless, we reached 50 m.p.h. from a standstill in less than 11 seconds, 0-60 m.p.h. took a little over 15 seconds, and the 0-70 m.p.h. figure was a fraction under 20 seconds!

The gear lever has a lovely, short and precise, movement. This, combined with the fact that the clutch behaves impeccably, makes it quite usual, when the road is clear and the driver is in the mood, for 0-80 m.p.h. to take less than 30 seconds without conscious hurrying.

All the gears are uncommonly quiet, and, with the exception of 70-75 m.p.h. when there is some slight exhaust resonance, high-speed travel is almost as remarkable for the low level of noise as for the high level of safety. Provided that the driver does not grasp the steering wheel ham-fistedly, inherent directional stability is almost uncannily good—the driver rests his hands gently on the steering wheel when cruising at, say, 85-90 m.p.h. on the open road, and the Sprint unerringly follows its straight course.

The range of the Sprint's equipment and the quality

The group of switches at the lower edge of the extreme left of the facia is non-standard, fitted to the test car to control the non-standard fog-lights.

of the car's external and internal finish and trim are fully in keeping with the price of more than £1,500.

Our adverse comments are few. We have already mentioned the hand-brake lever. Alas! six-footers will find it difficult to make themselves comfortable in the front of the car—and the rear is really suited only to the accommodation of children. We think that the nine turns of the handle required to lower or raise a window fully is excessive. We think that coat-hooks should be provided on the insides of the door pillars, and we would prefer the headlight-lever to give the main beams when in the upper, instead of the lower, position.

Thus—except for the matter of size, if you happen to be tall—our criticisms may fairly be considered to be of very minor character. Of course, high-quality gems—unless they are very expensive indeed—tend to be small, and the Alfa Romeo Guilietta Sprint is in very many ways a gem of a motor car. ●

SPECIFICATION AND PERFORMANCE

BRIEF SPECIFICATION

Make ... ... ALFA ROMEO
Model ... ... Sprint
Style of Engine Straight 4-cylinder. Water-cooled. Overhead valves; twin o.h.c. (chain-driven). Statically and dynamically balanced crankshaft running in 5 main bearings.
Bore ... ... 2·91 ins. (74·0 mm.)
Stroke ... ... 2·95 ins. (75 mm.)
Cubic Capacity 78·7 cu. ins (1,290 c.c.).
Maximum Horse-Power 80 b.h.p. at 6,000 r.p.m.
Brakes Hydraulic. Two-leading-shoe at front.
Front Suspension Independent. Coil springs and wishbones. Anti-roll torsion bar.
Rear Suspension Coil springs and radius arms.
Transmission System Clutch and four manually-engaged forward gears. All forward gears equipped with synchromesh.
Gear Ratios (overall) 1st 15·09 to 1
2nd 8·92 to 1
3rd 6·17 to 1
Top 4·56 to 1
Rev. 15·33 to 1
Final Drive Ratio ... 4·56 to 1
Steering Gear Worm and finger-roller.
Overall Length ... 13 ft. 0⅞ in.
Overall Width ... 5 ft. 0½ in.
Overall Height ... 4 ft. 4 ins.
Turning Circle ... 36 ft. 1 in.
Dry Weight ... 1,900 lbs. approx.
Price ... £1,531 at Coast Ports.

PERFORMANCE

Acceleration 0-30 m.p.h. 4·2 secs.
0-40 m.p.h. 7·1 secs.
0-50 m.p.h. 10·7 secs.
0-60 m.p.h. 15·2 secs.
0-70 m.p.h. 19·9 secs.
0-80 m.p.h. 27·3 secs.
0-90 m.p.h. 38·5 secs.
In top gear from a steady 40 m.p.h. to 60 m.p.h. ... ... 11·7 secs.
In top gear from a steady 50 m.p.h. to 70 m.p.h. ... ... 13·5 secs.
In top gear from a steady 60 m.p.h. to 80 m.p.h. ... ... 15·2 secs.
Maximum Speed Over 100 m.p.h.
Reasonable Maximum Speed in 3rd Gear ... ... 77 m.p.h.
Reasonable Maximum Speed in 2nd Gear ... ... 53 m.p.h.
Reasonable Maximum Speed in 1st Gear ... ... 31 m.p.h.
Fuel Consumption ... 32·4 m.p.g.
Test Conditions Sea level. Little wind. Dry road. 90-octane fuel.

BRIEF IMPRESSIONS OF A GIULIETTA SPYDER ON BRITISH ROADS

TOP MARKS FOR THIS ALFA

The diminutive grille retained today says Alfa Romeo as plainly as the familiar and massive angles from Crewe say Rolls-Royce

A RARE and charming visitor to these islands, as Peter Scott might say of some of his feathered friends—the Giulietta Spyder is just this. A chance came to try one that had been in private hands for the past year, but quite naturally the owner did not wish to spare it for the time and ordeal of a full test.

It was in the familiar bright Italian red with black interior; tiny, beautifully balanced and styled by Farina, and looking almost delicate. Only left-hand drive cars are made, and the long, very sturdy gear lever is a relic—not undesirable—of days gone by. Relatively the cockpit is enormous—bigger than that of some 3-litre sports cars. It is clean, functional and safe, with room for the tallest driver. There is a full complement of instruments and these and the secondary controls are conveniently arranged. Yet the Giulietta has an engine of only 1,290 c.c. capacity.

There is no more desirable small sports car; it sets standards of performance, handling and refinement that very few others can match. The steering is light, precise and the car responds to the slightest touch on the wheel. There is scarcely any roll or tyre squeal when cornering, and the driver at once feels confident about taking corners very fast indeed. The ride is an excellent compromise between firmness for high speeds and sufficient resilience to avoid bumps and shakes on poor road surfaces. Ground clearance is marginal, particularly for rally work. Protagonists of all-independent suspension must find it difficult to decry the Spyder's live rear axle and coil springing.

It is commendable on an open car with normal doors and no special height or thickness of sill, that the body is absolutely rigid, and the steering wheel is shock- and vibration-free. This must be achieved by good design rather than excessive structural strength, because the kerb weight of the car is only a moderate 16¼ cwt.

Alfa claim a top speed of 96 m.p.h., with 74 m.p.h. in third and 51 m.p.h. in second. These figures the car achieved without effort, although the last 5 m.p.h. of top speed takes time to realize. In fact, the little twin o.h.c. engine always seems to be working well within its capabilities. It feels completely smooth, and if it produces any vibration or mechanical noise, this is effectively excluded from the cockpit. Some care is needed not to exceed the 6,000 r.p.m. normal limit because the engine gives no indication of the high speed and would, apparently, be happy to turn faster still.

Italian manufacturers do not as yet fit disc brakes to their small sports cars. The Giuliettas have 10½in cast drums with cooling fins. Those on the Spyder were smooth in operation, reasonably light, and showed no signs of fading during bursts of fast driving. The balance, front and rear, seemed good, there being no premature rear-wheel locking. The 155×15 tyres were Pirelli Cinturatos.

One of the remarkable points of the Spyder's performance is its flexibility. In traffic it will potter, snatch-free, in top gear at under 20 m.p.h., and pull away without signs of distress. This speed represents no more than 1,100 r.p.m.

Apart from the performance and handling there are many little indications that this car has been designed and developed by a staff with real appreciation of the finer qualities of a sports car. The seats are carefully shaped, with built-up outer panels for the cushion and squab. Driver and passenger are located and supported unusually well, so that they are not thrown about when driving fast on winding roads, nor are they fatigued by tensing and bracing themselves against movements of the car.

This Alfa makes no pretence of being

Left: clean and roomy, the engine compartment quite dwarfs the 1,290 c.c. twin-cam engine, which has a single Solex carburettor. Below: the cockpit too is spacious. Instruments and controls are placed where a driver likes them: r.p.m. and speedo dials are central and right; combined on the left are water temperature, oil pressure and petrol gauge with three warning lights. Central knobs are for lamps, hand throttle, heater, fan and choke

Plenty of room in the boot. Right: sturdy wrap-round rear bumpers with chunky overriders protect the neat and elegant tail

more than a two-seater, but there is ample luggage space in the boot. The fuel tank holds 11½ gallons and normal consumption is about 30 m.p.g. Hard driving will pull this down to little more than 20 m.p.g.

A year's use and more than 8,000 miles driving have made little impression on this car. Without special preparation for our trial, it appeared in almost every respect to be in mint condition. New, the Spyder now costs £2,091, including U.K. purchase tax; the value of this one is estimated at between £1,700 and £1,800

Veloce

(Continued from page 43 *)*

around corners, the driver is allowed to safely meet any sudden emergency. Under certain circumstances, to take the power off in the corner would invite disaster for the car is literally steered by provoking rear end breakaway and then controlled by proper use of the throttle and steering wheel. The brakes have three inch wide drums, and are heavily ribbed in front and finned at the rear. Little pressure was required to bring the car to a quick stop, even though the car had just completed a race at Monza. Under sudden acceleration on rough surfaces, a certain amount of rear axle bounce is noticeable, but this does not seem to upset the handling qualities of the car at all. This is due undoubtedly to the location of the A-frame which is mounted on the upper left of the rigid axle. Shocks and springs on the Veloce are considerably stiffer than on the normal Sprint version. Thus the overall ride is considerably harsher. Tire noise was found to be excessive but this may have been due to the 15 inch Pirellis and the hot pavements encountered.

In summing up the Veloce, the most outstanding characteristic of the car is the flexibility factor. Second gear can be used almost continuously on short choppy race courses, and in normal driving is a wonderful gear for maneu-

vering in traffic Secondly, the flexibility in handling makes for complete security when driving hard and the driver of a Sprint Veloce will almost immediately have complete confidence in his car. Amazing things can be done with it and it's probably the most forgiving car on the road. A fast driver can make all kinds of mistakes in a corner and still come out ahead. A relatively large amount of luggage and child space is available, thus making the Veloce a really practical sports car which can be used on shopping expeditions the day before going off to a weekend race meeting where the car will take on all comers. In short, here is a true Italian competition machine with just enough comfort and flexibility left in the design to make it a really outstanding piece of equipment.

Jesse Alexander

Road Test– CONTINUED FROM PAGE 67

to the new gearbox. It is light and quick in action, yet the synchromesh is unbeatable. With this type of engine, one often wants to engage bottom gear at quite appreciable speeds. The perfection of the synchromesh encourages one to do this, rather than let the engine labour while pulling away on second. This box is so much better than the old one that it transforms the pleasure of driving the car.

The very large brakes have the situation well in hand at all times. They are smooth and progressive, and the wheels are pierced to assist the turbofins in their dissipation of heat. Regrettably, the hand-brake is of the umbrella-handle type, which hardly seems worthy of the car. Another small point of criticism is the absence of a high-beam indicator for the headlamps, though there is an indicator for the sidelamps.

The kind of driver who likes to stay in top gear all day will have no use whatever for this Alfa, but the keen motorist, who enjoys making intelligent use of the box, will love every moment of his journeys. It is a pity that punitive import duties render this a car for the very few, for it is as enjoyable to drive as it is beautiful to look at, which is praise indeed.

SOMETIMES one comes across a real jewel of a motor-car. The Alfa Romeo Giulietta Sprint Speciale, with bodywork by Bertone, is such a machine. Recently, owner Ron Atkinson lent me his pride and joy, with instructions to "see what she'll do". This beautiful little red coupé, supplied by Bolton of Leeds, was the one which excited so much admiration on the Alfa Romeo (Thomson and Taylor) stand at the 1959 Earls Court Show, and is the only one of its type in Great Britain.

The Speciale is, without question, an Alfa Romeo; competitions-bred and with the pedigree of a long line of racing and high-performance machines. Bertone has given the car a look of perfection, even improving on the normal SV, long acknowledged to be one of the prettiest small-capacity vehicles ever to be pro-

The Alfa Romeo Giulietta Sprint Speciale

Over 124 m.p.h. from Fully Equipped 1,300 c.c. Grand Touring Car

duced. There is a vague suggestion of "Disco Volante" in the low, swept-down bonnet line with its miniature "radiator" grille, and in the smooth contours of the bodywork. Undoubtedly the science of aerodynamics has been applied to the design, for rarely have I travelled in a car which has so little wind noise at high speeds. The exhaust note is almost inaudible from inside the car, for great care has been taken to avoid the bark which earlier SV models were apt to emit.

Driver and passenger comfort has been studied, and the Speciale has obviously been produced with long-distance, high-speed touring in mind. The seats are all that could be desired, and are adjustable for the longest legs. As for visibility, the car is a revelation; full-view, sharply-raked curved windscreen, wide rear window and slender door-pillars add up to a completely unobstructed view, not always an attribute of coupés. Although people could be accommodated in the rear, this should be regarded purely as luggage space, complementary to the boot. Equipment is comprehensive, and one feels that for modern motorway driving, such things as flick-switch headlamp controls ought to be made obligatory; they have, of course, been fitted on the majority of Italian-built cars for years.

Technically the Speciale differs from the SV in that the twin-o.h.c. camshaft

BY GREGOR GRANT

Photography by Francis Penn

engine is more highly tuned. With 9 to 1 compression ratio, special pistons, carefully gas-flowed cylinder head and strict attention to balancing of all moving parts, an output of 116 b.h.p. at 6,000 r.p.m. is claimed. It should be explained that this is an S.A.E. rating, probably achieved without dynamo and fan, and with a straight-through exhaust system. I should be inclined to give the true output as around 100 b.h.p., which, to judge from the performance figures obtained, must be near enough.

The transmission includes a competitions clutch, with cushioned centre hub, and a five-speed gearbox is used. This is all-synchromesh, giving very rapid changes. Fourth is a direct ratio (4.55 to 1 on the car tested), and fifth is an indirect gear of 0.84 to 1. In consequence, fourth gives 17 m.p.h. per 1,000 r.p.m., and fifth, 20.02 m.p.h. per 1,000 r.p.m. Changing from fourth to fifth at high r.p.m. requires care at first, for the gear lever position is fairly close to that of "third", being slightly to the right. However, the knack is soon achieved. On all gears, the synchromesh is so efficient that nothing whatsoever is gained by attempting to beat it. The clutch take-up is all that could be desired, and there is not the slightest suggestion of spin or slip even on the crucial first to second slick-change for maximum getaway.

The engine itself is audible, but not obtrusively so. With such an ultra-efficient 1,290 c.c. high-compression unit, there must, of necessity, be somewhat generous clearances for valve operation and so on. Nevertheless, the Alfa Romeo engineers have managed to avoid clatter. Piston-slap is evident when the engine is cold, but once the water temperature gauge rises this completely disappears. The indirect fifth gear is ever so slightly noisier than the direct fourth, but there is definitely no whine on any ratio.

Giulietta road-holding goes without saying, and is probably the ultimate obtainable with independent front and rigid-axle rear. Body-lean is not apparent when driving the car, but is evident when watching the machine being cornered at high speeds. As for understeer or oversteer, I feel that the car possesses something in between—in other words, it is just right! Oddly enough, worm-and-roller is retained; however, it is both light and positive, with absolutely no trace of kick-back even on the bumpiest of roads.

OFFICE: Well thought-out interior of the "Speciale", with anti-crash padding, hooded instrument panel, three-spoked sprung steering wheel and short lever for the five-speed gearbox.

Beautifully proportioned, the Alfa Romeo Speciale is one of the world's outstanding examples of Italian coachcraft.

Care must also be taken when using the extremely powerful brakes. If anything, the Giulietta is rather over-braked, with its enormous turbo-finned drums almost identical to those used on the Type 159 Grand Prix car, and with 2LS on the front. Still they can be made to fade, which is one more argument for disc-type brakes on cars which will obviously be used in competitions. With very large drums, and using racing-type linings to minimize fade, it is a fact that stopping efficiency at fairly low speeds is considerably reduced. The linings used on the Speciale offer a good compromise, but one foresees the time when the Milan technicians will go over to discs, like Ferrari and Maserati, and, it is said, Abarth.

Spring-damping could also be improved, the heat generated by the present type suggesting that they are somewhat overworked. Whilst sustained high speeds on M1 did not disclose any tendency to pitch, fast cruising on normal trunk roads did produce signs of "bumping", particularly at the rear. This was progressive, according to the speeds, and in view of the fact that, after a lengthy halt, it was some time before the symptoms reappeared, one must conclude that beyond a certain temperature damper efficiency begins to recede. I have found this to occur on cars which have been hotted up well beyond "standard", which supports John Bolster's theories that when performance is increased by any measurable amount on standard production machines, one must be prepared to strengthen and modify components to stand up to much higher speeds than the designers envisaged.

Although peak r.p.m. is stated to be 6,000 r.p.m., the engine runs quite happily up to 7 thou., without any indication of valve-crash. Some cars are actually faster in "overdrive" than in normal top, according to power-output and axle ratio, but the Speciale is very much quicker in indirect top with the standard 4.55 to 1 axle. Maximum (average of several each-way runs) was 112 m.p.h. in fourth, and 124.8 m.p.h. in fifth, denoting the absolute r.p.m. limit

★

(Right) The power comes from a highly developed, four-cylinder, two-o.h.c. engine developing over 100 b.h.p.

★

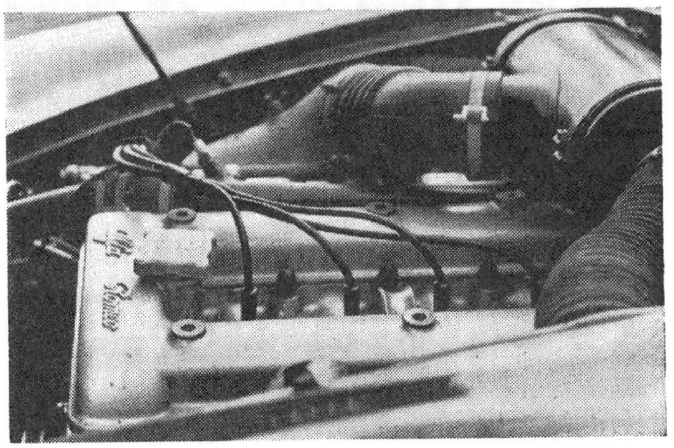

in both ratios. The speedometer was inclined to be optimistic after 160 k.p.h., and on quite a few occasions 230 k.p.h. came up when the tachometer indicated that under 200 k.p.h. was more reasonable. This meant that full allowance had to be made for discrepancies during acceleration tests, in addition to pasting an m.p.h. scale over the existing kilometre reading.

The best standing quarter-mile figures (17.2 secs.) were obtained using first, second and third gears, nearly 80 m.p.h. coming up at the end of the "quarter". A "straight-through" first to second was not attempted, and should not be practised with the all-synchro box. Usually when an SV is returned to the factory with gearbox failure, the Alfa people insist that it is due to "driver trouble".

Fuel consumption seldom worries people who run cars such as the Speciale, except for range purposes. I found the Alfa surprisingly light as regards petrol, and even driving hard, over 25 m.p.g. (100 octane) was the rule rather than the exception. With an 80-litre tank this gives the Speciale a very useful range, not only for touring, but for long-distance G.T. racing and rallies.

The headlamps were really excellent, giving range-cum-spread which was ideal for fast driving. However, despite the elaborate hooding of the instruments, the sharp rake-back of the windscreen produces reflective glare; the facia lighting ought to be more subdued and/or fitted with rheostat control.

During several hundred miles I had only one "incident". A front tyre burst at speed, and I was fortunate to escape without damaging the motor-car, as the cover more or less disintegrated. A nail

DESIRABLE PROPERTY: Three-quarters front view of the Speciale, emphasizing the sharp rake-back of the windscreen and the low, swept-down bonnet line.

A REVELATION in the matter of visibility is provided by the car. The sharply-raked rear window and slender door pillars combine with the full-view windscreen to give an unobstructed view.

caused sudden deflation, and may have been picked up much earlier on.

I was quite sorry to hand the Speciale back to Atkinson. It had been lots of fun to drive a small-capacity car of such refinement and performance, not to mention the enjoyment of being seen around in a car which must arouse the possessive instinct in anyone who cares something about automobile engineering.

SPECIFICATION AND PERFORMANCE DATA

Car Tested: Alfa Romeo Giulietta Sprint Speciale. 2-str. coupé Bertone.

Engine: Four cylinders, 74 mm x 75 mm. (1,290 c.c.). Twin overhead camshafts, 9 to 1 compression ratio. 116 b.h.p. at 6,000 r.p.m. Two twin-choke horizontal DC40 Weber carburetters. 12-volt coil and distributor ignition.

Transmission: Single dry-plate competition clutch. Five-speed all-synchromesh gearbox with central, ball change lever. Ratios 3.258, 1.985, 1.357, 1 and 0.84 to 1. Reverse, 3.255 to 1. Standard rear axle, 4.55 to 1. Open two-piece Hardy-Spicer propeller shaft.

Chassis: Monocoque steel main structure with aluminium alloy doors, bonnet top and locker lid. Independent front suspension by wishbones, helical springs and telescopic dampers with torsional anti-roll bar. Rear axle (rigid) connected to frame by an upper triangle and two lower struts with helical springs and telescopic dampers. Worm and roller steering with three-piece track rod. Girling hydraulic brakes (2LS at front) in bimetallic drums with herringbone fins. Bolt-on wheels with 155 x 15 Pirelli tyres.

Equipment: 12-volt lighting and starting, speedometer, revolution counter, oil pressure and temperature, water temperature and fuel gauges. Self-parking electric windscreen wipers with jet-type washers. Self-cancelling direction "flashers". Heating and ventilation system. 80-litre rear fuel tank.

Dimensions, etc.: Wheelbase, 7 ft. 5 ins. Overall length, 13 ft. 6 ins.; width, 5 ft. 5 ins.; track (front), 4 ft. 3 ins., (rear), 4 ft. 2 ins. Weight, (dry), 17 cwt., (as tested), 17 cwt. 62 lb.

Performance: Maximum speed (5th gear), 124.8 m.p.h. (201 k.p.h.). Speeds in gears, 1st 34 m.p.h., 2nd 58 m.p.h., 3rd 85 m.p.h., 4th 114 m.p.h. Acceleration, 0-30 m.p.h. 3.8 secs., 0-50 m.p.h. 7.2 secs, 0-60 m.p.h 10.2 secs, 0-80 m.p.h. 19.0 secs, 0-100 m.p.h. 32.4 secs. Standing quarter-mile, 17.2 secs. Fuel consumption (driven hard), 25-28 m.p.g.

ALFA

(Continued from page 55)

Giulietta Berlina, the TI version is fitted with a double-choke downdraft Solex 35 APAI-G with accelerator pump.

Once behind the wheel of the car and out on the open road, it's easy to feel at home. Our immediate reaction was that it would turn out to be a "fun car", and this proved to be true. The steering is light and quick: the gear lever, though situated on the column, is solid yet light, with the feel of the floor-mounted version found on the Sprint Veloce and Spyder. The engine is extremely responsive and the weight of the overall car has been kept down to a minimum. In traffic, one is laps ahead of the rest of the pack by merely using engine revs and the gear box as they ask to be used. Shifts can be rushed without gronching noises jumping out from under the floor boards.

On the *autostrada* we checked the speedo and found it to be hopelessly fast, but a maximum in third gear proved to be an honest 65 mph. As for top speed, we were able to clock only a 90 miles per hour two-way average, but with a bit of attention to the carburetor, the TI should reach its advertised 95 with ease.

At high revs engine noise was loud, but the wind noise was less than that of the Sprint Veloce. At high revs in high gear, there is the same feeling of extreme confidence that the Veloce inspires.

Brakes are taken directly from the Sprint, and no criticism is needed. Giulietta drivers know that their best weapon is their anchors, and this little four-door bomb is no exception. We had fade in our test, but recovery was almost immediate. Brake potency is amazing, even though the system is not servo-assisted.

Cruising speed of the TI is anything up to 90 mph. Stability is good at high speeds and fast cornering showed the car to have excellent road holding qualities on either Pirelli "Cinturato" or Michelin.

To determine the car's mannerisms on the track, we took our TI to the Modena *"autodromo"*. This is not an earthshaking circuit, but still it offered opportunity to discover how the car behaved when really trying. Fast corners offered no problems; but on slow to medium fast (such as the "S" corner at Modena) we found too much understeer for complete happiness. Adjusting tire pressures, both front and rear, altered the slip angles of the tires, putting the car in a better frame of mind to turn, and us just in a better frame of mind. Rear end breakaway could be provoked, but as the British say, in "an untidy manner". The high rev limit gives the car considerable flexibility.

Complaints center around the poor-quality and small dash and window controls, and the trunk latch is stiff and difficult to actuate without undue force. But aside from this, the TI is a first class sports-touring type car with near perfect brakes, and a lively, well-proven 4 cylinder dohc engine. For the frustrated family man who wants to put some fun in his motoring—real Italian style fun that is—don't walk, but run to your nearest Alfa dealer.

Jesse L. Alexander

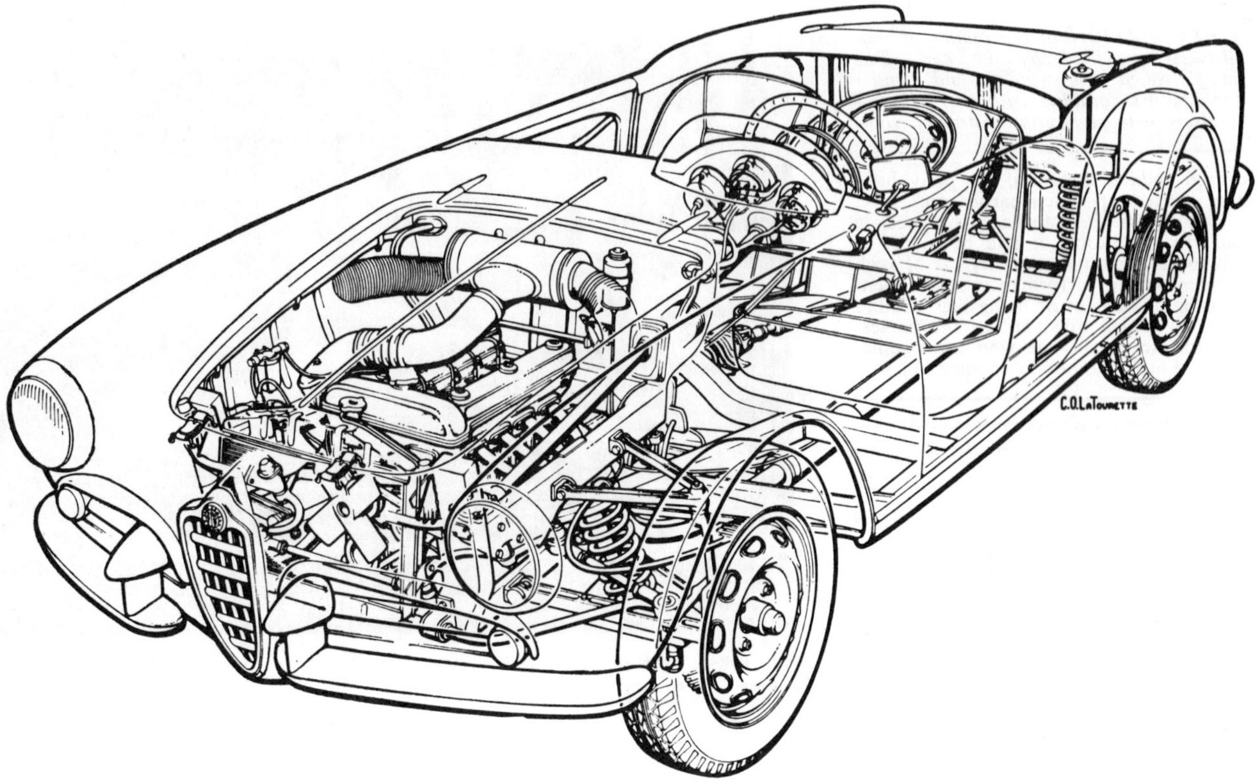

COMPETITION TUNING: ALFA SUPER SPIDER

Tom O'Brien

Are Alfas on their way out of the winners' lists thanks to the SCCA's new performance categories? The answer is a resounding "No," but you Veloce drivers may stand a better chance of dusting off the big-bore machinery by "prodifying" your cars.

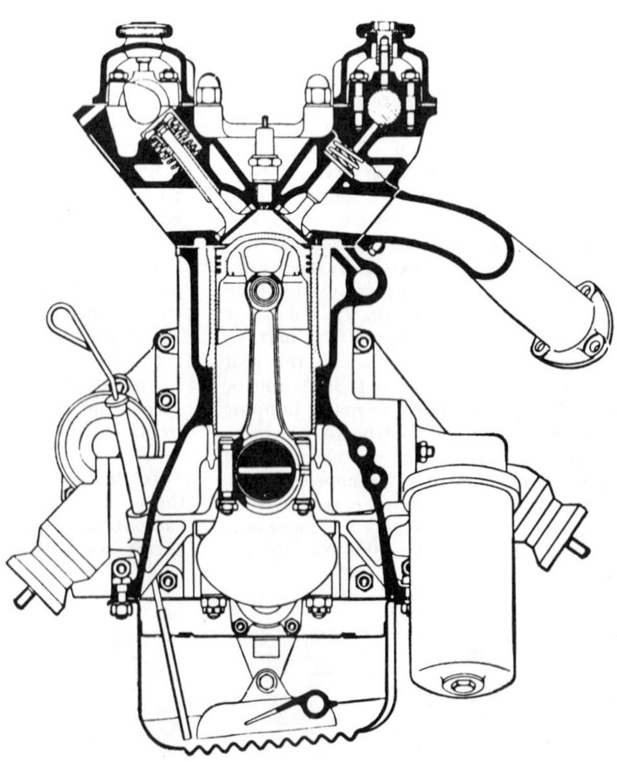

1. With the engine out of the car and disassembled, the first step is to take a light cut (.005 inches) on the head gasket sealing surface of the block. Removing any more will probably require machining of the cylinder liners to avoid distortion of the head. Positioned in the block, the liners may project above the block surface up to .005 inch. Check this without the O-rings in place. No other block renovations are required, but a word of caution: find a machine shop that's interested in doing a good job. This is essential since some effort must be expended to get the relatively short liners square with the surface to be ground. The next step is to hone the liners to give .006 inch piston skirt clearance at room temperature. Since Alfa pistons don't have expansion control inserts, you can count on the pistons swelling as the engine warms up. Don't be alarmed by the apparent loose fit of the pistons in the liners, but keep it in mind as you resist the temptation to rev high with a cold engine.

► How are the Alfa Veloces going to stack up against two-liter machinery in the SCCA's 1960 performance Class E when the record books are closed at the end of the year? At this point it's a moot question, but it's a tribute to the real performance of the Alfa that it's been placed in direct competition against such bigger-displacement machinery as the Porsche Supers, the Triumphs, the Healeys and the new Daimler 250's. Winning in an Alfa may no longer be the foregone conclusion it was under the old classification and the competition can be expected to be just as fierce next year.

There *is* a substitute for cubic inches however — at least as far as production class racing is concerned. To be in the running, Alfas have to be "prodified," that is, tuned and worked over to the hilt — within stock limitations. The 1290 cc Super Spider engine, built mostly of aluminum, is rated at 103 bhp in stock form. That's 1.3 horsepower per cubic inch, a respectable figure in anybody's book. However, the engine does have its faults and compromises which must be taken into account in preparing it for production racing, to ensure that all those horses are pulling for you, but without sacrificing reliability and flexibility.

Commonly known as "Sebring specifications," these Alfa urgings may be expected to result in about 110 or 112 bhp. Their main area of attention is in valve timing. The piston grooving, which results in a lowered compression ratio, is counteracted by milling of the block and head to regain a 9.5 to 1 compression ratio. Since engines vary, even off production lines, use caution in the milling and don't exceed cuts totaling .030 inches between the head and the block.

SHORT-COURSE TUNING

Tuning the Alfa for racing depends largely on the nature of the race. Each course may require different combinations of valve timing, carburetion and the like. On a short course, that is, one with a short straightaway, what's needed is a low gear ratio and good mid-range power. Try setting the ignition at 47° at 6000 rpm. Blip the engine to 7000 to be sure 47° is not exceeded. Small carburetor venturis for good low-speed characteristics are also in order.

The 4.1 rear-end ratio found on the stock Veloces is usually too high for most courses, while the Giulietta's 4.55 cogs are often just right. If after fitting a 4.55 set you find you need a higher rear-end ratio, tires of a larger circumference may be used. The 155 x 15 Pirelli Cinturatos that come with the car have a low profile so 5.00 x 15's with the 4.55 rear end may be just right, because of their greater rolling distance per revolution. The Cinturatos check

2. The pistons should be grooved as shown on the right, to keep them from striking the head in the event of con rod stretching at high revs. To prevent the top piston ring land from becoming the thrust surface when the piston "rocks" in the bore, the lands must be turned down at least .020 inch smaller than the bore. If this is not done, the top ring land will smear into the top ring groove, preventing the ring from effecting a good seal. After machining, radius all sharp edges. The final step in preparing the block assembly is to have the crankshaft, connecting rods and new connecting rod bolts magnafluxed. Along with the pistons, piston rings, flywheel and clutch pressure plate, they should be statically and dynamically balanced. The engine preparation described here was done by Competition Services in Paterson, New Jersey. The balancing and magnafluxing were carried out by Dick Simonek, East 29th Street in Paterson.

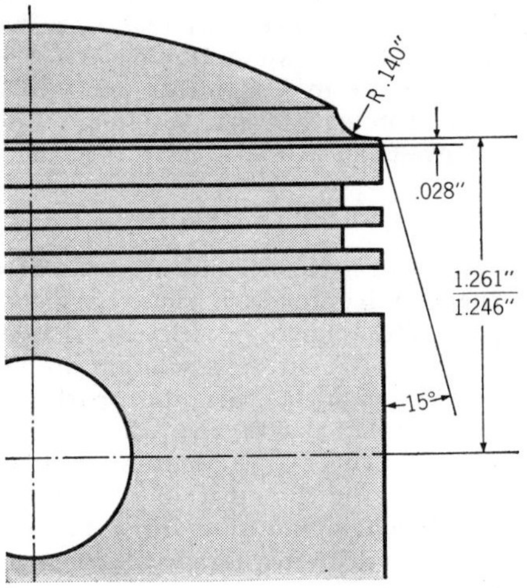

PHOTOGRAPHY: TYPOND

3. The Alfa's high horsepower-per-cubic-inch figure is for the most part due to its excellent breathing. Any improvements in this area should result in bettered performance, especially over 5000 rpm. First carefully match the ports in the intake manifold and the head. The ports should be polished, but not greatly enlarged. No major changes are called for in the carburetion-exhaust systems, but careful modifications may result in performance gains. A word to the wise: there is far more to be lost than gained. The Weber 40DCO3 carbs on the Veloce are equipped with 28 mm venturis. Larger (30 and 32 mm) venturis can be used, reducing venturi pressure loss at high speeds, provided you open up the exhaust system. Cut out the large muffler, replacing it with a piece of straight pipe, but leaving both resonators in place. The car can still be driven on the street if a "racket-buster" and a little discretion are used. Be sure your carbs are the latest type. There should be two 7 mm diameter holes drilled through the bowl cover, visible when the jet cover plate is removed. If you have the older type, the holes will be 4 mm, but they can be drilled out to 7 mm. The change was effected on engines starting with No. 30831 and on some between 30775 and 30821. The letter "M" may be found on the cover to indicate the carb was modified during production at the factory.

out at 75½ inches per revolution. Dunlop R-5 5.00 x 15's measure 77¾ inches per revolution while Dunlop R-3 5.50 x 15's come to 81¾ inches per revolution. Another 5.00 x 15 you might want to investigate is the Goodyear model.

LONGER TRACKS

On longer circuits and in cases where engine reliability is a factor, a higher gear ratio should be selected. Larger tires may be an easy solution to the gearing problem. Set the spark advance at 44° at 6000 rpm. If there are long straights, larger venturis may be tried, but be sure that the power loss coming out of the slow turns does not offset any high-speed gains. A point to remember is that you are looking for the lowest possible lap times, not just the fastest speed down the straights.

Carburetor venturis should be chosen to suit the course and jets chosen to suit both the venturis and the weather. Larger venturi areas require exactly proportionally larger main jets, at least over the narrow range we're concerned with. Since the Weber has a high-speed enrichening device operated by main venturi vacuum, it may be necessary to use a richer (smaller number) air correction jet to get the correct high-speed mixture when using larger venturis. Remember, colder weather requires richer jets; humid weather, leaner jets. An exhaust analyzer is undoubtedly the answer for on-the-money carburetion. At any rate, use one of the "super premium" fuels.

A word about ignition. The Marelli distributor is basically a good system; it can be spun to higher proportional rpm than the engine will deliver, without point bounce, and the coil voltage is sufficiently high, but there are two easily-cured drawbacks. The most serious is the lack of a bushing between the distributor shaft and body which in a short time permits the shaft to gall and wobble, causing erratic firing. If unchecked, the shaft may seize, bending the oil pickup drive in the process. Have the distributor bushed by a competent machine shop, preferably when the unit is new. The second problem is crossfire. Separate all high-tension leads from each other for their entire length and be sure the top surface of the coil is kept clean to prevent shorting at that point. Regarding spark plugs, try starting with a set of KLG 220's. Used with a set of warm-up plugs (such as Champion N-5's) we found they would almost never foul. A few hard practice laps should show if a colder plug is needed.

DRIVING AND MAINTENANCE

There are as many opinions on how high to rev the Veloce as there are drivers. The factory red-lines the tachometer at 6800 rpm. Audible valve float occurs at 7700 with anything less than brand-new valve springs. The maximum safe piston speed is generally agreed to be 4000 feet per minute and it occurs at 8000. Take your pick. Shifting out of second at 7250 puts you at 5000 in third – to go beyond this would seem foolish.

It's difficult to determine how often a competition engine should be overhauled. Rather than effect a complete tear-down and rebuilding after a given number of miles or hours, a four-point maintenance plan might be followed:

1. After each race, give the engine a complete and thorough tune. Check and record the compression reading of each cylinder and adjust the valve clearances.
2. After every third race, remove the oil pan and inspect the bearings. Replace any having a copper-like hue. If any bad bearings are found, check out the crankshaft with a micrometer. This latter inspection should be carried out after every exceptionally long or grueling race, or if anything is suspect in the lower end.
3. At least after every fifth race, and more often if possible, a complete valve job, including inspection of valve springs for weakness, should be undertaken. Decarbonizing can almost always be counted on to revitalize engine performance.
4. About midseason, install a new set of piston rings. Don't depend solely on the compression gauge to tell you when they're needed. If your engine just doesn't have it, even after a valve job, broken or tired rings are probably the reason. Piston ring failure occurs in a high-revving engine like the Alfa more often than you might expect. While the con rods are out of the engine, have them magnafluxed and replace the con rod bolts.

The Super Spider has an impressive competition record and is an excellent road car. To keep it winning and pleasurable to drive, however, three things are necessary: good preparation, thorough maintenance and sensible driving.

Tom O'Brien, the author of this story, will do his best to reply to any inquiries about "prodifying" the Alfa. You may write him at Competition Services, 99 Cross Street, Paterson, New Jersey. The estimated cost for the work described in the article is between $400 and $600, depending on the condition of the engine and the garage employed.

4. To prepare the head for surfacing, remove the lower row of exhaust manifold studs (shown projecting at the top of the photo). They will come out easily if the head is immersed in boiling water. Also remove the timing chain tensioning device. Inspect the chamfer on the periphery of each chamber; it will have to be recut if more than .015 inch is taken off the head. The valves should be lapped in to check seat width. The effective seat width, usually quite wide (about ⅛-inch), can be narrowed to about 3/32 inches by opening the port area of the seat. This can be done by using a 70° reamer piloted in the valve guide. In addition to aiding gas flow, this will increase contact pressure between the valve and seat and tend to discourage carbon build-up. Before installing the valves in the head, check all springs for free length and force compressed. The readings should be as follows:

Outer:	Free		43.0 44.6 mm
	Compressed	24-25 kg	22.5 mm
Inner:	Free		39.35—40.95 mm
	Compressed	14-15 kg	21.0 mm

Alfa springs may lose their strength quickly and need to be replaced often. Checking the free length usually uncovers the offenders. Don't try to increase pressure by putting shims under the springs; this will only result in coil bottoming.

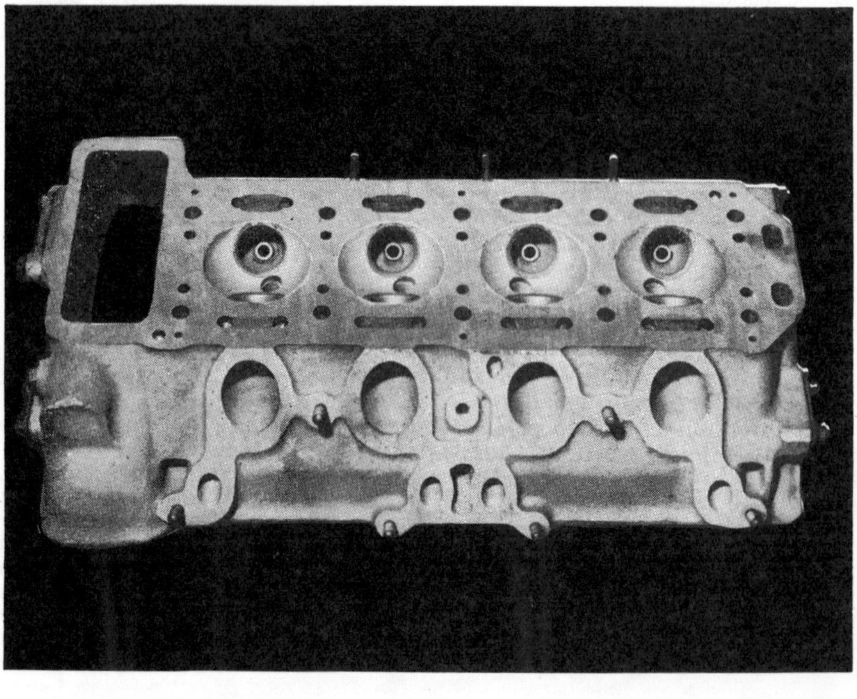

5. The straightforward Alfa cam and valve layout is shown right. After installing them in the head, check valve clearances with a feeler gauge. Exhaust valves should be set at .020—.021 inches and the intakes at .015 to .016 inches. By setting the valves .002 to .003 inches tighter, a noticeable increase can be had above 6000 rpm if the straight exhaust and larger venturis are used. But here is where it's possible to go astray, since the engine will fall flat on its face between 4000 and 4800 rpm. This isn't a problem in first or second gear since the car accelerates through this range rather quickly, but the lack of engine flexibility may be a drawback on some courses since it will compound transmission ratio faults. Fourth gear at 4800 rpm is equivalent to 6480 in third — no problem here. But 4800 in third equals 6960 in second, resulting in a problem if there are any corners in that rpm range, and there seems to be an uncommon number of them on long courses where changing to a lower rear-end ratio would do more harm than good. By using larger venturis and straight exhaust and tightening the valves that extra .002 or .003 inches, you may find yourself sitting in a corner in third gear with your foot to the floor waiting for things to happen or over-revving in second, the penalty for increased top-end go.

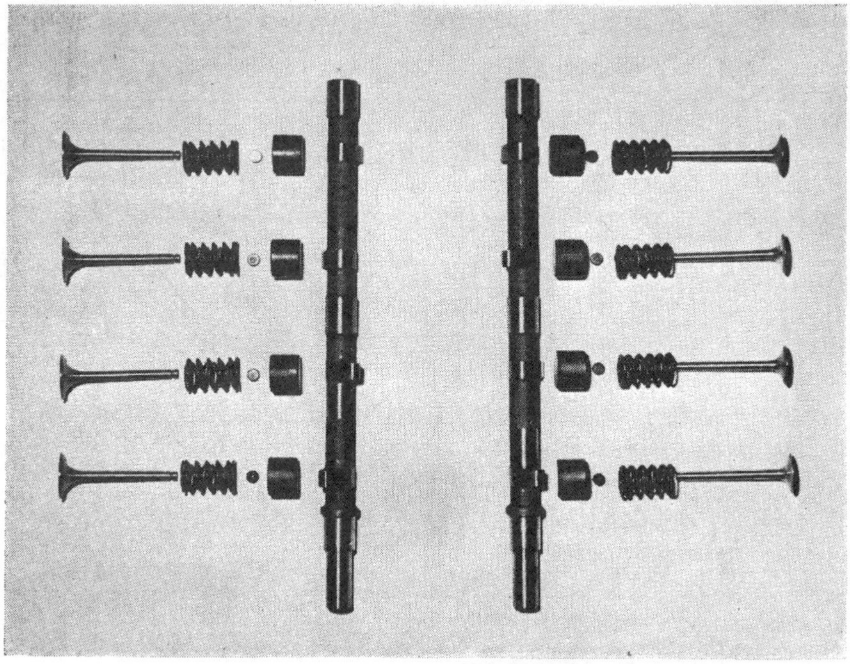

6. In reassembling the engine, place strips of clay on the piston heads to check clearance of the cylinder head, turning the engine by hand. Minimum clearance is .060 inch. Use the shop manual as a guide and install new gaskets and seals throughout. Follow recommended torque settings and check main and connecting rod clearances and crankshaft end play. Use new lock tabs, bolts and nuts on the connecting rods. Probably the only chance for error in assembling the engine is in installing the connecting rods, which are not numbered like other parts and are not symmetrical. As a check after assembly, be sure the small ends of the rods are centered on their wrist pins.

7. The first step in timing the cams is to tense the upper timing chain. Using a screwdriver behind the tensioning sprocket, draw the chain taut and lock it in this position. The spring behind the sprocket is insufficient. The next step is to degree the cams. Remove the bolts in the vernier adjustments (shown above) and loosen the large bolt on the front of each cam. This will permit the cams to be rotated independently of the sprockets and chain drive, facilitating the timing operation. Valve timing for stock settings is as follows:

	Intake	Exhaust
Opens	34° BTDC	63° BBDC
Closes	63° ABDC	30° ATDC

Make sure the clearances are set right, since minor variations here result in large changes in valve timing. For those who want to get blazing top-end performance, the Sebring timing specifications might be tried. They have a whopping 80° overlap compared with 64° for the stock settings, but they're very hard to set exactly right:

	Intake	Exhaust
Opens	46° BTDC	65° BBDC
Closes	65° ABDC	34° ATDC

GIULIETTA SPRINT

A really Grand Touring car for two (or occasionally four) people

ALFA ROMEO GIULIETTA SPRINT is a rather largish name designating a quite smallish coupe that was the first of the present crop of *Gran Turismo* cars, and even now, after nearly 6 years, the car fits the meaning of the term as well as anything since introduced. To be sure, there are faster cars and certainly more flamboyant ones, but there are precious few that can match the Giulietta for clean and classic styling, or for the refinement of its mechanical elements. Blood lines will tell, as our friends in the horsy set are so fond of saying, and in the Giulietta one can see the end results of many years of Alfa Romeo's sports and Grand Prix racing history.

The styling of the Sprint is so devoid of frills that it leaves one a trifle short of descriptive material. There are no fins, chromed spears or sculpturing—just smooth flowing lines. Naturally, such simple styling lacks the flash needed for mass appeal, but then so do a great many things connected with cars which have been designed with the enthusiast in mind.

For a car of such small overall dimensions, the Sprint has a marvellous amount of usable room inside. In fact, it is available as a 4-seater (which our test car was) and, although the rear seats are hardly spacious, 4 people can actually be transported about in the car. Large adults will not be too comfortable in the rear seat, but the car is perfectly suited to the man with a wife and no more than two children. Here though, as always seems to be the case, we encounter a slight snag. Trunk space in the Giulietta Sprint is rather limited and if one travels with more luggage than is required to contain the bare necessities for a week-end jaunt, there is sure to be some overflow. In the 2-seater version of the Sprint this overflow is dealt with by loading the luggage onto a shelf provided for that purpose behind the front seats, and in the space taken by the rear seats in the 4-seater. This is quite unhandy, but it also lends flexibility to the car and makes it more of a day-in, day-out proposition. One does not have to ponder long to bring to mind at least one or two popular sports cars with as little trunk space and no room at all for the overflow.

In the apportionment of interior space, the Giulietta is one of those rare examples of design in which the driver has been given primacy over all other occupants. For example, the engine and transmission have been offset to the right to provide more room for the driver's feet, and the fresh-air inlet is also placed where it will do the driver the most good. This, naturally, does not mean that the passenger has been left to suffer, but the driver—who in most cases is the person who paid for the car—has been given first consideration.

Making an entrance into the car is complicated only by the fact that it is very low; the door is large and opens wide. Also, it is high enough to clear most curbs. From behind the wheel one gets an excellent view in all directions—due to the large windows and very thin corner posts supporting the roof, and because one does not have to look under or through the steering wheel rim. The seats of our test car were leatherette covered, with bolstered edges that offered fine support against the side forces imposed by brisk cornering. This is wise indeed, for the basic controls—steering wheel, shift lever, clutch, throttle and brake—are so well arranged and perform their diverse functions with such willing precision that the most brisk type of cornering is surely inevitable.

The reporting-type instruments: tachometer, speedometer, oil pressure and temperature, etc., are in a hooded cluster immediately in front of the driver and can be read at a glance. Small toggle switches located along the lower edge of the panel operate the instrument lights and windshield wipers, while knobs control the choke, engine-idle, lights and heater. These knobs, instead of being marked with letters to indicate their function, have small picture-markings that mean the same thing in any language.

The mechanical elements of Alfa Romeo's Giulietta-series cars are, if anything, more delightful than the car's appearance. The engine is made almost entirely of light alloys—primarily aluminum—and has such features as a pair of camshafts above the inclined valves at the top of the engine and a 5-mainbearing crankshaft below. This engine is offered in two stages of tune: standard tune, which is the type in our test car and develops 92 bhp (SAE) at 6000 rpm; and the super-tuned version, which has more radical valve timing and a pair of double-throat Weber carburetors, along with various internal refinements, and pumps out about 103 bhp, also at 6000 rpm.

Actually, these ratings are somewhat misleading, for the Super version is *much* faster than its advantage over the standard, in terms of advertised bhp, would indicate. Of course, both cars are capable of very spirited performance, amazingly so, considering their small engine displacement, and the engines will, if pressed, deliver revolutions at a truly satisfying rate. The standard will get up to about 7000—though the power isn't with you much above 6500—and the super will crank out plenty of power even if pushed up to 8000, some 2000 rpm above the nominal power peak. Naturally, there is a "red-line" down lower, at 6800 rpm, but many of the people racing Giuliettas have exceeded 8000 repeatedly without bringing mechanical disaster down on their heads. This is, as goes without saying, not really very wise or necessary, for we have cut a 16.9 sec. standing ¼-mile with a super without going over 7000 rpm. In any case, the Giulietta engine is a marvel and is establishing a really fine record of reliability.

Unhappily, there is one area in which the Giulietta does not score so well; in the effectiveness and reliability of its electrical system. If we accept the evidence presented by experience with three staff-owned Alfa Romeos, and that provided by the many letters of complaint from the segment of our readership that owns Alfa Romeos, we must conclude that the car's electrical system just isn't up to the job. This is a great pity, for the rest of the car, especially in the later types (1959 and on), just never lets you down. Actually, once the individual owner becomes accustomed to the vagaries of the electrical system, it can be *made* to operate satisfactorily. But, in view of the overall excellence of the machine, which is as good as anything available today, it is silly to have such a marginal-quality electrical system. For those interested in the exact nature of the car's illness we can only say, at this time, that it centers around a generator, battery and starter that just aren't quite "heavy-duty" enough for North America. The margin is so slight that any small malfunction anywhere along the line causes a disproportionate amount of trouble.

Driving the Giulietta, whether in traffic or on the open road, is a pure pleasure for people who really like to drive. The car has tremendous agility and quickly translates the slightest movement of the controls into a change of pace or direction. Indeed, some first-time drivers of the car will find that they tend to jerk it about, before they become accustomed to its hyper-responsiveness. In the end though, when one is thoroughly familiar with the Giulietta, this responsiveness becomes a real delight. It should be stressed, too, that the car is not sensitive to

the point of being twitchy. Quite the opposite is true, and stability increases in direct proportion to road speed.

When pressed really hard, the Giulietta begins to lose the pronounced understeer that it shows while negotiating very slow, tight corners and becomes quite neutral in its steering response. Due to the very soft suspension, there is a considerable amount of tilting and leaning, but this does not upset the handling in the least. After the car is started into a fast corner and the initial roll-angle is taken, it seems to brace itself and never gives the driver any feeling of being tippy. Some complaints have been heard regarding the amount of roll—but this has come from people who haven't driven a Giulietta. The rolling doesn't affect the stability and one certainly can't have a soft ride without some roll.

The suspension used on the Giulietta is not in the least unusual; unequal-length A-arms are used in the front and a system of links at the rear, with coil springs all around. Unusual or not, this suspension gives very good results. Particularly impressive is the civilized behavior of the "live" rear axle, which functions at least as well (in quality of ride and cornering power) as a couple of independently-sprung machines that come to mind. Part of this is undoubtedly due to the lightness of the axle assembly, consisting of a light-alloy center section, which carries the gears, with thin, steel axle housings bolted on. The brakes, although very large, do not add much weight because they are also of aluminum, with iron liners.

Accessibility for service is quite good: the carburetor, distributor, spark plugs and fuel filter are all easily reached from the top, while the oil filter is not too badly hidden beside the sump, underneath the car. The only service problem is posed by the battery, which is tucked away in the right rear corner of the trunk where even the simple task of adding water becomes a real project. The spare tire is also in the trunk and there is a tie-down for the jack (which, in use, plugs into rectangular tubes located at four points along the lower sides of the body) and the tool kit is carried in the dished side of the spare wheel. And, as is the case with the spyder, there is a fuse-box located just under the dash on the left side of the steering wheel.

As a complete package, and in spite of its sometimes inadequate electrical system, the Giulietta Sprint is the kind of car that no enthusiast should ignore. It has a double overhead camshaft engine, an all-synchromesh transmission and unbelievably large, finned, aluminum-drum brakes, which give the kind of results that make one wonder if the disc-type brake is taking over after all. Add to this the panel-fit and the general finish of the various appointments and you have what is for many a nearly irresistible automobile.

ROAD TEST

GIULIETTA SPRINT

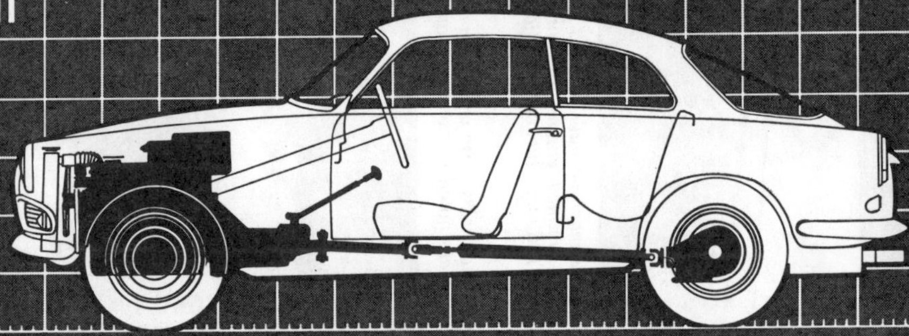

DIMENSIONS

Wheelbase, in.	93.7
Tread, f and r	50.6/50.0
Over-all length, in.	156.7
width	60.4
height	51.9
equivalent vol, cu ft	284
Frontal area, sq ft	17.4
Ground clearance, in.	7.0
Steering ratio, o/a	n.a.
turns, lock to lock	2.8
turning circle, ft.	36
Hip room, front	52
Hip room, rear	49
Pedal to seat back	43
Floor to ground	11

CALCULATED DATA

Lb/hp (test wt)	26.5
Cu ft/ton mile	66.8
Mph/1000 rpm (4th)	16.7
Engine revs/mile	3590
Piston travel, ft/mile	2120
Rpm @ 2500 ft/min	5090
equivalent mph	85.0
R&T wear index	76.1

SPECIFICATIONS

List price	$3878
Curb weight, lb.	2090
Test weight	2440
distribution, %	55/45
Tire size	155–15
Brake lining area	132
Engine type	4 cyl, dohc
Bore & stroke	2.91x2.95
Displacement, cc	1290
cu in.	78.7
Compression ratio	8.5
Bhp @ rpm	92 @ 6000
equivalent mph	100
Torque, lb-ft	79.6 @ 4000
equivalent mph	66.9

GEAR RATIOS

4th (1.00)	4.55
3rd (1.35)	6.17
2nd (1.96)	8.92
1st (3.31)	15.04

SPEEDOMETER ERROR

30 mph	actual, 27.3
60 mph	56.7

PERFORMANCE

Top speed (est), mph	103
best timed run	96
3rd (6500)	80
2nd (6500)	55.4
1st (6500)	32.8

FUEL CONSUMPTION

Normal range, mpg	25/35

ACCELERATION

0-30 mph, sec	3.9
0-40	6.1
0-50	9.1
0-60	13.2
0-70	18.1
0-80	25.8
0-100	
Standing ¼ mile	19.2
speed at end	71.8

TAPLEY DATA

4th, lb/ton @ mph	175 @ 44
3rd	250 @ 39
2nd	360 @ 35
Total drag at 60 mph, lb.	73

ENGINE SPEED IN GEARS

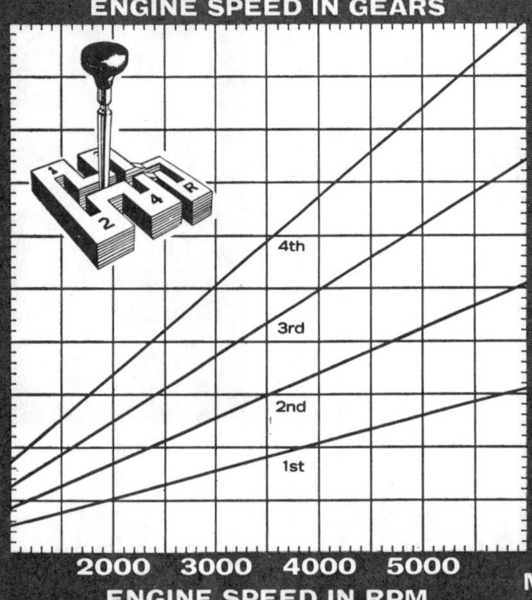

ACCELERATION & COASTING

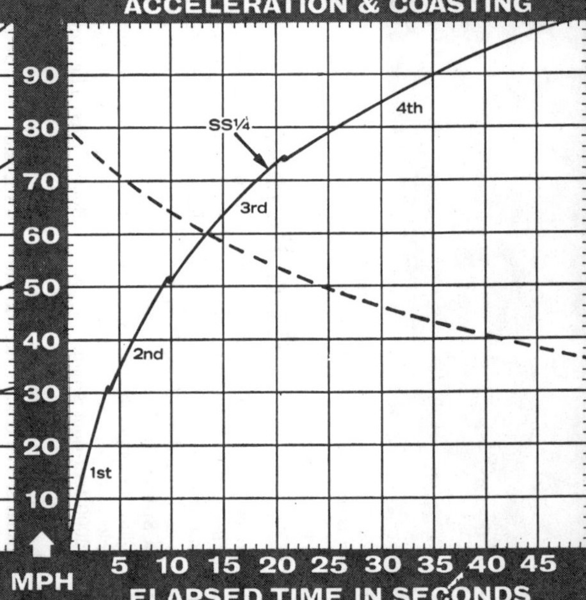

Alfa Romeo's Giulietta Sprint 1300 coupe is a popular import built on a 94-inch wheelbase. Power is supplied by double overhead-cammed engine of 78.65 inches displacement.

Alfa Romeo

Alfa Romeo of Milan, Italy, include six models in their 1959 line which, according to information available at press-time, remains unchanged for 1960. These are the Giulietta Spider roadster and Sprint coupe; the Giulietta Super Spider roadster and Sprint Veloce coupe; the two-liter roadster and the 2000 four-door sedan.

Engines used in these cars are similar in design. All have four cylinders and double overhead-camshafts. The engine fitted to the Giulietta Spider and Sprint displaces 78.65 cubic inches and utilizes a twin-barrel Solex carburetor; it has a power output of 91 bhp at 6000 rpm. The engine in the Super Spider and Sprint Veloce has the same displacement, two horizontal Weber carburetors and produces 103 bhp at 6000 rpm. These engines are coupled to a four-speed, full-synchromesh transmission.

The engine in the two-liter roadster and 2000 four-door sedan has a 120.5-cubic-inch displacement. Power output in the version fitted to the sedan is 120 bhp at 5300 rpm and 131 bhp at 5800 rpm in the version fitted to the roadster. These engines are coupled to a five-speed synchromesh transmission.

Suspension systems are similar on the Alfa Romeo cars. The front is independent with transverse linkage, coil springs, stabilizing bar, and telescopic shock absorbers. The rear suspension consists of a rigid axle housing with upper triangular thrust rods, lower radius rods, coil springs, and telescopic shock absorbers. ■

Alfa Romeo, Milano, Italy. Distributed in U. S. by Charles Rezzaghi, 815 Van Ness Ave., San Francisco, Calif.; Hoffman Motor Car Co., 487 Park Ave., New York 22, N. Y.; Hoffman of California, Inc., 9130 Wilshire Blvd., Beverly Hills, Calif.

The Giulietta-Berlina 1300 four-door sedan follows the styling of all Alfa Romeo cars.

Alfa Romeo's 2000 Spider is a slightly larger sports machine built on a 98-inch wheelbase. Also powered by a four-cylinder overhead-cammed engine, this unit is rated at 131 bhp at 5800 rpm. Top speed is a respectable 126 mph; fuel consumption: 20 mpg.

The smaller Alfa Romeo Spider is built on an 87-inch wheelbase with an overall length of 152 inches. Its engine is rated at 91 bhp. Alfas are noted as excellent handling cars.

The big Alfa Romeo 2000 four-door sedan only has a 107-inch wheebase.

ITALIANS BUILD SUCH EXCITING CARS

Alfa Romeo

ALFA ROMEO MILANO

New Low Prices!

DISCRIMINATING owners and competition drivers everywhere testify there's nothing to compare with Alfa Romeo performance and pride of ownership. For, built into every Alfa Romeo is the flair for great design and the technical superiority born of Italian passion for flawless craftsmanship that has been proved by countless victories on road and track. Plan now to see and test drive any of the great Alfa Romeos to experience the rocket acceleration, the silent power, the effortless handling and great roadability that will make this your proudest personal possession.

	OLD	NEW
GIULIETTA SPIDER 91 hp, 96 mph	~~$3515~~	**$3150***
GIULIETTA SUPER SPIDER 103 hp, 118 mph	~~$3885~~	**$3450***
GIULIETTA COUPE 91 hp, 103 mph	~~$3838~~	**$3495***
GIULIETTA VELOCE COUPE 103 hp, 118 mph	~~$4144~~	**$3795***
2000 SPIDER 131 hp, 126 mph	~~$5365~~	**$4650***

Giulietta Coupe $3495*

* Suggested POE Prices

IMPORTER

HOFFMAN MOTORS CORPORATION

NEW YORK: 443 PARK AVENUE, PLAZA 9-5120 • BEVERLY HILLS, CALIFORNIA, 9130 WILSHIRE, CR 4-8231 • CALL OR WRITE FOR NEAREST DEALER

COMPETITION TUNING

ALFA ROMEO

BY TOM O'BRIEN &
DON TYPOND

This ported and polished Alfa Veloce head has been surfaced to insure good gasket sealing, valve guides chopped.

Stock liner can be bored out to accept oversized piston. Liner for the 1500 piston (right) requires machining block.

Marks show where cylinder liner is to be milled away for clearance of con rod when stroker crank is to be used.

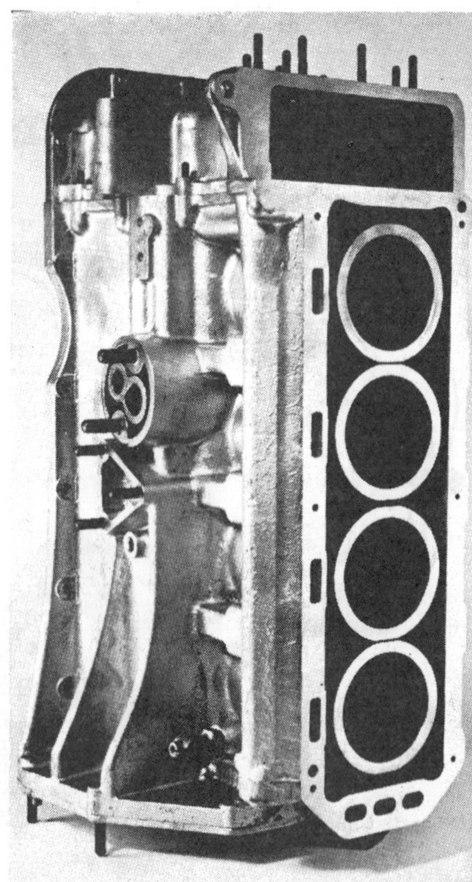

Alfa block with liners in place. The surface of the block has been ground true and the liners machined to fit.

What's On the Menu? POWER!

HUNGRY FOR MORE POWER? Maybe you can find something here to satisfy your appetite. Please notice that we call this a menu rather than a recipe. It is not intended to be a do-it-yourself instruction book for the home mechanic, but rather a list of parts and operations from which the discerning *aficionado* can select the dish he wants cooked. At best a recipe is only a guide. It lists ingredients and basic quantities, but it's up to the talent of the chef to combine them in the proper proportions necessary for the ultimate in cuisine.

The basic ingredient here is the Alfa Romeo Veloce engine — as it comes from the box, one of the finest production engines in the world. For this reason it isn't necessary to do very much to make the engine suitable for production class racing, and since most clubs forbid modifications anyway, we don't feel it's worthwhile to repeat the engine rebuilding procedure as outlined in the shop manual. Instead, we would like to show the various routes to more power, from which you can select the one that best suits your kind of driving.

To help you decide what modifications are in order for your needs, let's delve into some of the possibilities, their advantages and disadvantages, and exactly what is involved.

The crankshaft is the determining factor in planning the rest of the block components, so we'll start there. If you decide to retain the stock stroke of 75 mm (2.992 in.), check the shaft dimensionally and set it aside until later. Should you decide to increase the stroke, keep in mind the fact that any increase will require that the cylinder liners be notched to clear the rod bolts, and an increase of more than 1/4 inch will necessitate grinding work on the block to allow the rods to swing.

Generally speaking, reliability is not adversely affected to any great extent by stroking, but the choice of stroke will be affected by the choice of pistons. A variety of pistons are available, some of which can be used with a stroked crank. The first to be considered is the stock piston which is light and stands up extremely well under hard use. The largest oversize, 0.024 over standard, doesn't help much on cubic inches, but you can use it with a stroke increase of up to 1/8 inch. To get the necessary clearance between the piston and the head, the piston must be "grooved." To do this the piston is chucked in a lathe and the edge of the dome cut down, by half the stroke increase, with a radiused tool. This method of increasing displacement also results in a sizable boost in compression ratio. Incidentally, a badly scored crank can be saved by making it a stroker, thus killing three birds with one stone.

For a stroke increase of more than 1/8 inch, you must use special pistons in which the wrist-pin bore has been relocated. Jahns makes such pistons for any stroke in stock and oversize bores. We are currently modifying an engine using these pistons with a 3/16 stroked crank made by Crankshaft Company, Los Angeles.

If you want cubic inches without a stroke increase, the well-known "1500" piston and liner kit sold by Ram Engineering, Rome, Italy, is the answer. For anything less than out-and-out racing, the resulting 1470 cc and 10 1/2 to 1 compression ratio is hard to beat. Reliability at extremely high revs may be reduced due to the higher bearing loads, but this set up supplies what the Alfa needs for town driving . . . low end and mid-range torque. The block must be bored to accept the new, larger liners, but complete blueprints for this operation are supplied as part of the kit.

So far our changes have been somewhat extreme, but it doesn't mean that the engine won't respond to less expensive operations. Simply increasing the compression ratio without boring or stroking can do a lot. Don't get the idea that you can saw the head in half however . . . if you take off more than 0.020 you're pretty close to the valve seats. You can mill the head gasket surface of the block and timing chain cover, but the cylinder liners must be machined and the pistons "grooved" as well. This means removal and complete disassembly of the engine, with its accompanying cost.

Our recommendation is to install special high compression pistons. This has the additional advantage of avoiding any irrevocable machining. Compression ratios between 10 and 11 to one, with the use of super premium fuels, are possible without encountering detonation. There is little objection to raising the compression ratio, within reason, since bearing loads in a high revving engine are basically inertial.

The higher compression ratio will require a change in the spark advance curve. This is a cut-and-try process, but start with less spark advance in the mid-range, while keeping the static and maximum advance settings stock.

Although hemispheric combustion chambers, direct acting double overhead cams, straight through porting, and Weber carburetors are considered to be the ultimate in design, important changes to the Alfa head can result in big horsepower gains.

The place to start here is with the valves, since the head porting will depend upon what is done to the valves and valve seats. Close examination will reveal that the intake valve seats are about 1/16 inch larger in diameter than the intake valves. Larger valves, which take full advantage of this condition, are available from Ram Engineering. The throat of the valve seat should be opened up with a 70° reamer to leave a seat width of 3/64 to 1/16 inch. The head can now be ported to take advantage of the increased throat diameter at the valve seat.

The same basic process can be applied to the stock intake and exhaust valve and seat combination. The stock seat width is definitely on the high side, so much so that carbon deposits sometimes build up between the exhaust valves and their seats. Reduce the valve seat width to 1/16 inch by enlarging the throat with a 70° reamer. The effective discharge area can be increased by as much as 15 percent by this method using the stock parts.

We are currently installing cut-down Jaguar valves in a head into which large screw-in type American valve seats have been installed. This head will be used on a 1500 cc engine since the expense would probably not be justified on anything with less displacement.

When porting the intake passages, be sure to match the manifolds and head. There is generally severe mismatching at this point. A controversial subject is whether or not to chop off the ends of the valve guides which protrude into the port. Most of the objection centers around the reduction of guide length to support the valve. In our experience this causes no problem, but there is one that does exist pertaining to resurfacing the valve seats. Most seat refacing tools use

(continued on following page)

ALFA ROMEO

continued

tapered pilots which do not center properly when installed in cut down valve guides. Probably the only advantage in cutting the guides is the ease with which the porting can be done.

The dual 40mm Weber carburetors are fine for street use as they stand, even when used with larger than stock engines. Venturis larger than the stock 28 mm can be used if a sacrifice in flexibility can be tolerated. The 30 mm venturis are ideal for the racing driver who drives his car to the events, while 32 mm venturis can be used in all-out racing engines, such as those used in the factory "Sebring" cars. The fuel-air metering ability of anything larger than a 34 mm venturi is, in our opinion, questionable in the 40 mm carburetor.

Volumes could be written on the jetting of these carburetors, so let's just say that within the narrow range with which we're concerned, increases in the main jet *area* are proportional to increases in venturi *area*. Larger venturis also require richer air correction jets. The factory "Sebring" jetting is 1.30 mm main jet and 1.70 mm air correction jet, using a 32 mm venturi. Remember, air correction jets with smaller numbers are richer.

The camshafts chosen by Alfa Romeo for the Veloce engine combine acceptable idle and low end performance with good mid-range torque and very high revving ability. You can't ask for much more than that in a camshaft that has to suit everybody. But there are a few tricks you can try if you want more power in a particular range. By setting the intake cam six *crankshaft* degrees advanced from the stock position, the engine's low-end torque can be greatly enhanced. A definite increase in compression pressure can be measured which may require you to retard the spark advance a few degrees. Things stop happening over 6500 rpm with this set-up, but for town driving it really makes a difference.

At the other end of the performance scale is the factory "Sebring" timing. Setting the intake valve clearance at 0.012 and the exhaust valve clearance at 0.017 can produce timing of 46-65, 65-34 if the intake cam is advanced 5° crankshaft, and the exhaust cam is retarded 1° crankshaft. With nearly 80° of overlap, this setting is for top end. Under 4000 rpm the engine is flat, although strangely enough it is not erratic.

Special cams are available from several of the custom cam grinding companies, so take your pick. If you have special requirements in cam design, Ravesi Racing Cams, Brighton, Massachusetts, is familiar with the Alfa engine.

If possible, have the crankshaft, connecting rods, and connecting rod bolts magnafluxed, and the rotating and reciprocating parts balanced statically and dynamically. Before sending the parts to the balancing shop, dowel the pressure plate to the flywheel to prevent movement with resulting imbalance. Also mark the crankshaft and flywheel in their proper position, one to the other. In this way the flywheel will be balanced with the crankshaft in the right position, insuring that the timing marks will be visible after assembly.

In assembling the engine, carefully follow the Workshop Manual as to sequence, clearances and torques. Use standard Giulietta con rod lock tabs instead of cotter pins on the con rod bolts, and new lock tabs on the camshaft sprocket bolts, crank pulley nut, and flywheel bolts. Replace all split lockwashers with the closest American equivalent. Needless to say, all new gaskets, oil seals, and O-rings are in order. Tension the upper timing chain by forcing the idler sprocket out with a screwdriver as the spring pressure behind it is insufficient. Run in the engine as outlined in the manual.

The result? Power!

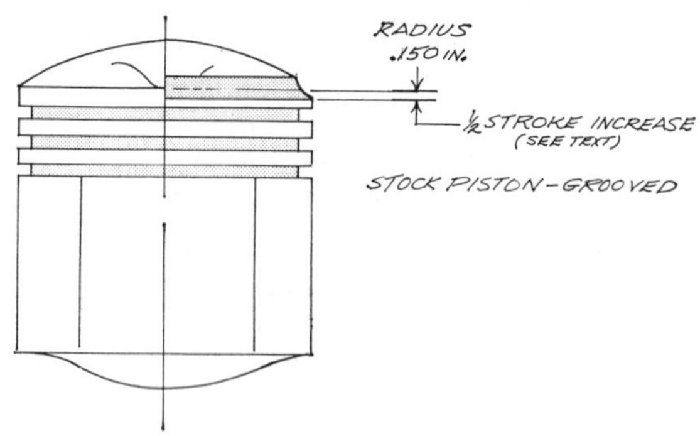

ALFA MENU . . . SELECT FROM COLUMNS A, B, C, AND D.

	A	B	C	D
Crankshaft	Stock	Stock balanced	Up to 1/8 stroke Balanced	1/4 stroke Balanced
Pistons	Stock Standard size	Stock oversize grooved	Special, oversize 1. High comp. 2. H/C for stroker	1500 Piston and liner kit
Liners	Stock Standard size	Stock Milled	Stock 1. Bored 2. Bored, notched	
Valves valve seats	Stock Stock	Stock Enlarged throat	Special 1/16 over Enlarged throat	Custom valves and seats
Intake ports	Stock	Match ports in manifold and head	Enlarge ports Polish	Enlarge ports, polish. Grind off valve guide ends.
Carburetor venturis	Stock 28 mm	30 mm	32 mm	32 or 34 mm depending on course
Cams cam timing (in **crankshaft** degrees) valve clearance	Stock Stock {34-63 63-30 Stock	Stock Advance intake 6° Stock	Stock, set at Intake 5° advanced Exhaust 1° retard Intake — .012 Exhaust — .017	Custom ground. See manufacturer's specifications.

Stock intake valve on right next to special 1/16-inch larger valve. The Jag valve (right) requires work on head.

Long ram tubes on Webers (right) are for increased mid-range torque. Use shorter tubes for high-range power.

Pistons available for the re-worked Alfa engine include (left to right) the stock piston, stock 0.024 oversize with groove, Jahns 0.060 oversize high compression piston and the "1500" piston.

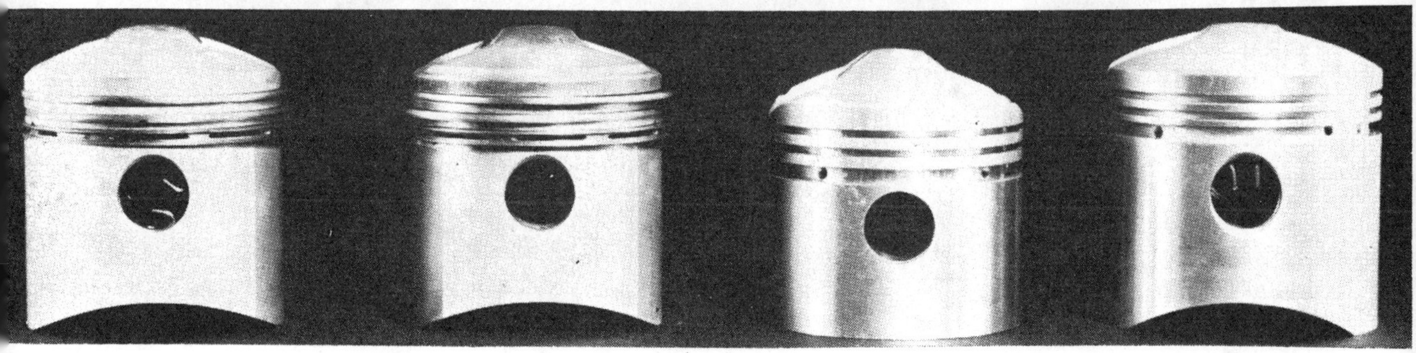

	NORMAL TOWN DRIVING	FAST HIGHWAY MODERATE COST	PRODUCTION CLASS RACING	HAIRY STREET MACHINE	MODIFIED CLASS RACING	FIRE-BREATHING TERROR DESIGNED TO BAFFLE OWNERS OF LARGER DISPLACEMENT MACHINERY
	A	A	B	C	C or D	B
	A	C-1	B	B	C-2	D
	A	C-1	B	C-2	C-2	D
	A	A	A	B	C	D
	A	B	B	C	D	D
	A	B	C	B	D	B or C
	B	A	C	A or B	C or D	A or D

ALFA SPRINT SPECIALE

Something special—for Alfa enthusiasts

ALMOST ANY CAR has its interesting features and some, naturally, are more exciting than others. Occasionally a car comes our way which is extra special. Unfortunately, the very interesting cars are usually very rare cars and we have to rely on friends to offer the family jewel (with, probably, a great deal of trepidation) for us to test. Our test car in this instance was furnished by Don Childers who purchased it new from Willie West Motors in Oceanside, Calif. West, one of the top production sports car racing drivers in the country, drove the car for the performance checks.

We've been eager to test the Alfa Romeo Giulietta Super Sprint Speciale (whew!) ever since its introduction at the 1958 Turin Auto Show. Going back to our first view of the car which was shown in the Turin show report (Feb. 1959 R&T) we said that the Giulietta SS Speciale was lighter, smaller and more streamlined than the standard model. We were right in one respect: It is more streamlined than the standard Giulietta. As we discovered, to our surprise, it is heavier (by 20 lb) and longer (by 10.3 in.) than a normal Giulietta Sprint.

The bodywork, by Bertone, who also builds the standard Alfa coupes, is flawless. The body construction is all steel with the exception of the aluminum deck lid, and every part of the Speciale that is visible is finished in excellent fashion. Doors, hood and trunk lid fit perfectly; the "fair lines" across the panels match perfectly and the highlights indicated good panel forming. In short, it exudes quality.

The interior is spacious and comfortable and has obviously been designed with driver ease and convenience in mind. The seats are comfortable, visibility is good, controls are well placed and easy to use.

Luggage capacity in the trunk compartment is limited by the space occupied by the spare tire and huge gas filler cap. Some of the new "soft" luggage will probably fit in quite readily but little else will. A great deal of space is available behind the seats, however, and in a

pinch this area can be used for two more passengers. No long trips should be attempted with back seat passengers and even confining children to this area is questionable. Interior space is much like the Giulietta Sprint in spite of the Speciale's shorter wheelbase, at 89 in. (compared to the Sprint's 93.7 in.), which is slightly longer than the Giulietta Spider at 86.6 in.

The Sprint Speciale's styling is slightly reminiscent of the Disco Volante and quite similar to the "Bat" series cars, toned down for production, and easily recognizable as Bertone bodywork. The appearance of the car suggests that wind tunnel tests were carried out and the ability of the diminutive powerplant to pull a high (low ratio) 5th gear with comparative ease backs this contention.

We were surprised to find the noise level to be lower than the standard model. Wind noise is low, which we expected, but engine noise level is extremely low considering the sporting nature of the double overhead camshaft design. Exhaust and road noise were virtually non-existent with the windows rolled up and created no cause for complaint with the windows down. We have two staff-owned Giuliettas—one Super Spider and one Sprint Veloce—and both are considerably noisier than the

Sprint Speciale, though these, too, are not out of line.

Handling is very much like the standard Giulietta's in all respects. This is to be expected because all the mechanical components are standard Giulietta, with the exception of the transmission which is like that of the 2000 Spider: 5 speeds forward (4th is direct) with all gears synchromesh.

The Porsche-type synchromesh transmission is almost unbeatable and quick, easy shifts seem natural even to the novice. There's little chance of getting the lever into the wrong slot, with the possible exception of hitting 3rd when the intention is to reach 5th (see shift pattern in the data panel).

Suspension is independent in front by unequal length A-arms and at the rear a live axle is positioned by a system of trailing links. Coil springs are used all around in conjunction with tubular hydraulic dampers.

The stability of the Alfa Giulietta chassis/suspension combination is well known throughout the automotive world and, we think, is about the ultimate possible with a live rear axle design. Like the other Giuliettas, the Speciale leans considerably during hard cornering, but this goes almost unnoticed by the occupants of the car. Steering is neutral under most conditions but oversteer can be induced if the driver presses to the limit. Because of the neutral characteristics (neither over- nor understeer) and almost vice-free handling quality of the Giulietta series, it is a car that would enable any driver to accustom himself (or herself) to it with a minimum of fuss and bother.

The fact that the Giulietta is so vice-free has probably kept a good many mediocre drivers out of trouble, because it invites hard driving by its very nature.

The performance, insofar as speed and acceleration are concerned, was disappointing to most of us, and especially to the Alfa owners on the staff. The car seemed to be running very well, and West is no novice to Alfa Romeo—having won 22 straight races (in class) with a standard Giulietta Spider—so we can't fault the driving. The car was new, having less than 2000 miles on the odometer, so there is a good possibility that it will loosen up and perform better as time goes on, as they usually do.

The brakes, like other Giuliettas (and the entire Alfa line), are large diameter, finned aluminum drums with cast-iron liners and the car would have to be driven to the brink of disaster before these brakes would fade.

Taken as a road car, the Speciale is ideal: fast, comfortable, safe, and economical insofar as gas mileage is concerned. As a town car, it has some drawbacks: Poor parking protection from any direction makes it almost impossible to park safely in either parallel parking or in the angle slots found in most market parking lots. The front overhang would also require that caution be used in angle parking near a curb of any height.

The large glass area affords excellent visibility for the occupants in all directions as long as the windshield and back window are clean. Like other cars with sharply raked front and rear glass, dust and dew will gather with alarming rapidity, turning the windshield almost opaque in the glare of oncoming headlights, or when driving into the sun.

The Sprint Speciale is not a car that will appeal to every sports car enthusiast. The close-to-$6000 price tag is pretty stiff when you consider that the main difference between the Speciale and the normal Sprint is the overdrive, a special body and the knowledge that you own a rare automobile from a highly regarded and reputable firm. In this latter statement we find the clue that helps justify the high cost.

The fact that these features are available to the buyer who wants something different, but *not* so different he can't get his car serviced on this side of the Atlantic Ocean, is a definite plus-factor for the car. No parts used on the Speciale (other than body) are unlike those used by some other currently made Alfa Romeos and, therefore, service and parts should be obtainable at any Alfa Romeo dealer or distributor.

Pride of ownership is also an important factor for a buyer who would be interested in something a little extra special in a car, and the Sprint Speciale is that type of car.

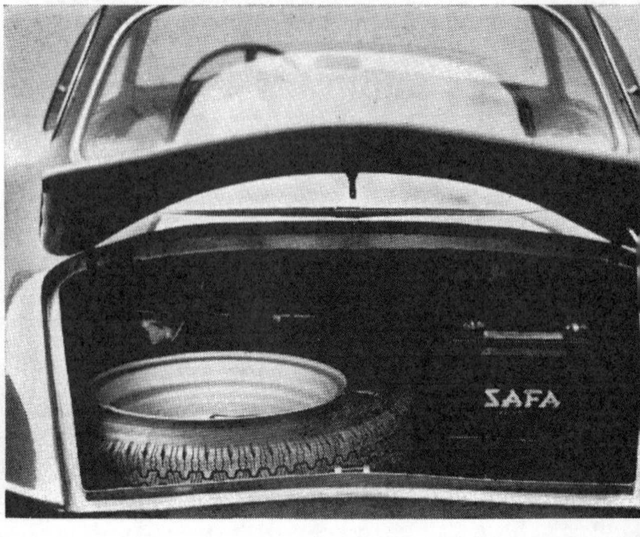

ROAD TEST

ALFA SPRINT SPECIALE

DIMENSIONS

Wheelbase, in	88.6
Tread, f and r	50.9/50.0
Over-all length, in	167
width	65.3
height	48.8
equivalent vol, cu ft	299
Frontal area, sq ft	17.7
Ground clearance, in	7.0
Steering ratio, o/a	n.a.
turns, lock to lock	2.8
turning circle, ft	34
Hip room, front	51
Hip room, rear	
Pedal to seat back	40
Floor to ground	13.5

CALCULATED DATA

Lb/hp (test wt)	21.1
Cu ft/ton mile	61.2
Mph/1000 rpm (5th)	18.3
Engine revs/mile	3290
Piston travel, ft/mile	1620
Rpm @ 2500 ft/min	5090
equivalent mph	92.9
R&T wear index	53.3

SPECIFICATIONS

List price	$5663
Curb weight, lb	2110
Test weight	2450
distribution, %	52/48
Tire size	155–15
Brake swept area	n.a.
Engine type	4 cyl, dohc
Bore & stroke	2.91 x 2.95
Displacement, cc	1290
cu in	78.7
Compression ratio	9.5
Bhp @ rpm	116 @ 6500
equivalent mph	118.5
Torque, lb-ft	96.2 @ 4500
equivalent mph	82.1

GEAR RATIOS

5th (0.85)	3.87
4th (1.00)	4.55
3rd (1.36)	6.19
2nd (1.99)	9.06
1st (3.26)	14.8

SPEEDOMETER ERROR

30 mph	actual, 28.4
60 mph	56.1

PERFORMANCE

Top speed (5th), mph	120
4th (7000)	108.5
3rd (7000)	79.9
2nd (7000)	51.7
1st (7000)	33.4

FUEL CONSUMPTION

Normal range, mpg	22/28

ACCELERATION

0-30 mph, sec	4.0
0-40	5.9
0-50	8.2
0-60	12.3
0-70	16.7
0-80	21.8
0-100	38.0
Standing ¼ mile	18.4
speed at end	73.9

TAPLEY DATA

4th, lb/ton @ mph	200 @ 70
3rd	300 @ 58
2nd	430 @ 42
Total drag at 60 mph, lb	70

ENGINE SPEED IN GEARS

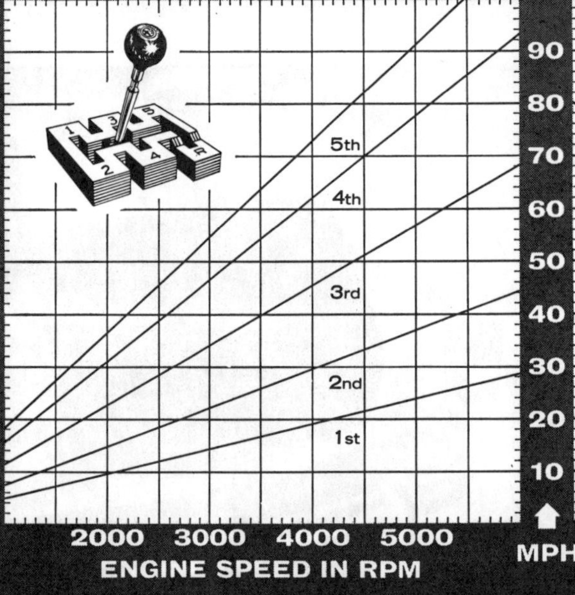

ACCELERATION & COASTING

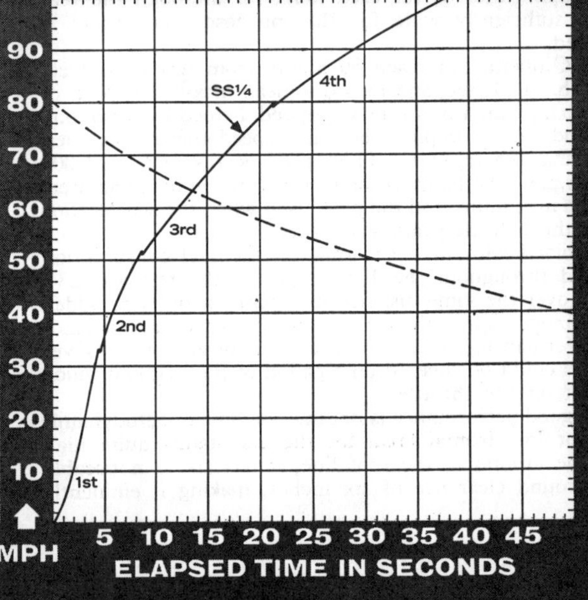

ALFA ROMEO GIULIETTA T.I.

Well balanced but simple lines are features of this small high performance family saloon. By modern standards it has a very generous ground clearance

TRADITION, it is said, dies hard. It is now roughly eight years since the Alfa Romeo Company of Milan withdrew from racing with factory-entered teams. They had competed successfully with a wide range of sports and Grand Prix racing cars almost from the inception of the company shortly before World War I. With such a background, therefore, it is not surprising that the present range of production cars, designed and developed by engineers trained during the racing era, should inherit rich legacies from their sporting lineage.

The four-cylinder, 1290 c.c. Giulietta Sprint coupé, introduced in 1954 with its exquisite body lines, has already become almost a legend. A four-seater and more angular saloon version, using the same basic mechanical components, was introduced in 1957, and is now available in this country with right-hand-drive. The subject of this road test is the more sporting version of this model, designated the T.I.

There are few four-seater cars of up to 1½ litres capacity which could match the performance of the Giulietta T.I.; some qualification is required here, for although the rear seat is sufficiently wide for this purpose, knee room is restricted.

The Giulietta can reach 50 m.p.h. from rest in 12·1sec, 70 m.p.h. in 24·5sec, and record a mean maximum speed of 94·6 m.p.h., with a best one-way speed, helped by a 5 m.p.h. tail wind, of 97 m.p.h. (the car's speedometer was then registering 105 m.p.h.). This speed corresponds to 6,000 engine r.p.m., the figure recommended by the manufacturers as a safe maximum, and the peak used when the acceleration figures through the gears were taken.

Combined with this performance, the fuel consumption recorded throughout the 1,088 miles of the test was 28·8 m.p.g.; over the same distance four pints of oil were added to the sump. Thus, the car not only possesses an outstanding performance, but is also very economical to run. Over 30 miles could be covered on a gallon of fuel without much restraint on the throttle.

This good performance is not the outcome of aerodynamic shape or low frontal area, for the car stands quite high from the ground to the roof line; it has also a noticeably good ground clearance of six inches, making it eminently suitable for operation over rough, non-metalled roads. Moreover, with a kerb weight of 18·5 cwt, and with 3 cwt more added for test crew and equipment, it is not particularly light. Finally, the engine is not so highly tuned as to be harsh or to give the impression that it might have only a short life. This five-bearing, twin o.h.c. unit is flexible and smooth throughout its range; although it has an 8·5 to 1 compression ratio, it will run without pinking when using premium grade fuel and will pull away well from low speeds.

The maximum installed power of the T.I. engine is 65 b.h.p. at 5,500 r.p.m., maximum torque being developed at 3,500 r.p.m. A more highly tuned version of this same power unit in the Giulietta Speciale produces 100 b.h.p. It is thus evident that the version tested operates well within its capabilities and should retain its stamina and reliability for long periods.

Compared with the Giulietta Sprint, which is a true coupé and intended to carry two people only, the T.I. has only 6in. more in the wheelbase. This undoubtedly explains the

The boot is more capacious than it looks for it is relatively deep and is rubber lined throughout. The battery is placed behind the spare wheel; adjacent to it are a well-equipped tool roll, screw jack and wheel brace

comparatively high build, for it has been necessary also to extend the dimensions upwards to accommodate the extra passengers. Thus on entering the driving seat the higher-than-usual cushions relative to the floor are apparent. However, the controls are well matched to this feature so that the bottom-hinged clutch and brake pedals, with their pads noticeably close to the rubber floor mats, have good actions and short movements, and are quite widely spaced. Forward vision is adequate for a driver of around 5ft 9in., but anyone over this height would find that the screen top rail limits forward vision—to the extent that the sun vizors would rarely, if ever, be required.

A bench seat with fixed backrest is fitted at the front. The seat cushion, with a length of 17in., is extremely short; this, combined with a pronounced roll on the leading edge, which pressed into the back of the thighs rather than lending support behind the knees, caused discomfort on long journeys. There is a similar roll on the top of the seat squab which presses against the shoulders. Adequate lateral support is offered, this being aided by a good-quality cloth trim.

The rear seats are more comfortable than the front ones, having longer cushions and greater inclination of the backrest. With the front seat set fully back—which is necessary to provide sufficient leg-room for a tall driver—knee-room for the rear seat passengers is severely restricted, despite deep recesses in the seatbacks.

All the instruments are in full view through the two-spoke steering wheel, and the controls, with functional code marks, are well set out. The rather optimistic ribbon-type speedometer did not suit all tastes, but it is supplemented by a circular tachometer, and the provision of a temperature gauge for the oil, as well as one for water is a commendable rarity. Another practical fitting is a hand throttle control which enables the engine to be warmed up quickly without much use of the choke; it is also useful for engine tuning.

The steering wheel is set at a comfortable angle and is conveniently small (14·75in. dia.) It is equipped with a full horn ring and a centre button for headlamp signalling. Beneath it is a short lever for the self-cancelling turn indicators, and below this a longer selector lever for side, dipped, or main beam driving lamps after they have been brought into circuit by a facia master switch.

Internal fittings and furnishings are good without being outstanding in the car's price class. There is a combined armrest and door pull on each door. For small articles and maps there are pockets in each front door and two in the backrests of the front seats; in addition, there is a very small facia locker.

If forward vision is not ideal for tall drivers, the high seating position is advantageous when looking to the side and rear. Side vision is the greater for the absence of hinged panels in the front doors. These, and the rear ones, are fitted with the popular Italian anti-draught transparent plastic panels set in the upper rear quarters of the windows. They are very effective, for the windows can be lowered half way, all at the same time, without any objectionable air buffeting occurring. Two minor objections to these fittings were noticeable distortions in the plastic material and the fact that care was needed when putting one's head through the open window for any purpose.

A Delaney Gallay heating and ventilating system is a standard fitting and very effective. A fresh air intake is located at the front of the car with the booster fan on this forward end of the trunking; it thus runs quietly, but is not required above 40 m.p.h. Cool or heated air can be directed through a flap on the heat exchanger unit above the gearbox tunnel; when the flap is closed the air passes up through three windscreen slots.

As with most Continental cars, the three indirect gear ratios are more evenly and widely spaced than on the average British car, and the T.I. peaks at 30 m.p.h., 48 m.p.h. and 70 m.p.h. respectively. These are considerably below the point of valve bounce, but there is no gain in performance through exceeding these recommended maxima. Although the engine is very flexible, it has a quick response and willing-

The five-bearing, 1,290 c.c. twin o.h.c. engine is very neat and remained clean throughout the test. Accessibility to the carburettor, distributor, fuel pump and sparking plugs is excellent, but the dipstick is difficult to replace

Seats are mounted high in relation to the carpet-covered floor. Leg room at the front is adequate but that at the rear is rather restricted. There are two elastic pockets in the backrest

ALFA ROMEO GIULIETTA T.I. . . .

There are no ventilators in the door windows, but plastic anti-draught panels permit them to be wound down without much buffeting or wind noise

ness to rev fast, with good torque in the higher ranges; these characteristics combined with a first-class gearbox, invite frequent use of third gear when the need arises to press on fast through a long, slow traffic stream.

Synchromesh is fitted to the four forward ratios and is unsurpassed in excellence, for it cannot be overridden however fast the changes are made. Movements of the sturdy central lever are very short indeed through and across the gate. The operating loads also are extremely light, so that a mere hand pressure is required for movement between ratios. This desirable feature applies also when selecting first from rest without any resistance to a first-time engagement. All gears were quiet, and the only minor criticism of an otherwise excellent transmission was a slight vibration around 60 m.p.h. in top gear; this may have emanated from the rubber mounting for the centre bearing of the two-piece propeller shaft.

Equal praise can be given to the smoothness of the engine. Throughout its wide range there was never a trace of roughness, although it became noticeably noisier at speeds above 5,000 r.p.m., mostly through increased air intake roar. For fast cruising on main roads, and third gear overtaking in such conditions, 5,000 r.p.m. was the normal maximum engine speed used. This rate also provides effortless cruising at nearly 80 m.p.h., and as the fuel consumption at this speed is about 24 m.p.g. it is, at the same time, economical. Starting was instantaneous, hot or cold, and only brief use of the choke was necessary. It was found that fuel surge starved the main jet on a fast and long right-hand bend and the engine would not continue to run on a 1 in 4 gradient.

It needs but a short distance in this car to appreciate the delicacy and precision of the steering. It is very quick, with 2.6 turns between extreme locks, light at low speeds and when cornering fast, with no impression of springiness in the linkage or sharp reaction from the road. On irregular surfaces with a series of fast opposite bends, the rear axle, despite its positive location by radius arms and an A-bracket in conjunction with coil springs, had some influence on stability. On such surfaces minor deviation occurred whenever one rear wheel was deflected, which resulted in a slight corkscrew motion. These characteristics, which were never serious or likely to give rise to a feeling of insecurity, are probably related to a noticeable degree of roll when cornering fast. The initial roll is more pronounced if a fast corner is taken late and hard, rather than attached early and smoothly. This minor criticism apart, adhesion was always good, and once the driver had assimilated the cornering technique, he usually drove much faster than in other cars of similar class or performance.

Straight ahead running qualities even when contending with a considerable side wind are excellent, indicating that the car has an understeer tendency, which proved to be true when this was investigated on an off-the-road steering pad. It is not excessive, and in normal road conditions strikes the balance which the majority of owners prefer.

Firm Suspension

Springing is quite firm, particularly at the rear, and with only the front seats occupied, this resulted in the passengers bouncing on the seats; the quick rate of the seat springs undoubtedly has a large influence in this respect. Fully laden, and with the tyres pumped up to the recommended higher pressures, the ride becomes much smoother, the rear passengers being more comfortable than those at the front. On roads with really deep potholes, the car could be driven at quite high speeds without any significant loss of steering control, and in these severe conditions the body structure proved to be very rigid.

As might be expected of a braking system having a very large effective area and using aluminium drums front and rear (the former having heavy radial ribs), it proved to be very stable and virtually immune from fade. These characteristics may not at first be apparent from the performance figures taken at 30 m.p.h. At these speeds quite heavy pressures were needed for moderate efficiency, but once the need for heavier-than-normal pedal loads at low speeds is appreciated, the system is above criticism, remaining balanced at all speeds no matter how frequently used. Such praise cannot be given to the umbrella-type hand brake mounted beneath the facia, for it failed completely to hold the car on a 1 in 4 incline.

Wind noise is very low even at the highest speeds the Alfa

The facia panel with ribbon speedometer and circular rev counter is sensibly arranged, the top deck having a crackle finished matt surface

Romeo can reach, indicating the clean form of external body details. Commendable also is the low level of transmitted road noise, except when passing over surface ridges or cats'-eyes, when the Pirelli Cinturato tyres thumped loudly.

Greased bearings are used throughout the suspension and steering linkage, and there are 20 points which need attention every 1,250 miles. There are eight separate fuses in the electrical circuits, and these are housed in an accessible box beneath the facia on the passenger's side; adjacent to it is a socket to take an inspection lamp.

British manufacturers do not produce a model comparable with the Guilietta, and few of its competitors in world markets can match it for performance, safety in handling, and running economy. It has established a reputation for reliability and sturdy service, and is well suited to meet the requirements of an enthusiastic driver who must have a small family saloon, and is also prepared to pay the extra burden of customs duty and additional tax which any imported car carries.

This car was kindly loaned by the London distributors of S. Morris & Co., 40, Conduit Street, London, W.1.

ALFA ROMEO GIULIETTA T.1.

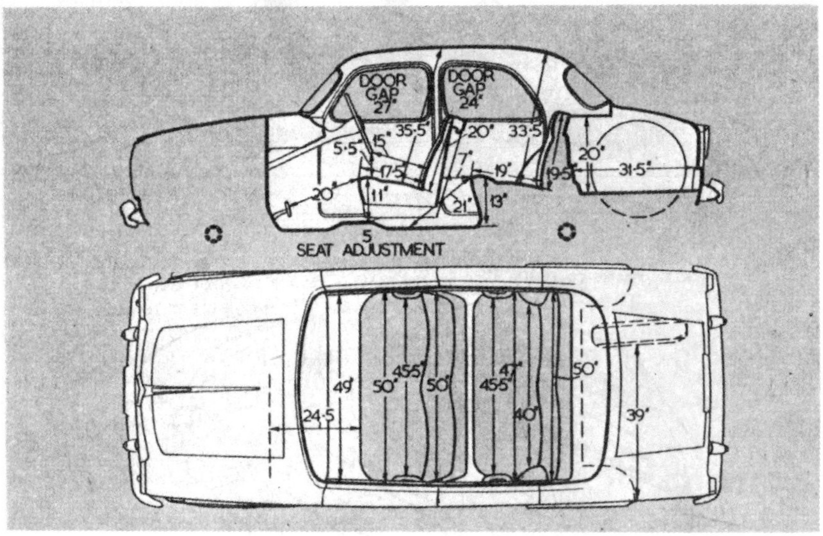

Scale ¼in. to 1ft. Driving seat in central position. Cushions uncompressed.

PERFORMANCE

ACCELERATION TIMES:

Speed range, Gear Ratios and Time in Sec.:

m.p.h.	4·55 to 1	6·18 to 1	9·05 to 1	14·84 to 1
10—30	13·9	10·0	6·2	4·3
20—40	14·6	9·3	6·3	—
30—50	12·7	9·0	—	—
40—60	14·7	10·6	—	—
50—70	18·4	12·1	—	—
60—80	21·3	—	—	—
70—90	30·3	—	—	—

From rest through gears to:

30 m.p.h.	..	**5·1 sec**
40 ,,	..	8·3 ,,
50 ,,	..	12·1 ,,
60 ,,	..	17·7 ,,
70 ,,	..	24·5 ,,
80 ,,	..	37·5 ,,
90 ,,	..	54·8 ,,

Standing quarter mile 20·8 sec.

MAXIMUM SPEEDS ON GEARS:

Gear		m.p.h.	k.p.h.
Top	**(mean)**	**94·6**	**152·2**
	(best)	**97·0**	**156·1**
3rd	.. ..	**70**	**112**
2nd	**at 6,000 r.p.m.**	**48**	**77**
1st	.. ..	**30**	**48**

TRACTIVE EFFORT (by Tapley meter):

	Pull (lb per ton)	Equivalent gradient
Top	180	1 in 12·4
Third	260	1 in 8·6
Second..	385	1 in 5·7

BRAKES (at 30 m.p.h. in neutral):

Pedal load in lb.	Retardation	Equiv. stopping distance in ft
25	0·30g	100
50	0·34g	89
75	0·49g	62
100	0·65g	46
125	0·76g	40
135	0·85g	35·5

FUEL CONSUMPTION: M.p.g. at steady speeds in top gear.

30 m.p.h.	..	40·4 m.p.g.
40 ,,	..	39·6 ,,
50 ,,	..	37·7 ,,
60 ,,	..	32·5 ,,
70 ,,	..	29·0 ,,
80 ,,	..	24·1 ,,
90 ,,	..	18·6 ,,

Overall fuel consumption for 1,088 miles, 28·8 m.p.g. (9·8 litres per 100 km.).

Approximate normal range 25-33 m.p.g. (11·3-8·6 litres per 100 km.).

Fuel: Premium grades.

TEST CONDITIONS: Weather: Dry with 0-5 m.p.h. wind.

Air temperature, 80 deg F.

STEERING: Turning circle,

Between kerbs, L, 34ft 5in. R, 37ft 11in.

Between walls, L, 35ft 1in. R, 38ft 7in.

Steering wheel turns, lock to lock, 2·6.

SPEEDOMETER: m.p.h.

Car speedometer	10	20	30	40	50	60	70	80	90	100
True speed	7	17	27	37	47	57	66	75	85	93

DATA

PRICE (basic), with four-door saloon body, (including heater), **£1,125.**
British purchase tax, **£515 10s.**
Total (in Great Britain), **£1,641 10s.**

ENGINE: Capacity, 1,290 c.c. (78·7 cu. in.).
Number of cylinders, 4 in line.
Bore and stroke, 74 × 75 mm (2·91 × 2·95in.).
Valve gear, Twin overhead camshafts with direct operating tappets.
Compression ratio, 8·5 to 1.
B.h.p. 65 (net) at 5,500 r.p.m. (b.h.p. per ton laden, 60).
Torque, 69 lb. ft. at 3,500 r.p.m.
M.p.h. per 1,000 r.p.m. in top gear, 15·7.

WEIGHT: (With 5 gals fuel), 18·5 cwt (2,079 lb).
Weight distribution (per cent): F, 54; R, 46.
Laden as tested, 21·5 cwt (2,415 lb).
Lb per c.c. (laden), 1·8.

BRAKES: Type, Girling, with light alloy drums.
Method of operation, Hydraulic non-servo.
Drum dimensions: F, 10·5in. diameter; 2·25in. wide. R, 10·0in. diameter; 1·75in. wide.
Swept area: F, 148 sq. in.; R, 110 sq. in. (240 sq. in. per ton laden).

TYRES: 155—15in. Pirelli Cinturato.
Pressures p.s.i.: F, 21; R, 23 (normal).
F, 23; R, 25 (fast driving).

TANK CAPACITY: 8·8 Imperial gallons.
Oil sump, 8·8 pints.
Cooling system, 13 pints.

DIMENSIONS: Wheelbase, 7ft 10in.
Track: F, 4ft 2in. R, 4ft 2·5in.
Length (overall), 13ft 1·5in.
Width, 5ft 1·5in.
Height, 4ft 8in.
Ground clearance, 5·5in.

ELECTRICAL SYSTEM: 12-volt; 38 ampère-hour battery.
Headlamps: 45-40 watt bulbs.

SUSPENSION: Front: Wishbones and coil springs with telescopic dampers and anti-roll bar.
Rear: Live axle with coil springs and telescopic dampers controlled by radius arms and upper "A" bracket.

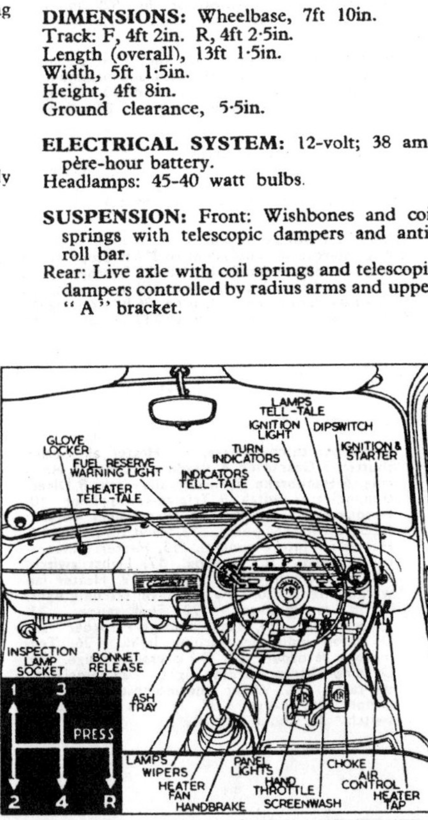

The Motor Road Test No. 28/61

Make: Alfa Romeo **Type:** Giulietta t.i.

Makers: Alfa Romeo & C., Milan, Italy

Concessionnaires: Thomson and Taylor (Brooklands) Ltd., Portsmouth Road, Cobham, Surrey

Test Car submitted by: K. N. Rudd (Engineers) Ltd., Worthing, Sussex

Test Data

CONDITIONS: *Weather: Warm and dry with light breeze. (Temperature 64°-72°F., Barometer 30.0 in. Hg.) Surface: Dry tarred macadam. Fuel: Premium-grade pump petrol (approx. 97 Research Method Octane Rating).*

INSTRUMENTS

Speedometer at 30 m.p.h.	9% fast
Speedometer at 60 m.p.h.	5% fast
Speedometer at 90 m.p.h.	4% fast
Distance recorder	1% fast

WEIGHT

Kerb weight (unladen, but with oil coolant and fuel for approx. 50 miles)	18¾ cwt.
Front/rear distribution of kerb weight	54/46
Weight laden as tested	22½ cwt.

MAXIMUM SPEEDS

Flying Mile

Mean of six opposite runs	96.3 m.p.h.
Best one-way time equals	99.4 m.p.h.

"Maximile" Speed. (Timed quarter mile after one mile accelerating from rest.)

Mean of four opposite runs	91.8 m.p.h.
Best one-way time equals	94.7 m.p.h.

Speed in gears (at 6,000 r.p.m.)

Speed in 3rd. gear	69 m.p.h.
Speed in 2nd gear	48 m.p.h.
Speed in 1st gear	28 m.p.h.

FUEL CONSUMPTION

40.0 m.p.g. at constant 30 m.p.h. on level
40.0 m.p.g. at constant 40 m.p.h. on level
37.5 m.p.g. at constant 50 m.p.h. on level
33.0 m.p.g. at constant 60 m.p.h. on level
29.5 m.p.g. at constant 70 m.p.h. on level
25.5 m.p.g. at constant 80 m.p.h. on level
21.5 m.p.g. at constant 90 m.p.h. on level

Overall Fuel Consumption for 929 miles, 37.8 gallons equals 24.6 m.p.g. (11.5 litres/100 km.).

Touring Fuel Consumption (m.p.g. at steady speed midway between 30 m.p.h. and maximum, less 5% allowance for acceleration). 30.3 m.p.g.
Fuel tank capacity (maker's figure) 8¾ gallons

STEERING

Turning circle between kerbs:

Left	33 feet
Right	34 feet
Turns of steering wheel from lock to lock	2¾

BRAKES from 30 m.p.h.

0.95 g retardation (equivalent to 31¾ ft. stopping distance) with 125 lb. pedal pressure.
0.76 g retardation (equivalent to 39¾ ft. stopping distance) with 100 lb. pedal pressure.
0.54 g retardation (equivalent to 55¾ ft. stopping distance) with 75 lb. pedal pressure.
0.35 g retardation (equivalent to 86 ft. stopping distance) with 50 lb. pedal pressure.
0.13 g retardation (equivalent to 230 ft. stopping distance) with 25 lb. pedal pressure.

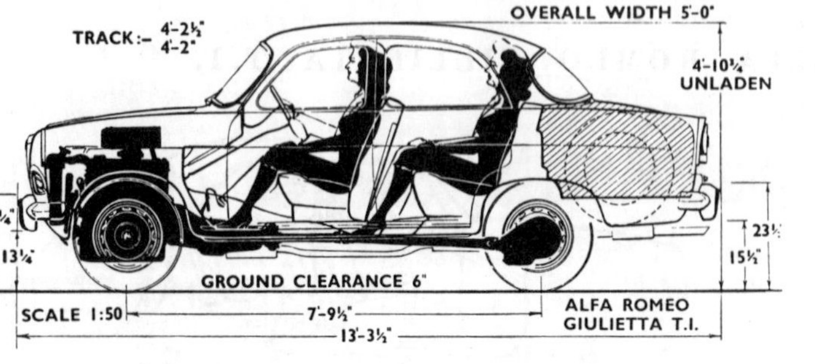

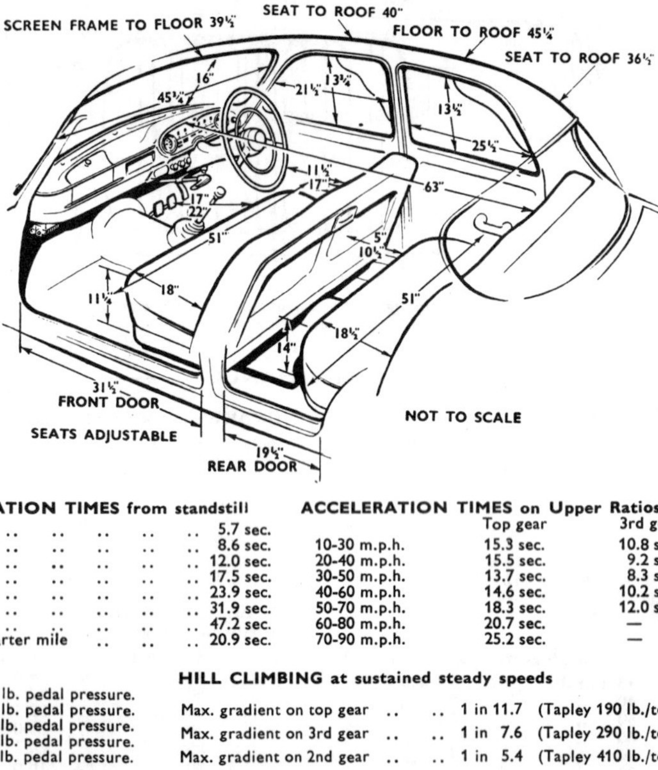

ACCELERATION TIMES from standstill

0-30 m.p.h.	5.7 sec.
0-40 m.p.h.	8.6 sec.
0-50 m.p.h.	12.0 sec.
0-60 m.p.h.	17.5 sec.
0-70 m.p.h.	23.9 sec.
0-80 m.p.h.	31.9 sec.
0-90 m.p.h.	47.2 sec.
Standing quarter mile	20.9 sec.

ACCELERATION TIMES on Upper Ratios

	Top gear	3rd gear
10-30 m.p.h.	15.3 sec.	10.8 sec.
20-40 m.p.h.	15.5 sec.	9.2 sec.
30-50 m.p.h.	13.7 sec.	8.3 sec.
40-60 m.p.h.	14.6 sec.	10.2 sec.
50-70 m.p.h.	18.3 sec.	12.0 sec.
60-80 m.p.h.	20.7 sec.	—
70-90 m.p.h.	25.2 sec.	—

HILL CLIMBING at sustained steady speeds

Max. gradient on top gear	1 in 11.7	(Tapley 190 lb./ton)
Max. gradient on 3rd gear	1 in 7.6	(Tapley 290 lb./ton)
Max. gradient on 2nd gear	1 in 5.4	(Tapley 410 lb./ton)

1, Bonnet catch release. 2, Heater air outlet shutter. 3, Gear lever. 4, Headlamp flasher. 5, Horn ring. 6, Handbrake. 7, Heater fan switch. 8, Direction indicator switch. 9, Trip re-setting knob. 10, Headlamp dip switch. 11, Panel light switch. 12, Screenwasher pump. 13, Hand throttle. 14, Heater air intake control. 15, Heater temperature control. 16, Fusebox. 17, Lights switch. 18, Windscreen wipers switch. 19, Heater fan warning light. 20, Low fuel level warning light. 21, R.p.m. indicator. 22, Fuel gauge. 23, Speedometer. 24, Total mileage recorder. 25, Direction indicator warning light. 26, Trip mileage recorder. 27, Oil pressure gauge. 28, Dynamo charge warning light. 29, Sidelamps tell-tale. 30, Water thermometer. 31, Oil temperature gauge. 32, Ignition and starter switch. 33, Choke control.

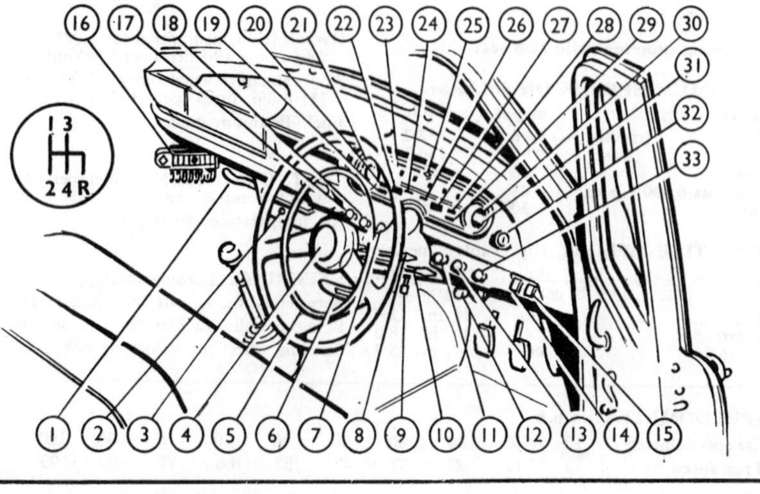

The Alfa Romeo Giulietta t.i.

Small-car Motoring with a Kick in It

WHEN a factory renowned for its successes in Grand Prix motor racing sets out to build a small family saloon, something unorthodox can be expected to result. Before introducing their Giulietta saloon model in the Spring of 1955 the Alfa Romeo factory used the same mechanical elements in Sprint Coupés for a season, and the Giulietta t.i. saloon which is now to be had with right-hand drive, is backed by a great mass of user experience as well as by racing ancestry. Unique and very Italian in its character, this model is astonishingly fast, reasonably quiet, and has a moderate thirst for fuel; it is rather uncomfortable for passengers but exceptionally controllable and gives every impression of being built for hard work. Whereas some carefully planned orthodox cars are quick to make a newcomer feel at home, the Giulietta t.i. has oddly-arranged controls with unconventional response which give some people a very poor first impression, but once familiar with the Giulietta drivers are often reluctant to go back to more normal kinds of motoring.

Most car manufacturers and importers are nowadays very anxious that journalists should sample their wares and tell the world about them, but the opportunity to drive an Alfa Romeo on British roads only arose when K. N. Rudd (Engineers) Ltd. added this make to the range of high-performance cars for which they are agents. Catalogues make the astonishing claim that this 1.3-litre saloon, which is roomy enough to accommodate four or five people and their luggage, has a top speed of 155 km. per hour, and our timed mean speed of 96.3 m.p.h. with a car that had run no more than 4,000 miles agrees precisely with the catalogue. As owners are asked to observe "running in" rules for 1,900 miles, a less new car would probably be even faster.

Nothing in the Giulietta's engine or body dimensions would lead one to expect a top speed higher than about 80 m.p.h., and whilst careful design of body details has obviously minimized wind resistance, the smallest Alfa Romeo engine clearly develops a lot of power. Twin overhead camshafts go with a cylinder head design that provides excellent breathing at 6,000 r.p.m., and although the engine of our test car had what sounded like a timing chain resonance at about 2,000 r.p.m., this unit compares very well with simpler and less potent designs for general quietness. A compression ratio of 8/1 does not fully exploit the anti-knock quality of British premium petrol, and the purchase of 100-octane fuel is quite unnecessary. Since we sampled a left-hand-drive example in Italy three years ago this model has been quietened very usefully, and although a certain amount of weight has been added in the refining process the astonishing performance remains virtually unimpaired. To our surprise, the test car did not have enough clearance to allow for the substantial movements of the engine on its rubber mountings: the exhaust manifold flange could rattle noisily against the left front shock absorber mounting, and on the right side damage to the starter's electrical connections occurred. Generally this car seems very sturdy, and despite widespread use of light alloys in the engine, gearbox, rear axle and body it weighs 18¾ cwt. as it stands at the kerb.

Small engines which are tuned to do the work of large engines sometimes become inflexible, but whilst this 1.3-litre Alfa Romeo unit with its progressive twin-choke downdraught carburetter does not develop a great deal of power at low r.p.m. it is perfectly tractable for town use. By the time 45 m.p.h. is reached, top gear acceleration has become unexpectedly good, and as the red column of a strip-type speedometer goes steadily on to the 90 m.p.h. mark and beyond it becomes increasingly difficult to believe that such a small engine lies under the bonnet. Wind noise at this sort of speed is commendably slight.

Pleasant though it is in main road driving to have the surge of top-gear power

In Brief

Price £1,125, plus purchase tax £520 12s. 5d., equals £1,645 12s. 5d.

Capacity	1,290 c.c.
Unladen kerb weight ...	18¾ cwt.
Acceleration:	
20-40 m.p.h. in top gear ...	15.5 sec.
0-50 m.p.h. through gears	12.0 sec.
Maximum top gear gradient	1 in 11.7
Maximum speed	96.3 m.p.h.
"Maximile" speed	91.8 m.p.h.
Touring fuel consumption ...	30.3 m.p.g.

Gearing: 15.5 m.p.h. in top gear at 1,000 r.p.m.; 31.5 m.p.h. at 1,000 ft./min. piston speed.

BENCH seating is used in this saloon model of the Alfa Romeo, and a central gearchange goes with right-hand drive. The speedometer is a very clear strip-type instrument, supplemented by a rev. counter. Hollows in the back of the front seat provide rear-seat kneeroom and also form map pockets.

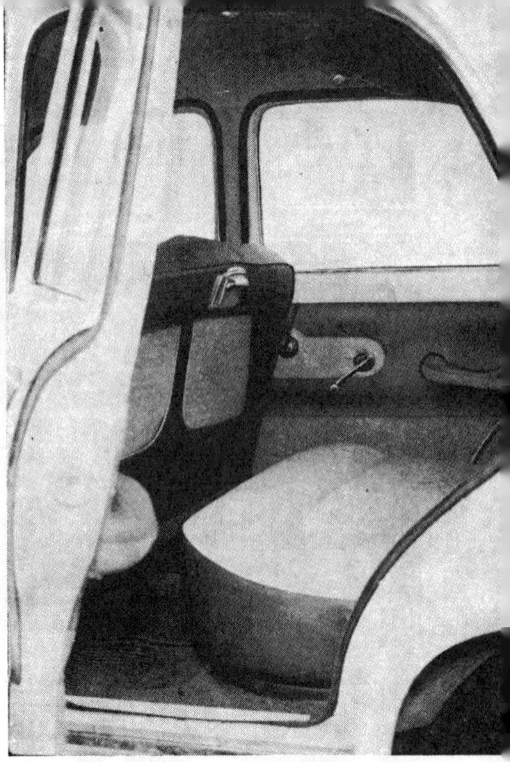

The Alfa Romeo Giulietta t.i.

that this little engine provides, lively-performance in the gears is an even more attractive feature for much British motoring. Cars with right-hand drive have a floor-mounted central gear lever, immensely sturdy yet needing only a very light touch to operate—so gentle is the lateral spring loading of the gear lever towards the central top-3rd gear plane of its gate that a driver needs practice not to over-ride the reverse catch. All four gears have synchromesh, light in action and not hindering engagement of a gear from rest yet thoroughly effective, and they are all pleasantly quiet. Using 5,000 r.p.m. on the rev. counter, the indirect gears provide speeds of 23, 40 and 57 m.p.h. with at least 20% more speed in reserve for more hurried driving.

By 1961 standards the Giulietta saloon is quite a high-built car, and the driver sits upright on a bench seat, there being just enough headroom for a tall man although the thickening upper corners of the windscreen pillars tend to limit his vision. There is enough kneeroom in the back for adult passengers, thanks to a hollowed-out front seat backrest, but certainly no excess of headroom or kneeroom. The luggage locker is flat-floored, conveniently cubic in its proportions and of very useful size.

In character with their racing background, Alfa Romeo have sprung this rather high-built little car so that it corners very fast without any side-sway and with no more than a moderate amount of roll, using an anti-roll torsion bar to help the coil front springs, and placing the triangulated link which provides lateral location of the coil-sprung rear axle high up beside the top of the differential casing. These measures are thoroughly effective in providing much flatter cornering characteristics than appearances would suggest.

Moderately firm springs do a good all-round job of cushioning direct road shocks, and although on an ordinary British road the Giulietta t.i. gives a livelier ride than do most present-day touring cars, extreme conditions of potholed going or of "switch-back" undulations do not trouble it much. Probably because of its rather high build and an exceptionally high rear roll centre however, bumps rock the car sideways to an unusual extent, somewhat to the discomfort of passengers. Sitting rather upright on a bench seat which, although not slippery, is not shaped to give any lateral support, a passenger is acutely conscious of the fast cornering and hard braking in which a Giulietta driver almost inevitably indulges.

At first acquaintance, not all drivers like the handling qualities of this saloon. Our test car had the optional Michelin "X" tyres, and from being heavy at the lowest speeds the steering becomes light and immensely sensitive for small deviations from the straight at anything over 30 m.p.h. The combination of this sensitivity to fractions of an inch of wheel movement (despite two universal joints in the steering column, no trace of lost motion can be detected in the steering mechanism when the front wheels are pointing straight ahead) with the car's sideways motion over bumps can give the newcomer to this model a feeling of instability, but familiarity proves this to have been an entirely false impression. It is possible for a driver who has not yet acquired the appropriate delicacy of touch to over-steer this car when entering a corner, but one soon learns not to do this unintentionally, enjoying instead a quite exceptional ability to make a sudden swerve should the actions of some other road user make this necessary.

When put into a corner, the Giulietta t.i. shows a degree of understeer which increases with the driver's cornering verve: sampling the car's extremes of cornering power on a circular skid pan, one can apply more and more steering until the front wheels are on full lock, without any sudden loss of tyre adhesion taking place. A marked characteristic, presumably associated with changed inclination of radius arms which locate the rear axle, is that increasing the load of passengers or luggage in the back of the car diminishes the initial steering sensitivity and emphasizes its stable understeering tendency. Whilst it may take time to explain the unorthodox handling characteristics of the Giulietta t.i., a driver who has "played himself in" with it enjoys quite exceptionally accurate and responsive control.

Whereas the steering has an almost exaggerated degree of initial response, the brakes give little return for moderate

TWIN CAMSHAFTS are used to operate the inclined valves of an exceptionally neat looking 4-cylinder engine. On t.i. saloons, a twin-choke downdraught carburetter supplies the mixture.

AMPLE ground clearance for rough going is provided under the compact four-door saloon body. A lift-up lid gives access to the flat-floored luggage locker, at one side of which the spare wheel is clamped in an upright position.

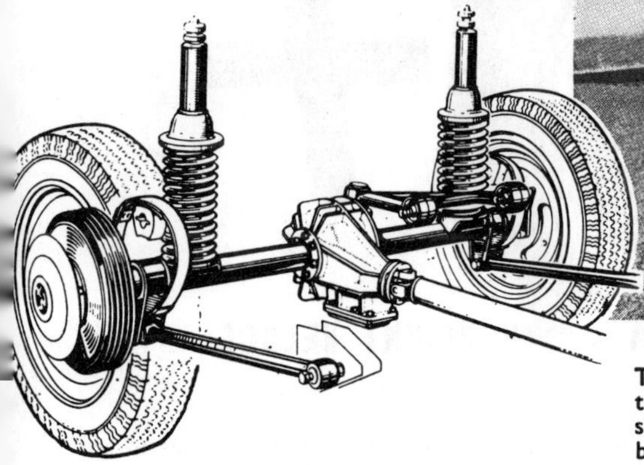

THREE-LINK positioning of the coil-sprung rear axle is shown in this sketch, the wishbone which takes lateral loads minimizing cornering roll by its high position.

pressure on the pedal but their effectiveness then builds up quickly. With rather more than 100 lb. pedal pressure really excellent stopping power is produced, and goes on being produced after a quick series of hard stops such as might overheat many braking systems. Both the brake and clutch pedals move upwards and forwards as they are pressed, their pivots being below and ahead of the toeboard, and whilst initially this direction of movement feels thoroughly unnatural a driver comes to take it for granted. The long pendant accelerator pedal has (in the case of an Italian car one might add "of course") been arranged to permit a form of toe-and-heel operation for downward gear-changes whilst braking.

Far from ostentatious, the Giulietta interior furnishing is quite pleasing to British eyes, a crackle black top to the plain facia panel avoiding any windscreen reflections. Whilst few strip-type speedometers have earned our praise, this one is certainly clear in its readings, showed about 3 m.p.h. error through most of the range, and is supplemented by a rev. counter. There is a low level warning lamp as well as a fuel gauge, an oil pressure gauge, and an oil thermometer as well as a water thermometer. In summer, the choke was not needed for starting the engine from cold, the hand throttle provided serving to ensure a reliable fast tick-over during a brief warm-up period. The fresh-air heating and screen demisting system fed plenty of cold, warm or hot air into the body.

Corner panels of transparent plastic proved effective in checking draughts from partially-opened windows, although looking back when turning into a major road a driver could suffer distorted vision from the slightly curved plastic. Supplementing the ring-controlled horn, a button at the steering wheel centre flashes dipped headlamps, nocturnal motoring proving the shallow but quite long and broad headlamp beams to be excellent. Repeaters for the flashing turn indicators mounted on the body sides are a sensible provision for town traffic.

Whilst the Giulietta "turismo internazionale" model is a thoroughly practical small saloon for everyday use, it is a car about which one cannot have other than strong feelings. It is possible to dislike quite strongly the rather jerky way in which it gets from place to place, regarding its talent for reaching any desired destination in an exceptionally short time as no more than a slight extenuating circumstance. It is equally possible to find that the speed and controllability of this vivacious saloon have put the fun back into motoring.

Specification

Engine

Cylinders 4
Bore ... ... 74 mm.
Stroke ... ... 75 mm.
Cubic capacity ... ... 1,290 c.c.
Piston area ... ... 26.65 sq. in.
Valves ... 2 overhead camshafts, chain driven
Compression ratio 8/1
Carburetter ... Solex 35APAI-G twin-choke downdraught
Fuel pump ... ... Fispa mechanical
Ignition timing control Centrifugal and vacuum
Oil filter ... . Fispa full-flow
Max. power (gross) 75 b.h.p. (65 b.h.p. net) at ... ... 6,000 r.p.m.
Piston speed at max. b.h.p. 2,950 ft./min.

Transmission

Clutch ... 7.9-in. dia. single dry plate
Top gear (s/m) 4.556
3rd gear (s/m) 6.17
2nd gear (s/m) 8.92
1st gear (s/m) 15.09
Reverse 15.32
Propeller shaft ... ... Divided open
Final drive ... 9/41 hypoid bevel
Top gear m.p.h. at 1,000 r.p.m. 15.5
Top gear m.p.h. at 1,000 ft./min. piston speed **31.5**

Chassis

Brakes: Hydraulic with ribbed alloy drums, 2 l.s. at front.
Brake dimensions: Front drums $10\frac{1}{2}$ in. dia. by $2\frac{1}{4}$ in. wide; rear drums 10 in. dia. by $1\frac{3}{4}$ in. wide.
Friction areas: 139 sq. in. of lining area working on 258 sq. in. rubbed area of drums.
Suspension:
Front: Independent by coil springs, transverse wishbones and anti-roll torsion bar.
Rear: Rigid axle with coil springs, trailing lower radius arms and upper A-bracket.
Shock absorbers ... Girling telescopic
Steering gear ... ZF worm and roller
Tyres: 155-15 Michelin "X" on test car (or Pirelli Cintura), tubed.

Coachwork and Equipment

Starting handle ... ... None
Battery mounting ... In luggage locker
Jack ... ... Pillar type
Jacking points ... 2 external under each side of body
Standard tool kit: Jack, wheelbrace, tool roll containing 4 open-jaw spanners, sump drain plug spanner, box spanner and tommy bar, screwdriver, pliers, hub cap lever, hammer.
Exterior lights: 2 headlamps, 2 sidelamps/flashers, 2 stop/tail lamps, number plate lamp, reversing lamp.
Number of electrical fuses ... 8
Direction indicators: Self-cancelling flashers (white front, amber rear, amber repeaters on body sides).
Windscreen wipers: Electrical self-parking two-blade.
Windscreen washers ... Hand pump type
Sun visors 2
Instruments: Speedometer with total and decimal trip distance recorders, r.p.m. indicator, fuel contents gauge, oil pressure gauge, oil thermometer, coolant thermometer.
Warning lights: Dynamo charge, low fuel level, sidelamps, heater fan, turn indicators.

Locks:
With ignition key: Ignition/starter switch.
With other key: Either front door, luggage locker, glove box, fuel filler cover.
Glove lockers One on facia with lockable lid.
Map pockets... Two in front, two behind front seat
Parcel shelf ... Behind rear seat
Ashtrays One on facia, one behind front seat
Cigar lighters ... ... None
Interior lights: One, with courtesy switches on front doors.
Interior heater: Fresh air heater and screen de-mister as standard.
Car radio. Accommodation on facia for any standard-size control unit.
Extras available ... None specifically for this model
Upholstery material ... Cloth and P.V.C.
Floor covering Carpet with rubber insets
Exterior colours standardized ... 4
Alternative body styles: None with identical mechanical specification. Giulietta 1.3-litre range includes lower-performance saloon, and sports models on shorter wheelbase.

Maintenance

Sump ... ... $8\frac{3}{4}$ pints, S.A.E. 10W/30
Gearbox ... $2\frac{3}{4}$ pints, S.A.E. 90 gear oil
Rear axle $2\frac{1}{2}$ pints, S.A.E. 90 hypoid gear oil
Steering gear lubricant S.A.E. 90 gear oil
Cooling system capacity ... 14 pints (2 drain taps)
Chassis lubrication: By grease gun every 1,250 miles to 23 points.
Ignition timing ... ... 8° before t.d.c.
Contact-breaker gap ... 0.014-0.016 in.
Sparking plug type Lodge HLN or Marelli CW225G
Sparking plug gap ... ... 0.024 in.

Valve timing: Inlet opens 22° before t.d.c. and closes 65° after b.d.c.; exhaust opens 55° before b.d.c. and closes 12° after t.d.c.
Tappet clearances (cold):
Inlet ... ... 0.017 in.
Exhaust ... ... 0.019 in.
Front wheel toe-in ... ... 0.12 in.
Camber angle ... ... 0°
Castor angle ... ... $\frac{1}{2}$°
Steering swivel pin inclination ... 8° 35′
Tyre pressures: Michelin "X" 21 lb. at front and rear. Pirelli $19\frac{1}{2}$ lb. at front and 21 lb. at rear.
Battery type ... ... S.A.F.A. 12-volt

ALFA  +

—OR HOW TO GET MORE URGE FROM A GIULIETTA SPRINT VELOCE

SOME people are never satisfied. Most of us would be content to own just an Alfa Romeo Giulietta of the most standard variety, but here's a man who isn't even content with his Sprint Veloce coupé. He must stretch its 1,290 c.c. engine to 1,470 c.c. and drag some extra Roman horses from it. This particular chariot is of 1957 vintage, when its advertised engine output was 90 b.h.p. at 6,000 r.p.m., which isn't exactly mean for 1·3 litres.

It was in June of last year that the engine was removed from the car, and for the next six weeks or so Jack Mount spent practically every evening and week-end at work on it. It meant sending to Italy for special 79mm bore cylinder liners of centrifugally cast Mehanite, and boring and counter-boring the crankcase to take them and their sealing rings. Special Mondial pistons to give a 10-to-1 compression were also obtained from Italy to fill the bigger new holes. A real worry was the cylinder head, as the new liners overlapped its waterways and these therefore had to be welded up locally and re-shaped. Despite welding by the relatively low-temperature Argonarc method, there was considerable distortion, which put the camshaft bearings out of line. To overcome this the centre bearings were bored 0·025in. oversize and the shaft journals built up by electro-plating to suit.

Then followed two dummy assemblies without liner seals or piston rings to see that all was well, which it wasn't quite. For one thing, the larger inlet valves (up from 37 to 39mm diameter) were found to foul the recesses already provided in the piston crowns; for another, the flats on adjacent pairs of liners had to be pared down before they would condescend to be good neighbours. On the subject of neighbours, those in Corkran Road, Surbiton, seem to have uttered no protest when the souped engine was bench-run for its first 20 hours after assembly in modified form. A special head gasket obtained from Italy is the jam in the head-block sandwich.

In its new form the Giulietta engine draws its fuel through the two standard

The Alfa Romeo has a slightly nose-down attitude on its larger tyres

double-choke Weber carburettors, type 40 DC03, but their main jets are two sizes up—120 instead of 110—and the chokes are now of 30mm dia. instead of 28mm. It breathes out through a simplified exhaust system with one of those Alextracter megaphone devices sheathing the tail pipe.

A word, now, about valve timing. The standard tappet clearances are 0·40mm for the inlets, 0·56mm for the exhausts, and the diagram then reads as follows:

Inlet opens at 34deg BTDC (Before Top Dead Centre, in case you are unfamiliar with the jargon)
Inlet closes at 64deg ABDC (After Bottom Dead Centre)
Exhaust opens at 64deg BBDC
Exhaust closes at 30deg ATDC.

One can get away with slightly reduced clearances, every 0·5mm augmenting the open duration by 6deg, and by doing this the owner claims a notable delay in power fall-off beyond 6,000 r.p.m. Now he has two American camshafts to play with, but they were not fitted at the time of this brief test. These so-called Sebring shafts give this timing:

Inlet opens 46deg BTDC
Inlet closes 65deg ABDC
Exhaust opens 65deg BBDC
Exhaust closes 34deg ATDC.

With these it should be better furnished in the upper storeys than at present, but rather drab in the basement. It would be interesting to know what the differences in performance are, and we might persuade Mr. Mount to let us try the car again with these camshafts.

Among other factors which might affect performance are the fitting of a Kenlowe electrically driven cooling fan for the radiator, with the usual thermostatic switch supplemented by an overriding switch. This fan is fitted in front of the radiator, blowing back through it. For sprints and really fast runs Lodge R47 plugs are used, for ordinary road work 2HLNs, and a Bosch coil is fitted. There is a radiator blind to help the engine to warm quickly in cool weather, the normal running temperature now lying in the range 80-90deg.C.

Brakes and Tyres Changed

With something approaching 130 b.h.p. (at 6,500 r.p.m.), so he says, under the bonnet, Mount thought it wise to fit Mintex M14 fade-resistant brake linings, despite the rather heavy pedal pressures which they entail. At present Pirelli Cintura tyres are fitted to the front wheels, and 5·50-15in. Michelin X at the rear, and the owner runs these at 28-30 and 35 p.s.i. respectively for ordinary road use.

So much for the preamble—does it really go? The answer is a very definite yes, but the pity is that since this journal has never tested such a car in standard form, the actual improvement cannot be assessed exactly.

Some guidance might be drawn from a test of an 80 b.h.p. Sprint (*not* Sprint Veloce) tested by the Swiss journal *Revue Automobile* in 1959. This reached 80 k.p.h. from a standstill in 9·5sec, whereas the Mount machine leapt to 50 m.p.h. (80·5 k.p.h.) in 7·7sec. The Swiss car's time from rest to 87 m.p.h. (140 k.p.h.) was 35·5sec, as compared with the hot 1,470's 23·3sec to 90 m.p.h. With all the stops pulled the standard Sprint recorded just short of 102 m.p.h. as a mean, while the subject of this report achieved a mean of 120·2 m.p.h., with its best run in one direction at 122 m.p.h.

There is a slight catch here; when most of the acceleration figures had been taken during our initial test runs, the radiator sprang a leak which grew quickly into a water spout and we had to abandon the car to have this sorted out. As the rev counter was reading 6,000 r.p.m. at almost exactly 100 m.p.h. on *The Autocar's* electric speedometer during those first runs, which would have meant only 116·7 m.p.h. theoretical at 7,000 r.p.m., the owner decided to raise the gearing for its high-speed runs by fitting larger section tyres. Perhaps he overdid this, for on the new 5·90-15in. Dunlop R55s the top gear limit became just over 6,500 r.p.m. Moreover, with these brand new big tyres, still with the "shine" on their treads, the car lost some of its previously admirable handling qualities.

This radiator trouble also confused our attempts to gauge the car's fuel consumption accurately in the short time we had it, but we suspect that most people who tune their cars as sharply as this are not too worried about how much wind goes into the trumpet.

PERFORMANCE

Acceleration from rest through gears to:

M.p.h.	sec
50	7·7
60	10·0
70	13·7
80	17·5
90	23·3
100	38·0

Standing start quarter-mile	17·6
20-40 m.p.h. in 2nd	4·0
30-50 m.p.h. in 2nd	4·4
30-50 m.p.h. in 3rd	6·8
40-60 m.p.h. in 2nd	4·6
40-60 m.p.h. in 3rd	6·5
50-70 m.p.h. in 3rd	6·8
50-70 m.p.h. in top	10·1
60-80 m.p.h. in 3rd	7·5
60-80 m.p.h. in top	11·4

Maximum speed:

mean m.p.h.	120·2
best m.p.h.	122·0

Still Tractable

In my experience one cannot increase the cylinder capacity of any standard engine and retain all its customary smoothness and mechanical refinement, but as the Giulietta's crankshaft runs in five main bearings this unit has broader shoulders than most for carrying extra loads. In this case it proved easy to start and quite tractable at low speeds, although the power muscles don't really bulge until over 3,000 r.p.m. are indicated on the tachometer. These early Giuliettas suffer from a miserable gearbox, which might even be better without the little synchromesh that remains when the car has covered a big mileage. One has to grit one's teeth and snatch it from first to second, the least co-operative step, so quickly that the box is caught unawares and hasn't time to protest. The only other snag, which applies still to the current Sprint and Sprint Veloce coupés, is that none is made with right-hand drive. If all other vehicles on the road were as low-built as this it wouldn't matter much; as it is, one rarely has a through-view beyond the rear of the car in front.

In every other respect the Giulietta pleased greatly, for its quietness and comfort on the road despite the high tyre pressures used, its lightness and precision of control, and, above all, perhaps, its freedom from any signs of deterioration in its structure. For these and other reasons it provides a fine basis for personal attention such as Jack Mount has bestowed upon it. R. B.

Externally the engine looks unchanged, but 1,470 c.c. are under the head

ALFA GUILIETTA SZ

". . . It responds quick as thought to every touch on the controls." Gordon Wilkins gives his impressions of the Sprint Zagato, one of the world's most desirable small cars and the newest variant of Alfa's classic Guilietta coupe.

ZAGATO'S LITTLE JEWEL

FEW high performance cars have been offered in so many different versions as the Alfa Guilietta. Sedan, TI sedan, Sprint and Sprint Veloce coupes by Bertone, Spider and Spider Veloce with detachable hard top by Farina, SS coupe by Bertone and SZ coupe by Zagato—there is something for everyone.

The SS began as a streamlined lightweight racing machine to designs by Scaglione but in course of time it acquired extra avoirdupois in the form of bumpers, a higher roof line, and a revised front with a token radiator grille.

It is still quite a respectable Gran Turismo machine but a lot seem to be bought as play cars by the gilded youth of Roman cafe society. This left the way open for a new competition car, and Zagato produced it—the flyweight snub-nosed SZ coupe. After trying it out on Swiss roads, I cannot imagine a better machine for fast mountain motoring and it is easy to see why it won the 1960 Alpine Rally on its first time out. It responds quick as thought to every touch on the controls, hurtles from a standstill to 100 mph without using more than four of its five speeds and it has a gearbox which gives a keen driver the kind of pleasure one gets from playing a good musical instrument.

Power unit of all the Guiliettas is a four-cylinder engine of classic design with twin chain driven overhead camshafts and valves at an angle of 80° in hemispherical combustion chambers. Bore and stroke are 74 X 75 mm giving 1290 cc and on a compression ratio of 9.7 to 1 the SZ engine delivers 99 bhp installed at 6000 rpm (116 hp SAE) with peak torque of 83 lb ft at 4500 rpm (91.3 SAE). With the highest axle ratio of 4.1 to 1 piston speed at 60 mph in top gear is only 31.7 ft/sec. Aluminium alloy is used for block and head, with wet cylinder liners. The crankshaft runs in five bearings, thin lead indium shells being used for main and big ends. An electric pump fuels the two twin choke Weber carburettors.

Transmission is via a single plate dry clutch to a five speed all-synchromesh gearbox with direct fourth and geared up fifth, the gearbox ratios being 3.258, 1.985, 1.357, 1.0, and 0.854 to 1. There is a choice of three axle ratios, 5.125, 4.555, and 4.1 to 1. Although it is only a short car with a wheelbase of 88.5 in the propeller shaft is in two sections, with flexibly mounted centre bearing.

Front suspension is conventional with wishbones, coil springs and an anti-roll bar. Rear suspension is one of the most successful rigid axle designs with two single radius arms which take the drive and central A bracket above the differential to resist brake torque and locate the axle laterally. Coil springs are the suspension medium with concentric telescopic dampers. Body and chassis form a unit and kerb weight without fuel is 1685 lb. Though shorter and apparently less well streamlined than the SS, the SZ is about three mph faster and has more luggage space; you can even get a small suitcase in the trunk on top of the spare wheel.

Zagato's single minded pursuit of lightness is apparent in the tiny finger grips and press buttons on the doors, the plastic windows and the lightweight seats with pierced backrests and slim padded side bars to hold the occupants against centrifugal force.

Interior finish is mainly in practical matt black. Under a cowl in front of the driver are three dials; one contains oil and water thermometers plus fuel gauge with reverse warning lamp; in the centre is the tachometer reading up to 8000 rpm, with inset oil pressure gauge but no warning about rev limits and on the right is the speedometer with warning light for headlamp high beam and generator charge. Arrangement of the controls reflects Alfa's long racing experience, except for a rather nasty pull-out handbrake. The short central gear lever works with the utmost precision and the gearbox is such a joy to handle that you soon find yourself singing up and down through the gears, changing far more often than is necessary for the sheer pleasure of it. There are no indications of gear positions but they are exactly where one would instinctively expect to find them, so there is nothing to learn. Having been told that I could go round to 7000 rpm in the gears, I saw nearly 78 mph in third and over 100 mph in fourth, before snatching fifth speed which continued the acceleration in a most inspiring way. With the 4.55 axle which I was using, limiting speeds on the first four gears are 18, 33, 78 and 106 mph, with a theoretical 128 mph available in fifth. Quick acceleration checks showed 0-50 mph in about 9 secs and 0-60 in 12.5 secs.

The clutch handles racing starts and full throttle changes, but don't drop it in at low revs or you may hear a bump from the propeller shaft, and a vibration in the drive line. Yet this is not a fierce machine suitable only for the race track, the engine is quite flexible and fourth speed can be used for quiet driving around town provided you are ready to drop a couple of ratios for quick acceleration.

On dry roads the handling is terrific but those who have driven it in the wet say that you have to use some discretion because the combination of short wheelbase, high power and lack of overhang at front and rear makes it easy to twitch it into a slide. The brakes are drums with helical cooling fins in light alloy. They are very powerful but do fade in hard competition use and owners who race the SZ are now being offered the option of a special new front brake with three leading shoes which raise the friction lining area of the front brakes by 84 percent and reduce the pedal effort by 20 percent.

The SZ steers to a hair line on corners. It needs three turns of the wheel from lock to lock but feels more direct and no more than flexing of the wrists is needed to hold it on the straight. Basically there is slight understeer but it is easy to steer with the throttle. It is a nimble versatile little bomb which doesn't seem to be worried by rough roads. The structure feels taut and rigid and the ride, though firm is never harsh. It is certainly one of the most exciting small cars I have ever driven.

Its great rival is, of course, the Lotus Elite and although the SZ has scored an impressive number of competition successes, the Elite usually comes off best on fast circuits. Trying to match the Elite's top speed Zagato has now come up with yet another Guilietta; it has a longer lower nose than the SZ and a longer streamline tail but is at present unnamed.

This is not the first Alfa to which Zagato has lent his magic touch, but he admits it is his favorite. There are dozens of SZ's in Italy and they come one-two-three in their class with almost monotonous regularity in their own country.

In this year's gruelling Targa Florio the little Alfas came in to the finish line straight after the works Porsches which were sixth and seventh. #

ALFA SPRINT ZAGATO

BY HANSJOERG BENDEL

THE ALFA Giulietta Sprint Zagato is the lightest, smallest, best streamlined, most powerful and therefore also the fastest of the entire Alfa range. The engine is the well known dohc 4-cyl fitted in this instance with two twin-choke Weber carburetors and a U.S.-made Bendix electric fuel pump, as on the Veloce model; front suspension with wishbones and a rigid rear axle with light-alloy differential casing, located by two lateral and one triangulated center link, belong to well-established Alfa practice; springs are coil all around; brakes feature heavily finned aluminum drums. In common with the Bertone-bodied Speciale, the Zagato has a wheelbase of 88.5 in., shorter by 5 in. than on the standard coupes, and is fitted with a 5-speed gearbox with Porsche ring-type synchromesh.

Although these mechanical specifications are very close to those of the other Giulietta models, the Zagato nevertheless has a distinct personality, for which the body, built by one of the most famous Italian coachbuilders (compare Salon, May 1961 issue, and see pages 32-35 of Dec. 1961), is primarily responsible. It is strikingly low, giving a very small frontal area, well streamlined, constructed of aluminum and fitted out in a businesslike way —no unnecessary chromium anywhere and only the lightest seats and interior furnishings.

Our test car was handed to us at Milan by Alfa's amiable Press Relations manager, Mario Bernasconi. The odometer showed almost exactly 10,000 miles. We soon felt at home and left Italy to cross the Alps into Switzerland, from where we continued toward the south of France and into Spain. On our return, we followed the Mediterranean along the famous Costa Brava where—in addition to some cork-screwing previously accomplished—we sampled another collection of tight corners, bumpy surfaces and gravel roads with bumps up to two inches high. The whole trip took roughly 2000 miles and provided us with a very complete seminar on The Behaviour of a Fast Car under All Conditions Imaginable.

Now—what kind of car is this? When we first walked around it, we made a note of the complete absence of bumpers front and rear; obviously, low weight was more important to Zagato than considerations of shunting on a parking lot or mishaps at the hands of a beginner. Compactness in length is matched by an over-all height of only 48.5 in. and in the interest of generous elbow room coupled with small frontal area, the body tapers strongly from the window sills upward. This entails a roof line rather far inside, so that rain water can drip onto one's elbow. Remarkably thin doors house curved wind-down windows in order to achieve maximum width utilization.

Two lightweight seats are built on steel tube "space frames." These hold the occupants firmly and with reasonable comfort, but for long distances, a little more padding—particularly along the lateral struts—and a more solid base cushion would be desirable. This car, of course, is intended to be driven with the straight-arm technique. Even so, the range of lengthwise seat adjustment is exceptional. Others please follow—you too, E-type!

Driving position is dead right, slightly reclining yet high enough in relation to the low engine hood. Instruments are grouped in a cluster in front of the driver and contain everything one can possibly want, including an amusing polarizing lens on the headlamp warning light, which turns to adjust brightness in a most unusual fashion. As on all Alfas, pedal arrangement reflects racing experience, although with synchromesh as fast and unbeatable as this one, the heel-and-toe method is great fun but certainly no necessity. The curved windshield has remarkably little distortion over its entire width and features slender corner posts. Side windows are made of plexiglass; this material is more easily scratched than real glass, so that replacement may become necessary.

Equipment is—surprisingly—as complete as on the Bertone coupes: Heater-defroster with electric fan (the hot water tap is still in the engine compartment; relocation of this control, to make it accessible without stopping, would certainly be an advantage); a separate fresh air inlet on the driver's side and opening rear side windows combine to provide efficient ventilation in hot climate; there is a foot-operated windshield washer with automatic wiper engagement; a map reading light is located underneath the instrument panel on the passenger's side—too low for good reading. There are no sun visors, but a pair could easily be installed and would not really add much weight. A rear mirror is suspended from the windscreen top and leaves vision ahead quite unobstructed, but offers totally inadequate rear vision; it would be better to mount it atop the instrument panel as on the Bertone coupes.

The steering wheel is beautifully made, with polished aluminum spokes; but a low sun shining in from behind causes unpleasant reflection—proving again that beauty must be paid for. The horn—much too weak on the test car—is actuated by a central button, surrounded by a ring-type switch for headlamp flashing. The ignition key has a safety device which allows only one actuation of the starter motor. If a second start is necessary, ignition must first be switched off so that accidental starter operation with the engine still running is impossible. An interesting body detail is a rear edge of the engine hood raised by about half an inch, allowing hot air to escape from the engine compartment and impinge on the windscreen, helping to defrost in winter.

Sensibly, no attempt has been made to provide "occasional seats," and the rear compartment thus offers a very useful space for additional luggage. Large suitcases are not easy to stow in here, because the seat backs do not tilt—this ensures that the back of an empty seat cannot come forward at an inopportune moment even when braking under race conditions. The tail contains a lockable luggage compartment which is sufficient for one medium size suitcase and a bag, and also houses the fuel filler.

Interior furnishings and floor mats are more notable for lightness than for robustness; this is one instance where the heavier Bertone bodies display superior finish and are likely to provide longer repair-free life.

We have commented on these "static" aspects at some length because they show that Zagato is indeed a fully-equipped road car. In order to complete the description we note that no safety or crash-padding has been installed; Alfa is obviously of the opinion that it is better to prevent accidents from happening, by developing chassis with superb controllability, rather than by mitigating their effects after its too late.

Starting the engine is an operation which, as on all Giuliettas we know, is needed only once—even this hottest version was an instant starter, when cold after a rainy night in the open as well as when hot after a fast run. This is particularly praiseworthy because the engine is equipped with cold Lodge RL-47 racing plugs, which could easily be oiled in an engine of lesser balance.

The Zagato's engine proved very satisfactory. It warmed up quickly, and even without the use of a radiator blind, oil and water temperatures of 185° were maintained with astonishing certainty. Considering the size—1290 cc, the four large carburetor throats and the tune applied to it, the engine displays outstanding flexibility. Reasonable acceleration starts shortly above 2000 rpm, and from about 3000 rpm upward opening the throttle is accepted with real eagerness, accompanied by a highly delightful tire scream when using 1st and 2nd gear. Racing starts are easy, the rear tires leaving nice grey traces of rubber.

When driving in really spirited fashion, one almost automatically shifts at 6500 rpm which coincides with maximum power output; on the open road, in 5th gear (giving an overdrive ratio) the car chooses something like 100 mph as its natural cruising speed. This represents 5500 rpm and a piston speed of 2750 fpm—well within the limits of the 5-bearing short-stroke engine.

This kind of performance could not be exhilarating and would even be dangerous on a chassis unable to match the power of the engine—we are pleased to report that the Zagato was found to have an extremely good balance. The steering—with Pirelli Cinturato tires inflated to 24 and 26 lb/sq in. front and rear—was light, sensitive but shockproof, steering gear ratio about right with just below 3 turns from lock to lock. Roadholding merits high marks; whereas the standard Bertone coupe has unmistakable understeer and feels a bit heavy in some corners, the Zagato strikes a good compromise between the ability to keep straight on bumpy roads and a very unusual degree of controllability. For immediate response in emergencies, this is one of the best cars we have driven.

ALFA SPRINT ZAGATO

As on most streamlined cars, sensitivity to side winds is pronounced, but even at high speed, corrections necessary to keep the nose pointing in the right direction are made almost without conscious effort. Springing is rather firm, more so than on the heavier Bertone coupe; when driven fast on bumpy roads, occasionally there was some sideways hopping of the rear axle. This characteristic must probably be attributed to two distinct reasons. First of all, a glance at the speedometer always proves that one has been going quite a bit faster than it felt; secondly, since the car as a whole is lighter but the unsprung weight practically the same as on the standard models—in spite of the use of very beautiful Borrani aluminum wheels—the ratio of sprung/unsprung weight must be slightly worse.

We said that the car always goes faster than it feels; this can be attributed to the free-revving engine, the unmistakably sure-footed chassis and a really low level of wind noise. Added to this is the knowledge of having brakes able to cope with any situation—even repeated high-speed braking during performance tests failed to produce fade, and there was never any trace of roughness or side pull.

The performance figures speak for themselves, particularly as they were recorded after our 2000-mile trip to Spain, without any intermediate tune-up. Our best high-speed run, timed over a quarter mile, produced 121.8 mph; it might have been faster if a tune-up had been made, because the plugs indicated correct mixture strength for cylinders 1 and 2, definitely over-rich mixture for 3 and 4.

To us, the most interesting result of the test is the confirmation that the Zagato is also practical for fast touring. It is comfortable enough, appears almost as difficult to over-strain as its slower sisters, and its moderate fuel consumption and exemplary gas tank capacity of 20.5 gallons eliminate the need to look for gas stations every 200 miles, as with some other GT cars.

This enables us also, we believe, to answer the question as to what Giulietta model one should buy. If your car must have race-winning potential in addition to the features required for fast touring, the Zagato is the obvious choice. If fast touring is the only aim, then the steel-bodied Veloce or Speciale models, very little slower but less expensive, seem preferable.

R&T ROAD TEST

ALFA SPRINT ZAGATO

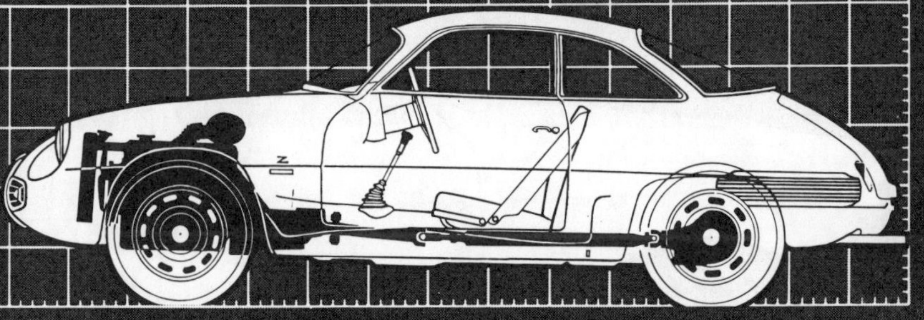

DIMENSIONS

Wheelbase, in	88.5
Tread, f and r	50.8/50.0
Over-all length, in	151.5
width	60.2
height	48.5
equivalent vol, cu ft	256
Frontal area, sq ft	16.2
Ground clearance, in	5.1
Steering ratio, o/a	n.a.
turns, lock to lock	2.6
turning circle, ft	33.5
Hip room, front	2 x 19
Hip room, rear	n.a.
Pedal to seat back, max	45
Floor to ground	11

CALCULATED DATA

Lb/hp (test wt)	19.1
Cu ft/ton mile	68.3
Mph/1000 rpm (5th)	18.2
Engine revs/mile	3300
Piston travel, ft/mile	1620
Rpm @ 2500 ft/min	5090
equivalent mph	92.5
R&T wear index	53.4

SPECIFICATIONS

List price	n.a.
Curb weight, lb	1890
Test weight	2220
distribution, %	52/48
Tire size	155-15
Brake swept area	259
Engine type	4 cyl, dohc
Bore & stroke	2.91 x 2.95
Displacement, cc	1290
cu in	78.7
Compression ratio	9.7
Bhp @ rpm	116 @ 6500
equivalent mph	118.1
Torque, lb-ft	96 @ 4500
equivalent mph	81.8

GEAR RATIOS

5th (0.854)	3.88
4th (1.00)	4.55
3rd (1.36)	6.17
2nd (1.98)	9.03
1st (3.26)	14.8

SPEEDOMETER ERROR

30 mph	actual, 26.1
60 mph	53.7

PERFORMANCE

Top speed (5th), mph	120
best timed run	121.8
4th (6900)	107
3rd (7000)	80
2nd (7050)	75

FUEL CONSUMPTION

Normal range, mpg	19/24

ACCELERATION

0-30 mph, sec	3.6
0-40	5.8
0-50	8.1
0-60	11.2
0-70	14.4
0-80	19.1
0-100	33.4
Standing ¼ mile	17.8
speed at end	77.0

TAPLEY DATA

4th, lb/ton @ mph	215 @ 74
3rd	340 @ 57
2nd	510 @ 40
Total drag at 60 mph, lb	80

ENGINE SPEED IN GEARS

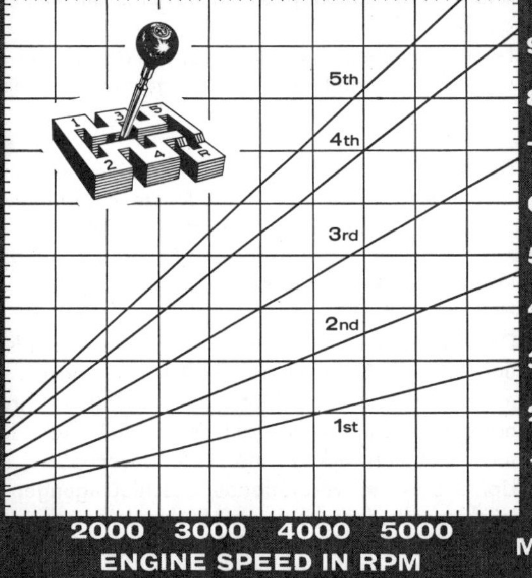

ACCELERATION & COASTING

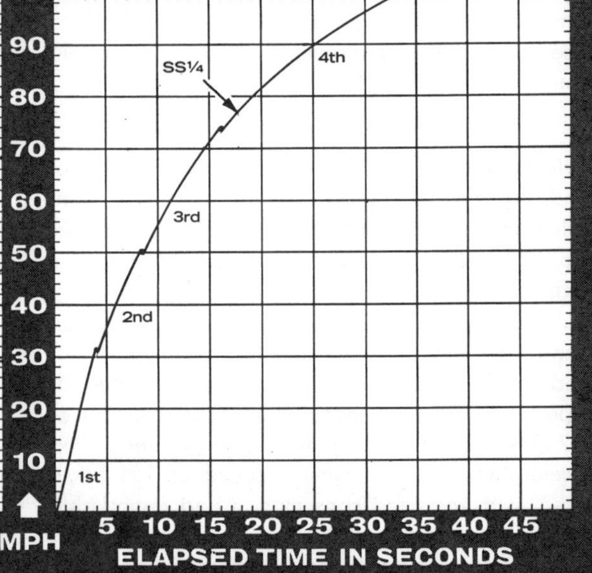

giulietta berlina e t. i.

L'Alfa Romeo **Giulietta berlina** è una vettura a 5 posti, 4 porte, adatta per il lavoro e per il turismo, elegante, pratica, d'ingombro limitato e di basso costo di esercizio. Essa ha linea piacevole, carrozzeria confortevole, eccezionale tenuta di strada e frenata oltremodo sicura.
La « **Giulietta t. i.** » possiede tutte le caratteristiche di resistenza e di praticità della Giulietta berlina, ma è munita di un motore più spinto; è quindi una vettura di turismo normale avente prestazioni di gran turismo. Ha una potenza di 65 CV (75 CV-SAE) a 5500 giri/1', che le consente una maggiore ripresa ed una velocità massima di 155 km/h, con un consumo assai limitato e notevole silenziosità.
Il sistema di frenatura, già largamente sperimentato, rappresenta la migliore garanzia di sicurezza della Giulietta berlina e t. i.

La **Giulietta berline** est une conduite intérieure, à 5 places, convenant au travail comme au loisir, élégante, pratique et en même temps d'entretien économique et de dimensions raisonables. Elle a un profil agréable, une carrosserie confortable, une tenue de route suprenante et un freinage extrêmement sûr.
La **Giulietta t. i.** a toutes les caractéristiques de résistance et de praticité de la Giulietta berline, mais possède un moteur plus puissant elle est portant une voiture de tourisme normale offrant les performances du grand tourisme. Elle bénéficie d'une puissance de 65 CV a 5500 t/m qui lui permet des reprises et des accélérations foudroyantes et une vitesse maximum de 155 Km/h., avec une consommation très réduite et... un silence remarquable.
Le système spécial de freinage, qui a déjà fait ses preuves, offre en effet la meilleure garantie de sécurité.

The **Giulietta berlina** is a 5-seat, 4-door car suitable for work and pleasure, very handy, of small scale dimensions and of low consumption. The car has an agreeable line, a comfortable interior, exceptional road-holding and a very safe braking action.
The **Giulietta t.i.** retains all the characteristics of endurance and handiness of the Giulietta berlina, but it is equipped with a hotter engine which confers to the car higher performances. This model has a rated output of 65 HP (75 HP SAE) at 5500 r.p.m. allowing better pickup and acceleration and a maximum speed of 155 km/h (96 h.p.h.), with restricted consumption and remarkable noiseless.
The braking system, already long experienced, represents the best safety guaranty for both "Giulietta berlina e t. i.".

Der **Giulietta Berlina** ist eine 5-plätzige Limousine. Er ist gleich geeignet für Arbeit und Musse; elegant, zweckmassig und zugleich wirtschaftlich im Unterhalt und vernünftig in seinen Ausmassen. Er besitzt ein gefälliges Profil, eine bequeme Karosserie, erstaunliche Strassenhaltung und überaus zuverlässige Bremsen.
Der **Giulietta t. i.** weist die ganze Widerstandsfähigkeit und Zwerckmässigkeit des Giulietta Berlina auf, ist aber mit einem kräftigeren Motor ausgestattet: so wird er zum normalen Tourenwagen mit der Schnelligkeit des Granturismo. Er entwickelt eine Leistung von 65 PS bei 5500 UpM, die ihm verblüffende Beschleunigungen bis zur Höchstgeschwindigkeit von 155 km/h bei geringen Benzinverbrauch und bemerkenswert geräuschlosem Lauf ermöglicht.
Das lang erprobte Spezialbremssystem bietet tatsächlich die beste Sicherheitsgarantie.

I presenti dati hanno solamente valore indicativo.

Caratteristiche

	berlina		t. i.
Cilindri		n. 4	
Alesaggio		mm. 74	
Corsa		mm. 75	
Cilindrata		cmc. 1290	
Potenza a 5500 giri/1'	CV 53 CV-SAE 61		
Potenza a 6100 giri/1'			CV 65 CV-SAE 75
Potenza fiscale		CV 15	
Passo		mm. 2380	
Carreggiata anteriore		mm. 1292	
Carreggiata posteriore		mm. 1270	
Lunghezza massima		mm. 4106	
Larghezza massima		mm. 1555	
Peso a secco	kg. 915		kg. 920
Velocità massima	km/h 140		km/h 155
Pneumatici		155 × 15	
Numero posti		5	
Impianto elettrico		Volt 12	
Capacità serbatoio		l. 40	

Distribuzione: camere di scoppio emisferiche; due valvole in testa comandate direttamente da due alberi a camme in testa azionate da due catene silenziose.

Frizione: monodisco a secco con mozzo elastico.

Cambio di velocità: 4 marce sincronizzate e retromarcia. Comando sotto il volante.

Ponte posteriore: con coppia ipoide.

Sospensione anteriore: a ruote indipendenti; a quadrilateri trasversali con flessibilità variabile lungo lo scuotimento; molle ad elica; barra stabilizzatrice ed ammortizzatori telescopici.

Sospensione posteriore: a ponte rigido collegato al telaio con triangolo superiore e puntoni inferiori; molle ad elica ed ammortizzatori telescopici. Flessibilità variabile lungo lo scuotimento.

Guida: a vite globoidale e rullo.

Freni: idraulici sulle quattro ruote; gli anteriori a due ceppi avvolgenti; ceppi flottanti; tamburi bimetallici a grandi dimensioni; anteriormente con alettatura elicoidale.

Caractéristiques techniques

	berlina		t. i.
Cylindres		n. 4	
Alésage		mm. 74	
Course		mm. 75	
Cylindrée		cmc. 1290	
Puissance à 5500 tr/mn	CV 53 CV-SAE 61		
Puissance à 6100 tr/mn			CV 65 CV-SAE 75
Empattement		mm. 2380	
Voie avant		mm. 1292	
Voie arrière		mm. 1270	
Longuer max		mm. 4106	
Largeur max		mm. 1555	
Poids à sec	kg. 915		kg. 920
Vitesse max	km/h 140		km/h 155
Pneus		155 × 15	
Nombre de Places		5	
Equipment électrique		Volt 12	
Capacité du réservoir		l. 40	

Distribution: soupapes inclinées dans la culasse hémisphérique; commande directe au plateau avec deux arbres à cames en tête; distribution à collets des soupapes à bain d'huile.

Embrayage: monodisque à sec.

Boîtes de vitesse: 4 vitesses synchronisées avec levier de commande sous le volant.

Pont arrière: à couple hypoïde conique.

Suspension avant: à roues independantes et quadrilatères transversaux, ressorts hélicoidaux, barre stabilisatrice et amortisseurs hydrauliques télescopiques.

Suspension arrière: à pont rigide avec supports longitudinaux inférieurs et triangle supérieur; ressorts hélicoidaux, amortisseurs hydrauliques télescopiques.

Direction: à vis globoidale et rouleau.

Freins: hydrauliques; frein avant à deux mâchoires enveloppantes; tambours avant munis d'ailettes hélicoidales de refroidissement.

Technical features

	berlina		t. i.
Cylinders		4	
Bore		mm. 74	
Stroke		mm. 75	
Cylinder capacity		cc. 1290	
Power at 5500 r.p.m.	HP 53 HP-SAE 61		
Power at 6100 r.p.m.			HP 65 HP-SAE 75
Wheel base		7'10"	
Front track		4' 2"1/4	
Rear track		4' 4"	
Overall length		13' 1"	
Overall width		5' 1"/2	
Dry weight	lbs 2013		lbs 2024
Maximum speed	m.p.h. 88		m.p.h. 96
Tires		155 × 15	
Number of seats		5	
Electrical system		V 12	
Tank capacity		l 40	

Valve gear: inclined valves in the hemispherical cylinder head, cams acting directly on valve stems, springs cap oil cups, double overhead camshafts.

Clutch: single dry-plate.

Gear-box: 4 synchronized forward speeds and 1 reverse. Lever under the steering wheel.

Differential: with hypoid bevel gears.

Front suspension: independent with transverse linkage, coil springs, stabilizing bar and telescopic hydraulic shock absorbers.

Rear suspension: rigid axle housing with upper triangular thrust rods and lower radius rods, coil springs, hydraulic and telescopic shock absorbers.

Steering: by worm and roller.

Brakes: hydraulic, the front with two leading shoes; helical cooling fins on the drums.

Technische Daten

	berlina		t. i.
Zylinderzahl		4	
Bohrung		mm 74	
Hub		mm 75	
Hubraum		mm 1290	
Leistung bei 5500 U/min	CV 53 CV-SAE 61		
Leistung bei 6100 U/min			CV 65 CV-SAE 75
Radstand		mm 2380	
Spurweite vorn		mm 1292	
Spurweite hinten		mm 1270	
Gesamtlänge		mm 4106	
Gesamtbreite		mm 1555	
Leergewicht	kg. 915		kg. 920
Höchstgeschwindigkeit		155 × 15	
Reifen		5	
Platzzahl	km/h 140		km/h 155
Elektrische Anlage		Volt 12	
Tankinhalt		l. 40	

Ventilsteuerung: schräghängende Ventile in halbkugelförmigem Zylinderkopf; direkte Steuerung durch zwei obenliegende Nockenwellen; die Ventilführungen laufen im Oelbad.

Kupplung: Einscheibentrockenkupplung.

Getriebe: 4 synchronisierte Vorwärtsgänge und Rückwärtsgang. Schaltung unter dem Lenkrad.

Hinterachse: Konisches Hypoid.

Federung vorn: Unabhängige Räder; viereck - Querlenker; Schrauben federn mit Kurvenstabilisator; hydraulische Teleskop-Stossdämpfer.

Federung hinten: durchgehender Achsträger; Dreieck oben; längsgehende Stützen unten; Schraubenfedern und hydraulische Teleskop-Stossdämpfer.

Lenkung: mit Schnecke und Rolle.

Bremsen: hydraulisch; vorn mit zwei Gleitbacken; die vier Trommeln mit schraubenförmigen Flügeln zur Belüftung.

ALFA ROMEO GIULIETTA SPRINT

We always wondered why Alfa owners wore a big smile as they rasped past—now we know.

● Italians have been in love with wheeled vehicles since the days of the chariots. This deep affection is still very much alive, as witness the Alfa Romeo Giulietta Sprint. A small, almost delicate-looking coupe, the body is drawn in metal with such economy of line that you have to sit and look at it in repose for a goodly length of time before realizing just how handsome a piece of automotive sculpture it is. Finding it at rest for any great length of time was a problem once its fun-producing performance had been sampled by the members of the C/D crew.

Don't be misled by the word fun. The Italians, being a

PHOTOGRAPHY: DOLIN

An Alfa in full cry. In touring trim there's quite a bit of body lean, which doesn't seem to disturb cornering power.

people steeped in ancient wisdoms, are serious about important things such as women, food, and automobiles. And the Alfa coupe is a serious answer to the problem of transporting two people and their luggage far and fast on a modicum of gasoline. Can they help it if the solution comes up fun?

Many things about the Alfa come to the fore even before you set sail down the highway. The car is low, even though the basic design is seven years old, but the door is wide and opens at almost right angles to the body so that slipping behind the wheel might only be a problem for people who don't bend at the waist. The unoccupied Alfa seat looks a little skimpy around the cushions, but once aboard it is comfortable with good support and ample wiggle room for long trips. In the *conduttore's* position behind the wheel all is made clear as to the purpose of this machine. The angle of the wheel, in relation to the driver, is fairly straight up and down, which lends itself to continental-style straight arm driving. The wheel itself is small in diameter, of black plastic, with a man-sized rim that gets right away from the bent knitting-needle feel of many contemporary steering wheels. The pedals, a touchy point with one slew-footed staff member, just can't be faulted in regards to room and their spatial relationship. Heel-and-toe work (it never hurts to show off once in awhile) comes almost naturally with the Alfa controls, and engine response to the throttle is so quick and accurate that even a few clutchless downshifts can be assayed, although this is not recommended.

Starting up is quick and simple. A twist of the key and the gem-like double overhead camshaft engine whirs into life with the typical Alfa rasp, and then settles down to about a 750 rpm idle. With arms outstretched, clutch foot down, and the whole wide road in view the first-time Alfa driver is ready for a pleasant automotive experience. Initial acceleration is not of the cheek-sagging variety, but as the revs mount and the gears in the Porsche-type box are selected, used, then rejected for the next higher ratio, things get more and more interesting. Shifts are smooth and fast with a possible increase in both qualities in a box with more miles on it than the one we had. One wonderful aid to picking off the right cog at the right time was the gearshift lever itself. This was a really healthy polished steel shaft topped with a generous handful of hard black plastic. Even if the gearbox wasn't up to snuff—which it was, although some complaints have been leveled at early Alfa ZF boxes—you felt that with this surgeon's instrument in your right fist you just couldn't mangle a shift. After a few standing starts it becomes apparent that this little car demands to be driven with precision and a high degree of enthusiasm. The clutch, almost slip-free after repeated timing runs, takes up low in the pedal travel and is nearer to the right-now type than many fitted to cars billed as real road-burners. Driven with brio, using all the gears including the synchromesh first and a six grand rev limit, the Alfa has a sparkling road performance. Acceleration up to 60 mph is in the 13-second bracket (not earth-shaking but a handy bit of go to have under your right toe), and an effortless cruise in the 80s makes this an automobile that can stand on its own four wheels in any country and in any company.

Alfa roadholding, among the best of any car utilizing

Low seat, tiny wheel, hooded instruments plus Alfa agility combine to make even cloddish drivers look and feel good.

a solid rear axle, puts it head and shoulders above 90 percent of the passenger vehicles available today. Hilly, twisty, two-lane roads are Alfa country—outright. Roadside observers will comment on the amount of body lean, but this is hardly noticeable in the car. Handling can be described as marked understeer up to the point of a strangled gulp on the part of the driver when it turns—in an easy transition period—into slight oversteer. Braking? What can we say, except that it is superb. Four of the biggest finned drums in the business stop the Giulietta come hell or high speeds. Steering? Well, the shaft that runs from the wheel to the worm and roller steering box does so in a straight line, and is made of solid steel. 'Nuff said? The Alfa, in a few words then, is a small car with a rev-happy d.o.h.c. engine that can carry two people from point A to point B over all types of roads quicker than most cars twice its size. It does this not with blinding speed, but with a wonderful combination of roadholding, compact size, and sheer willingness. As a plus factor to all this you can't budge the gas mileage below about 20 mpg. Admittedly few Alfa drivers in this country think about the cost of *benzina*, but it would provide grist for the cocktail party chatter mill to be able to say that you own a 90-plus mph sports car that's a real miser with fuel.

Still another nice thing about the Alfa coupe is the quality and quantity of cockpit gadgets. Most builders of "sports" cars seem to feel that creature comforts should be on the Spartan side. The coupe has roll-up windows, ash trays, window washers, water and oil temperature gauges, heater, choke, hand throttle, back-up lights, trafficators, and generator warning light. A nice touch is the pedal for the windshield washers that gives you one squish and one stroke of the wipers for each toe tap.

Up till now this has been one paean of praise—probably because we've saved the barbs for the end. It must

The secret of the Alfa is under the hood. Small (1290 cc) dohc unit has race-car responsiveness and auto show finish.

be realized that this car is powered by a small engine. This is half its charm. It is the David and Goliath concept demonstrated whenever Alfa Veloces whip bigger-engined Porsches and Sunbeam Alpines. This is a drawback for transcontinental work. After two or three hours in the 70s the engine begins to make its presence felt. One way around this is to get the four-speed gearbox converted into a five-speeder. This can be done for about $149 plus the cost of labor. The window winders take 10½ turns from open to close. This is too lengthy a distraction for the driver of a *(Continued on page 147)*

ALFA ROMEO GIULIETTA SPRINT

Price as tested: $3495
Manufacturer/importer:
Alfa Romeo Inc.
231 Johnson Avenue
Newark, New Jersey

ENGINE

Displacement....78.8 cu in, 1290 cc
Dimensions....4-cyl, 2.91 in bore, 2.95 in stroke
Valve gear: Double overhead camshafts, with valves disposed in 80° included angle
Compression ratio....8.0 to one
Power (SAE)....92 bhp @ 6000 rpm
Torque....83 lb-ft @ 4250 rpm
Usable range of engine speeds....750-6000 rpm
Corrected piston speed @ 6000 rpm....2940 fpm
Fuel recommended....Premium
Mileage....22-30 mpg
Range on 11-gallon tank....230-330 miles

CHASSIS

Wheelbase....93 in
Tread....F 51 in, R 50 in
Length....157 in
Ground clearance....5 in
Suspension: F, ind., Wishbones and coil springs enclosing telescopic shock absorbers, anti-roll bar; R, Live axle, left-mounted A-bracket and radius rods, coil springs enclosing telescopic shock absorbers
Turns, lock to lock....2½
Turning circle diameter, between curbs: L 34½, R 38 ft
Tire and rim size....5.50 x 15, 15 x 4½J
Pressures recommended....Normal, F 19, R 21 psi
High-speed, F 22, R 24 psi
Brakes, type, swept area: Girling hydraulic 2 LS front, finned light alloy drums, 10½-inch front, 10-inch rear, 264 sq in
Curb weight (full tank)....2088 lb
Percentage on the driving wheels....45.5

DRIVE TRAIN

Gear	Synchro?	Ratio	Step	Overall	Mph per 1000 rpm
Rev	No	3.25	—	14.80	−6.0
1st	Yes	3.26	64%	14.84	5.9
2nd	Yes	1.99	46%	9.05	9.6
3rd	Yes	1.36	36%	6.18	11.6
4th	Yes	1.00	—	4.56	15.7

Final drive ratio: 4.56 to one

ACCELERATION

Zero to	Seconds
30 mph	4.2
40 mph	6.5
50 mph	9.4
60 mph	13.7
70 mph	18.7
80 mph	24.0
90 mph	31.4
Standing ¼ mile	19.1

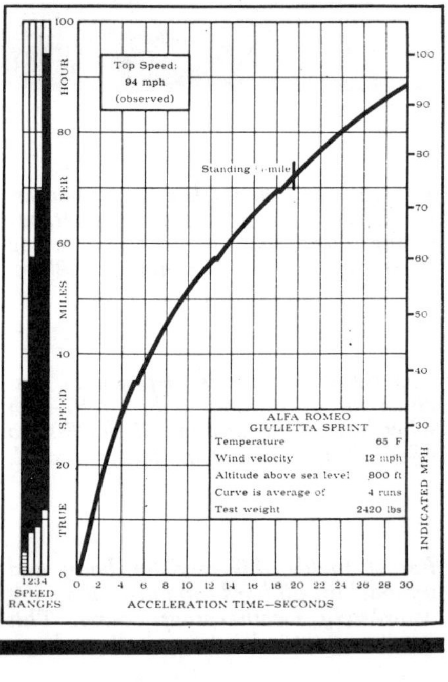

BEST OF THE REST *Italy*

Top: Alfa's Giulietta has long since established its supremacy in sportscar racing, due largely to the high-revving twin-cam engine displacing 1290cc.

Top right: Interior is surprisingly spacious for a car with only an 87-inch wheelbase. Alfa was one of the first to offer roll-up windows in roadster model.

Farina builds this roadster body and optional lift-off hard top makes it a snug completely weatherproof coupe. This car is a great value in a small package.

ALFA ROMEO GIULIETTA SS

Even though the cars appear virtually identical on the surface, Alfas have come a long way since the first Giuliettas were introduced several years ago. Horsepower has been nearly doubled from the original Spyder to the Super Spyder which now cranks out 103 at 6000 rpm from a remarkable little 1290cc twin-cam engine packing a pair of business-like Weber carburetors.

The Giulietta looks small and *is* small with its 87-inch wheelbase and 152-inch overall length. But slide the bucket driver's seat way back and even a long-legged Texan can sit comfortably. This was one of the first cars to offer roll-up windows in an inexpensive sports roadster—a feature that many other small sportscars are just now copying. And the all-synchro four-speed gearbox has never hurt its sales appeal either.

Suspension is conventional — independent at front with coil springs and stabilizer bar; a solid axle at the rear with coil springs and tubular shock absorbers. But that does not explain the Super Spyder's tenacious cornering ability. And we almost forgot to mention those huge finned drum brakes. All in all, quite an amazing little car.

JOHN BOLSTER TRIES

An ALFA ROMEO with RIGHT-HAND DRIVE

John Bolster tries the Ken Rudd-converted, right-hand drive Alfa Romeo Giulietta Sprint Veloce. With R.H.D., this beautiful car is now perfect in every respect.

This picture shows the rebuilt instrument panel and the repositioned steering wheel and foot pedals. The centrally-mounted gear-lever works just as well in the left hand!

FOR many people, the Alfa Romeo Giulietta is the perfect small car. Of superb appearance, it is beautifully appointed and finished and its almost unbreakable twin-camshaft engine gives it a really splendid performance. To the man who doesn't have to count the pennies, the little Alfa may well represent his heart's desire.

Yet, there has been a snag. These delectable machines have not been available with right-hand drive, and steering on the left is a serious handicap in this country. Many prospective buyers must have turned away with regret for this one reason, and so the announcement that Ken Rudd has successfully achieved a conversion is exciting news for many of us.

This conversion was always considered impossible, for so many things "got in the way". However, the persevering Ken has found that the Sprint converts relatively easily and that the Sprint Veloce, although representing a job of considerable complexity, can be tackled successfully. I have recently spent some memorable days at the wheel of a Ruddspeed Veloce, and the wheel was on the proper side, too.

Of course, it is the two big twin-choke Weber carburetters that fill up the right side of the bonnet. The steering wheel is repositioned and a new universally jointed column goes just over the top of the Webers. New ducts had to be made for the air cleaner, and the pipes to the heating and ventilation system were rerouted. The foot pedals all had to be moved to the right side of the car—a major undertaking in itself.

The instrument panel has been rebuilt, and a fair amount of re-upholstering has been necessary. The finished car shows no signs of having been worked upon, however. At the front of the column, a new steering box is mounted high up behind the radiator header tank, and its output shaft leads downwards to the point where it is connected to the former slave arm on the three-piece track rod. The former steering box on the left is "gutted" and acts as the slave arm bearing. In the case of the Sprint, which has a downdraught carburetter, a simpler installation suffices.

I found that the converted Alfa handles just like a standard one. The central gear lever works just as well in the left hand, and what a gearbox it is! With close ratios, really light action, and excellent synchromesh on all gears, it never fails to delight the enthusiastic driver. All the pedals work normally, and nobody could tell that they have been transplanted. The instruments are easy to read at speed, and although the angle of the steering wheel has been slightly altered, the driving position is just as good as on a left-hand drive car, if not better.

I thoroughly enjoyed driving the Veloce and the new location of the driving seat made me appreciably faster through traffic. The cornering power of the Giulietta must be the highest of any car with a rigid axle, yet the ride is comfortable and one covers the miles in effortless ease. I found no disadvantage in the converted car and I understand that the Alfa Romeo engineers approve of the work—it should certainly bring them some new customers.

The price of the complete operation is £175 for the Sprint, £200 for the "2000" and £225 for the much more complicated Sprint Veloce. For further details, telephone Worthing 7773/4.

"Streepie's" lowered suspension shows up i this picture, which also illustrates how th front valance has been cut away to improv brake cooling. The silencer was fitted for th test.

'STREEPIE' Modified Alfa-Romeo TI

Report: L. SYMONS **Pictures: A. L. BERNSTEIN**

IT WAS IN THE SECOND HALF OF 1960 that a much-modified and lightened Alfa T1, painted silver with a broad white stripe bordered with red — from which came the name "Streepie" — first appeared on South African circuits, and soon proceeded to motor so rapidly that the term "modified touring car" took on quite a new meaning to racegoers..

In its first form it had the 105 b.h.p. version of the 1,290 c.c. Giulietta engine, as fitted to the Alfa Sprint Speciale, and a four-speed gearbox. Since then it has, at one time or another, embodied practically very possible variation of the Alfa except for a "stroker" crankshaft.

Andre Pieterse of Pretoria was usually the conductor in the earlier races, and the combination of his dash and the car's performance produced results that really were

(*Right*) **From the front "Streepie" is businesslike.**

something — and made lifting the inside front wheel on corners quite the thing to do.

Few who saw it will ever be likely to forget the performance of Andre and "Streepie" in the "curtain-raiser" to the first post-war S.A. Grand Prix at East London in December, 1960, when they worked their way through the field to win their class and finish second overall to an Austin-Healey 3000 equipped with three twin-choke Webers — and were closing rapidly on the Healey at the finish. At that time the car still had the SS engine, but in July last year, when Andre put up another terrific performance at Lourenco Marques, its engine had been enlarged to the present 1,473 c.c. capacity.

On this occasion the car was an "unofficial" runner in the G.T. race, being too modified to be acceptable under F.I.A. rules for that category, but Andre did not allow that to deter him from leading Ian Fraser-Jones in his Porsche Carrera Abarth for the first lap and finishing close behind him in second place. Running officially in the event for sports cars, Streepie was fifth overall, led only by such machinery as Neville Austin in the D-type Jag., Gene Bosman's Lotus-Alfa XV, Alvaro Lopes of Luanda in a Maserati 300S, and "Frones" once again.

In the One-Hour race at the Pietermaritzburg Royal Show meeting last year the car was second to Bosman's Lotus-Alfa. In November Jan Hettema scored a win with it at the opening meeting at the Zwartkop circuit near Pretoria and followed up by a win at some 74 m.p.h. at the S.A.G.P. meeting at East London and another at Kyalami on February 10. In between Doug. Serrurier scored a second, behind a Dart-Climax, in a handicap race at the Fairfield meeting at Westmead — and had an additional handicap in the form of a slipping clutch.

These are only some of the high spots in a racing career which has taken the car to races on every circuit in South Africa — and made it the most outstanding vehicle in saloon car racing in the country.

At present the engine has the 1,500 c.c. sleeves used by the many drivers here who power their G.P. machinery with the Alfa motor, but has the standard crank, giving a capacity of 1,473 c.c. The compression ratio has been raised considerably from the standard 8 : 1 it was made with, and there are Sprint Veloce cams, two twin-choke Webers and a completely redesigned exhaust system, but valves are standard, though the ports have been enlarged, matched and polished. The whole shooting match has, of course, been carefully balanced and the engine will rev. to 8,000 — but I did not take it over 7,500 (I hope — anyway, there is no tell-tale on the tachometer). A Sprint Veloce clutch copes with the extra power, and a five-speed SS gearbox is fitted.

The mods. do not stop here, of course. The two-leading-shoe brakes and turbo-finned drums normally fitted on the front have been moved to the back wheels, and the front is fitted with similarly-finned, but wider, drums and three-leading-shoes. Both front and rear suspension have been lowered, and there are Koni

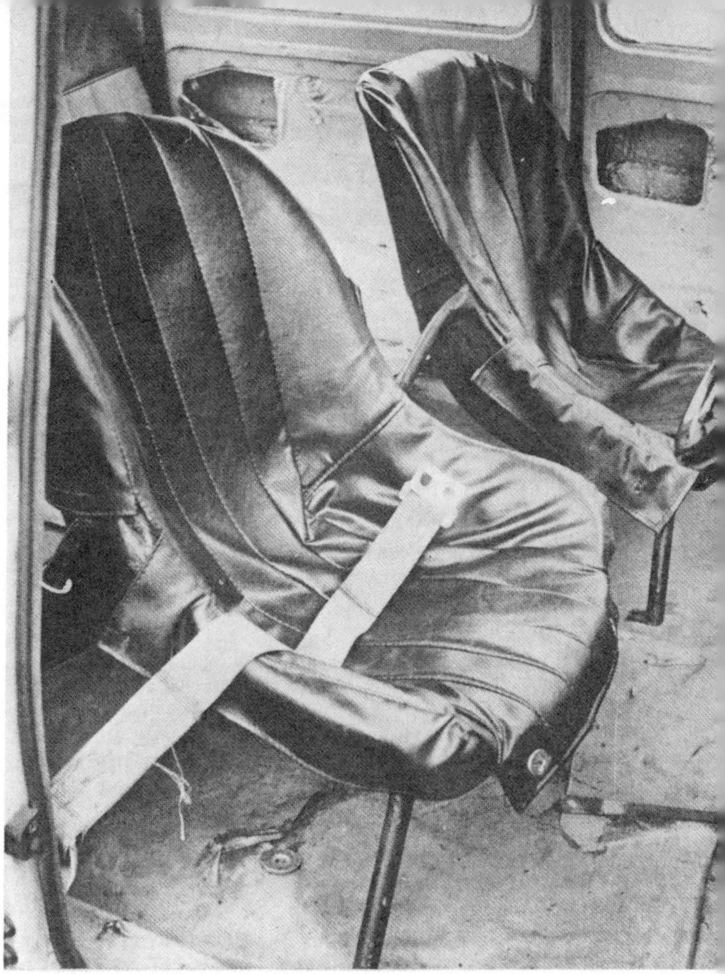

(Above) **The bucket seats are stark and sparingly upholstered, but comfortable and excellently shaped.**

(Below) **A detailed view of one of the front brakes, showing the massive proportions of the three leading shoes and the finned drum.**

'STREEPIE'
Modified Alfa-Romeo TI

SPECIFICATION

MAKE AND MODEL: Alfa Romeo TI "Streepie".

ENGINE: 4-cylinder, in-line, water-cooled, twin overhead cams, two twin-choke Weber 40 DCO E2 carbs., compression ratio 10 : 1.

BORE: 79 mm. **STROKE:** 75 mm. **CAPACITY:** 1,473 c.c.

MAXIMUM B.H.P.: Approx. 135 at about 6,200 r.p.m.

ROAD SPEEDS AT 1,000 R.P.M.: 4th (direct top) — 15·9 m.p.h. 5th — 18·34 m.p.h.

BRAKES: 3-leading-shoe front, two-leading-shoe rear. Turbo-finned drums.

SUSPENSION: (Front) Independent, wishbone and coils (shortened). (Rear) Live axle, located by A-bracket and radius rods. Coil springs (shortened). Koni telescopic dampers all round.

TRANSMISSION: Manually operated, five-speed, all synchromesh gearbox. Lever in floor. Single, dry-plate clutch, mechanically operated.

OVERALL GEAR RATIOS:

1st: 15·07	3rd: 6·16	5th: 3·886
2nd: 8·92	4th: 4·55	Rev.: 15·31

FINAL DRIVE RATIO: 4·55 : 1. **TYRES:** Michelin "X" (or Pirelli Cinturato "S"), 155 x 380.

GROUND CLEARANCE: 3 in. **FUEL CAPACITY:** 17 gals. **OIL CAP:** (Incl. filter) — 13 pints. **WEIGHT AS TESTED:** Approximately 2,150 lbs.

STEERING: ZF worm-and-roller. Three turns lock to lock. Turning circle 30 ft.

COST: Considerable.

PERFORMANCE

(Figures in parenthesis are those obtained by CAR on a road test of a standard TI, published in February, 1961.)

ACCELERATION THROUGH GEARS:

M.P.H.	Secs.	M.P.H.	Secs.
0—50	8·4 (12·3)	0—80	19·4 (37·5)
0—60	10·8 (18·0)	0—90	25·6 (56·8)
0—70	15·6 (24·0)	0—100	36·7 —

ACCELERATION IN HIGHER RATIOS IN SECONDS:

M.P.H.	3rd	4th	5th
50—70	7·0 (12·0)	—	—
60—80	8·0 —	11·6 (19·7)	—
70—90	—	12·6 —	21·5
80—100	—	17·1 —	30·9

STANDING QUARTER MILE: 18·05 secs.

REASONABLE MAXIMUM SPEEDS IN GEARS: (Approx. 7,500 r.p.m. — see text re. speeds in 4th and 5th), 1st — 36. 2nd — 60. 3rd — 86. 4th (actual) — Approx. 115. (Recorded) 109·7. 5th (actual) — Approx. 120.

ACCELERATION AND BRAKING: 0—90 m.p.h. — 0: 30·5 secs.

FUEL CONSUMPTION (racing): Approx. 12 m.p.g.

TEST CONDITIONS: Cloudy, cool, moist with intermittent rain later. Tarmac surface damp to wet. Altitude 4,500 ft. Wind about 8 m.p.h., 30 degrees to test strip.

Car lent by: Continental Cars, Andries Street, Pretoria.

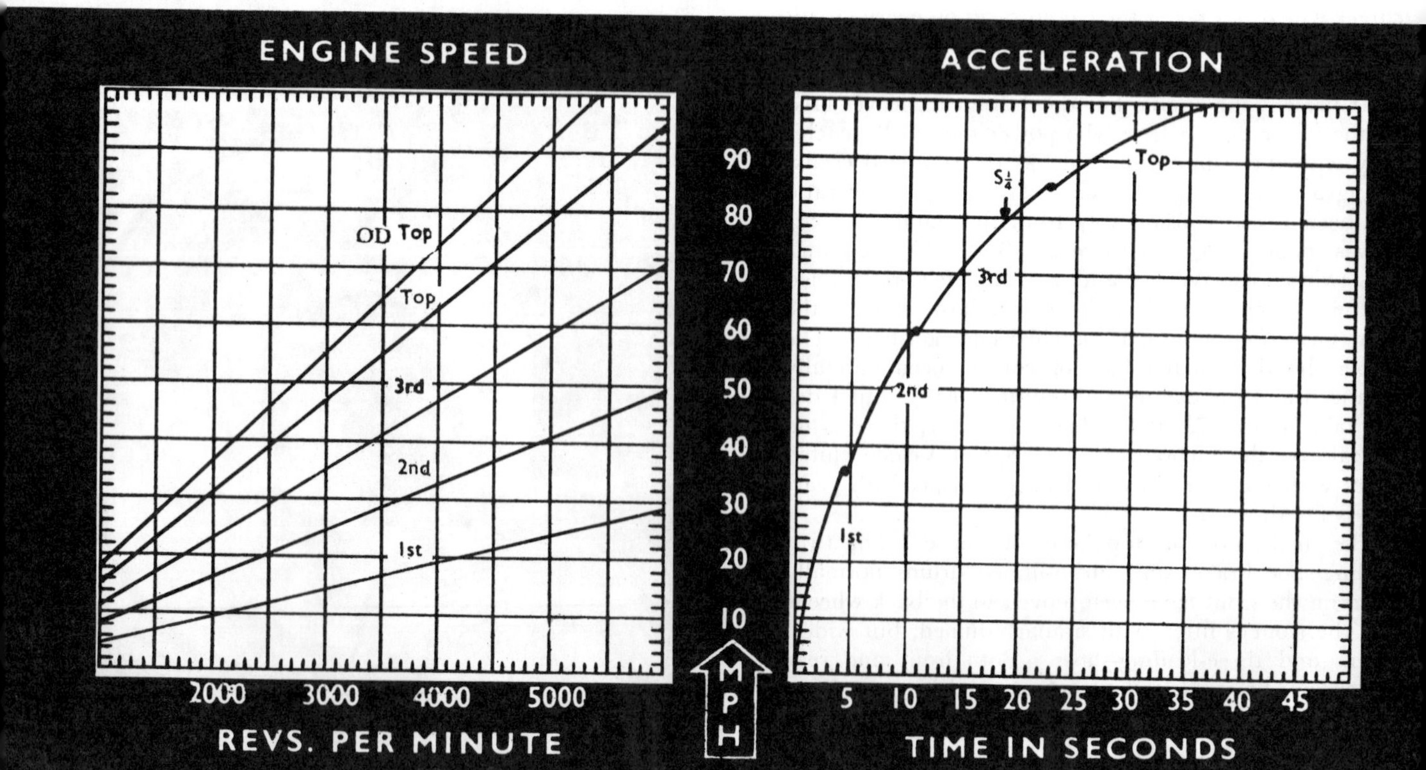

A bird's-eye-view of the engine, showing the lack of a fan and the extra header tank. The side panels have been cut away to allow a free flow of air to the Webers.

shock absorbers all round. There is no cooling fan, but an extra header tank has been fitted and the bonnet top is extensively louvred. For brake cooling the front valance has been cut away and vents have been cut in the rear valance.

Lightening has been extensive, reducing the weight to nearly 200 lb. below the standard model. This has involved getting rid of all such items as sound-damping and dust-proofing material, window-winders and interior door handles, badges, trim and even the glove box and its lid. All glass, except in the windscreen, has been replaced by plastic sheet, and the bumpers have gone. The rear "seat" consists of the covering material only, without "innards", and the front seats are two fibreglass buckets, fixed in position and with a thin covering of sponge-rubber and leather.

Thanks to cams, carbs. and the rest, nothing much does happen below about 4,500 r.p.m. The car will run — after a fashion — but to the accompaniment of a noise rather like a giant concrete-mixer at work against the background of an artillery barrage. So driving through town in traffic is rather hard on the nerves.

There is a tale that Jan Hettema, taking the car down to East London with the normal straight-through fitted, was stopped in every town between Pretoria and East London. He managed to talk his way out of it — but firmly delegated to someone else the task of driving the car back again. It may not be true — but easily could be.

Between 4,500 and 5,000 r.p.m. the noise settles down to that of strong canvas being ripped — very loudly. That is outside the car. Inside there are also the general thrashing noises of high-efficiency machinery getting down to business. My only regret, after the test was over, was that I had not been warned that ear plugs were required.

Getting going from a standing start is rather similar to what one imagines being launched into orbit is like, and not only for the noise. It is a fairly tricky exercise to keep the revs. up at a figure which means real power from the engine and at the same time to avoid wasting too much time through wheelspin.

Reasonable success at this produces results which are also rather rocket-like, for a "family saloon". The figures speak for themselves. As a matter of interest the times recorded in a standard T1. and published in a road test in CAR in February, 1961, are given to provide a comparison. It is also worth mentioning that the standard car would not have reached 80 by the time Streepie is doing 100.

Streepie has also been timed over the same standing quarter as I used, though in dry instead of damp conditions, at some 0·2 secs. better than the best I could manage.

Maximum speed was limited by the roads available for the test. It is obvious that, when pulling an overdrive fifth gear, some distance is needed to get up to maximum in that gear. The only suitable straight stretches which could be reached in the time available are rather too busy for this sort of caper to be popular — let alone safe — on them.

So the straights available, while adequate for getting something like the D-type up to near maximum, were on the short side for Streepie.

The brakes may be outstanding — as a time of 30·5 seconds to accelerate a 1,500 c.c. saloon from 0 to 90 m.p.h. and stop it again testifies — but even so stopping from over the "ton" takes some distance.

So the maximum recorded reflects the need, as far as I was concerned, to "tap off" when the car was still accelerating in fourth gear. In these circumstances fifth gear naturally gave no better results and in fact the main use for this gear, in racing, is to save the engine rather than to give more speed. Given the room, however, I would put Streepie's maximum in fifth, at 5,000 feet altitude, at about 120 m.p.h.

The gearbox needs little further comment. It is in effect the normal four-speed box with the overdrive fifth added, and is certainly at least as good as the normal one, which is good enough for most people.

The lowering job has made the suspension a good deal harder than that of the normal T1 but the "Alfa lean", though limited, is still there. This roll, combined with the very much above-normal power available, at first produced an impression of very slight oversteer. Increasing familiarity showed, however, that the roadholding and steering characteristics are first-class — as near neutral as makes no matter.

Going into a bend too fast with moderate power produces understeer, but the driver can get out of a lot of trouble by putting his foot firmly on the loud pedal. Of course, he can get into it in the same way.

The bucket seat, though hard, is very comfortable. It gives excellent support to the driver, without interfering with his freedom of movement or needing help from the lap-type safety belt which is fitted — and which I used. The seat is fixed in just the right spot, for me at any rate, to hold the wheel at arms' length — but I was disappointed to find that it was awkward to "heel-and-toe" the brakes and throttle.

That is "Streepie". Certainly no car for town use or even for the open road, but quite something for saloon car racing, and nothing less than a phenomenon to drive. ●

ALFA ROMEO

"COMPETITION 1600 GIULIETTA GT"

BY BERNARD CAHIER

The prototype Alfa Romeo "Competition 1600 Giulietta GT" is shown here during its tests on the Italian Monza circuit, where it turned laps five seconds faster than the record for the 1300 cc Giuliettas. Alfa Romeo is quickly producing 100 of this model, with less restricted vision to the rear, it is hoped, so that the car can qualify in the Grand Tourismo category with the F.I.A.

The multi-tubular chassis and body are very light, some 220 pounds less than the lightest Giulietta SS, and the car should tip the scales at about 1600 pounds. The wheelbase is 2.5 inches shorter than the standard Giulietta, at 86.5 inches.

According to body-designer Zagato, the engine to be used in the 100 cars to be produced for FIA homologation will not be the 1600 cc type, which is still experimental, but rather a hot, 125 horsepower version of the present 1300 cc Giulietta engine. It seems likely, though, that the larger engine will soon be used, as Grand Touring Class II runs from 1001 to 2000 cubic centimeters and there is no point in giving away any more displacement than necessary, particularly with 1600 and 2000 Porsche GTs in the same Class.

In roadster trim, the GT Giulietta looks very much a competition car. Large finned brake drums show through slots in front wheels.

GT GIULIETTA

Alfa Romeo brings out a long-awaited answer to Porsche 1600 supremacy

Rumors of a bigger Grand Touring version of the Giulietta which Alfa Romeo was said to be planning, with the intention of fielding a car to compete on even terms with the Porsche 1,600-cc models, have been borne out by the appearance of the car shown above and at right.

The new car, which has been tested in both open and closed versions, is obviously intended as a competition machine, though tamer models will probably be developed from it.

Very different from the rest of the current Alfa Romeo line, the new GT Giulietta is highly streamlined, with its tail cut off sharply much like that of the current sports-racing Ferrari models.

Its engine is a 1,600-cc version of the typical Giulietta powerplant, with twin overhead camshafts and two double-choke carburetors. Surprisingly, the stopping power is supplied by drum brakes. The rear brakes are fitted near the differential casing, which means independent suspension or at least a de Dion arrangement.

The chassis is tubular and extra light. The car is very low, and weight distribution is said to be excellent. Its body line has been very carefully worked out according to Kamm's theories of streamlining (see CAR AND DRIVER, May, 1962, p. 12). The very large sloping hood directs a flow of engine-warmed air to the windshield. Instead of the familiar Alfa Romeo grille, there is a low racing-style airscoop.

What is known of the new Giulietta's performance is impressive. On the Monza road course the 1,600-cc car has lapped many times in two minutes flat, compared with the best previous GT Giulietta time of 2:04.6, set last September by a highly modified 1,300-cc model using the new coupé body. The 4.6-second gain indicates considerable improvement in both chassis and engine since then.

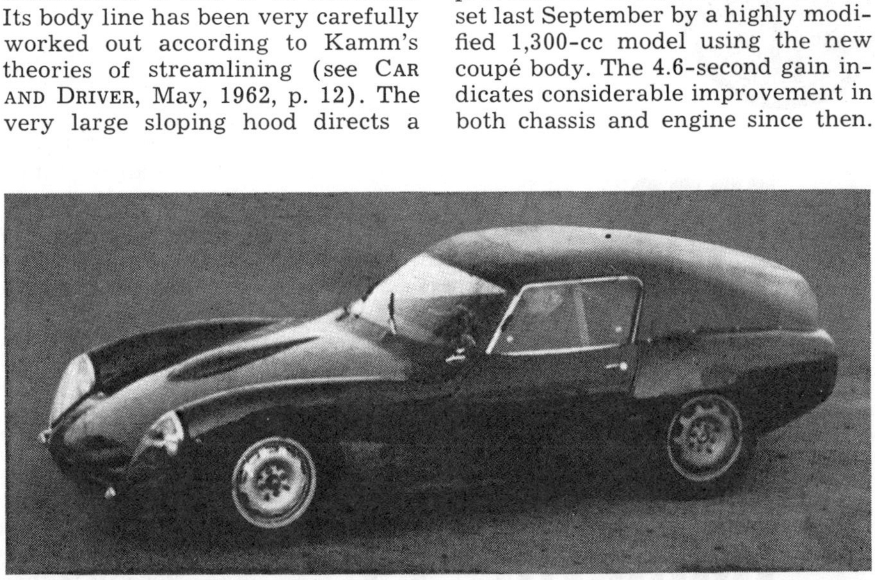

Closed version which has also been tested is the same car with hard top attached. Sharply chopped-off rear end follows current theories of boundary-layer control.

The Alfa is a neat car, even to the "built-in" number plate on the front bumper.

1962 Alfa Romeo Giulietta T.I.

THE CAR THAT SHAKESPEARE NAMED? According to fairly well-authenticated legend, it is true. Some years back, when Alfa Romeo were casting around for a name for their latest model, someone with an apt sense of fun linked Alfa's "Romeo" with the Bard's "Juliet", and the Alfa-Romeo Giulietta (pronounced "jew-li-etta", the Italian version of the girl's name) came into being. We do not doubt that Shakespeare himself would have approved of this touch — and probably of the car as well.

For the Alfa T1, as it is widely known (the initials are for Turismo Internazionale — "international touring") is an elegant and energetic means of transport which has won itself prestige all over the world, in both the fields of motor sport and private motoring.

When local assembly of the Alfa T1 was commenced at the CDA plant at East London about two years back a substantial cash saving was passed on to the purchaser.

Now it has come even more firmly within the reach of the salaried family man at R1,996 (£998). These drops in retail price have meant economies in some directions. For instance, the new Alfa does not have a fancy hooter or automatic windscreen washers, but every essential (and a considerable amount of luxury) goes in to it.

The interior of the Giulietta T1 is above the average for cars in the under–R2,000 price bracket. The front seat is split in two, and both are fully reclining and adjust fore-and-aft for a distance of 7½ in. There is ample leg room and shoulder room, and both doors are fitted with padded armrests.

Above the passenger door a useful grab-handle is mounted in the roof. Both sunvisors are padded and the one on the passenger side has a vanity mirror built into it. There are roof lights on either side.

The rear seat will hold three adults in reasonable comfort, and both rear doors have both armrests and deep chromium ashtrays. Even with the front seats set fully back, knee room at the back is adequate because of the hollow backs of the front seats. The floor, front and back, is covered with carpets and integral rubber mats, which extend over the gearbox hump and up the bulkhead.

The interior, showing the pull-out handbrake (below steering column) and the gearlever, which has a non-standard knob.

The dash has an impressive array of instruments, which include a rev. counter and oil temperature gauge, besides the usual range such as water temperature gauge and oil pressure. Speedometer is of the ribbon type. The hooter ring obstructs clear reading of some of the instruments, but not seriously.

Working readings during test were 160 deg. F. oil and 170 deg. F. water temperature, and 70 lb. oil pressure.

An unusual item (and very much an Alfa feature!) is that pressing the hooter button itself produces no noise, but gives warning flashes of the headlights. This is something to be appreciated not only by competition types, but by traffic officers and city motorists.

Instead of irritating other drivers and risking traffic fines by hooting at an obstructing car in front, the Alfa owner merely presses his hooter button to give a silent "I-want-to-overtake" signal — and does not have to fumble for the light switch to do so.

Manually operated windscreen washers, an electrically operated heater unit, a dipping rear-view mirror and rubber-cushioned bumper overriders are standard fittings.

The boot has a press-button opening mechanism and is held open by a rubber-tipped rod. The interior is rubber-matted and deep, having the rather square shape which holds a lot. The capacity is not stated in the car's specification sheet, but we took measurements and calculated it at a total of 15 cubic feet. Part of this space — but not much — is taken up by the spare wheel mounted vertically on the starboard side. Behind the wheel are stored the battery, jack, wheelbrace and a serviceable tool kit in a roll-up holder.

The boot was found to be weather-tight in two days of testing in damp conditions, but by reason of the weather the dust-proofing could not be assessed.

The bonnet lid is light and fairly small, opening alligator-wise and being held up by a semi-automatic catch. A working light, with switch, is mounted under the lid. The engine compartment has a tidy layout, but with a tendency towards cramping on the starboard side, where the air cleaner and twin-choke carburettor make access to such components as the fuel pump and distributor — both mounted low — difficult.

But the twin overhead cam covers and the carburettor itself are very get-at-able. The engine compartment is well finished, with tight seals between it and the car interior and all of the plastic-covered wiring neatly enclosed.

The lack of superfluous and stray wiring is noticeable, and the gleaming bronze-finished header-tank on the radiator is a serviceable and attractive feature. The fan is six-bladed and made of nylon, for quietness, cleanliness and durability.

Altogether it is a most business-like power unit, in keeping with the character of the car.

The solid, floor-mounted gearlever is short and extremely smooth in operation. All four forward ratios are synchromeshed, and absorb ultra-quick changes up or down without any hesitation.

The 15-cu. ft. boot. Battery and tool kit are behind the spare wheel.

1962 Alfa-Romeo Giulietta T.I.

SPECIFICATION

MAKE AND MODEL: Alfa-Romeo Giulietta TI Saloon, 1962.

ENGINE: 4-cylinder, in-line, water-cooled, twin o.h.c.

BORE AND STROKE: 74 x 75 mm.

COMPRESSION RATIO: 8·5 to 1.

CUBIC CAPACITY: 1,290 c.c.

MAXIMUM HORSE-POWER: 84 at 6,200 r.p.m.

ROAD SPEED IN DIRECT TOP GEAR AT 1,000 R.P.M.: 16·0 m.p.h.

PISTON SPEED AT MAXIMUM HORSE-POWER: 2,950 ft./min.

BRAKES: 4-wheel hydraulic, two leading shoes at front.

SUSPENSION: (Front) Independent, coil springs and transverse wishbones. (Rear) Rigid axle, coil springs with A-bracket and radius arms.

TRANSMISSION: 4 forward gears, all synchromeshed, floor shift.

GEAR RATIOS: 1st: 3·258 to 1; 2nd: 1·985 to 1; 3rd: 1·357 to 1; Top: Direct; Reverse: 3·252 to 1

FINAL DRIVE RATIO: 4·55 to 1. **TYRE SIZE:** 155 x 15.

LENGTH: 162 in. **WIDTH:** 61 in. **HEIGHT:** 55 in.

GROUND CLEARANCE (laden: 5·5 in. **STEERING:** Worm and roller, $3\frac{1}{4}$ turns lock-to-lock. Turning circle: 36 ft.

FUEL TANK CAPACITY: 8·75 gl. **BOOT CAPACITY:** 15 cu. ft.

LICENSING WEIGHT: 2,084 lb. **WEIGHT AS TESTED:** 2,409 lb.

ANNUAL LICENCE: R16·00. **PRICE ALL ROUND:** R1,996·00.

INTERIOR DIMENSIONS:

Width of front seat(s): 23·5 in.
Driver's seat squab to clutch pedal: (Max.) 41 in., (Min.) 33/5 in.
*Front seat headroom: 4 in.
Width of rear seat: 49 in.
Rear seat kneeroom: (Max.) 14 in., (Min.) 6·5 in.
*Rear seat headroom: 1·25 in.
(*Measurements taken with 6 ft. man seated, no hat.)

PERFORMANCE

ACCELERATION THROUGH GEARS:

M.P.H.	Sec.	M.P.H.	Sec.
0—30	4·9	0—60	15·7
0—40	7·9	0—70	22·1
0—50	10·3	0—80	35·4

ACCELERATION IN HIGHER RATIOS IN SECONDS:

M.P.H.	Top	Third	Second
20—40	13·4	8·6	5·7
30—50	12·4	8·0	—
40—60	13·9	9·2	—
50—70	17·6	11·0	—
60—80	22·4	—	—

STANDING QUARTER MILE: 19·6 sec.

REASONABLE MAXIMUM SPEEDS IN GEARS:
1st: 30 2nd: 47 3rd: 71

MAXIMUM SPEED IN TOP: 93·7.

EMERGENCY STOPS (10 at 30 sec. intervals, from 50 m.p.h., in sec.): 3·4, 3·5, 3·0, 2·8, 3·1, 3·5, 3·4, 2·9, 2·9, 3·2. (Average 3·17 sec., no fade).

FUEL CONSUMPTION AT STEADY SPEEDS IN TOP:

30 m.p.h.:	46·0	60 m.p.h.:	33·9
40 m.p.h.:	42·5	70 m.p.h.:	30·6
50 m.p.h.:	39·8	80 m.p.h.:	26·3

FUEL CONSUMPTION AT FULL THROTTLE: 18·0 m.p.g.

TOURING CONSUMPTION: 32·7 m.p.g.

SPEEDOMETER CORRECTION:

Indicated:	20	30	40	50	60	70	80	90
Actual:	20	30	40	49·5	59·5	69	79	89

TEST CONDITIONS: Barometer reading 30·02; at sea level; air temperature 81 deg. F.; wind 0–5 m.p.h. at 80 deg. angle to test strip; dry new tarmac surface; weather sunny and mild; 93-octane fuel.

CAR SUPPLIED BY: Wolman Motors (Pty.) Ltd., Cape Town.

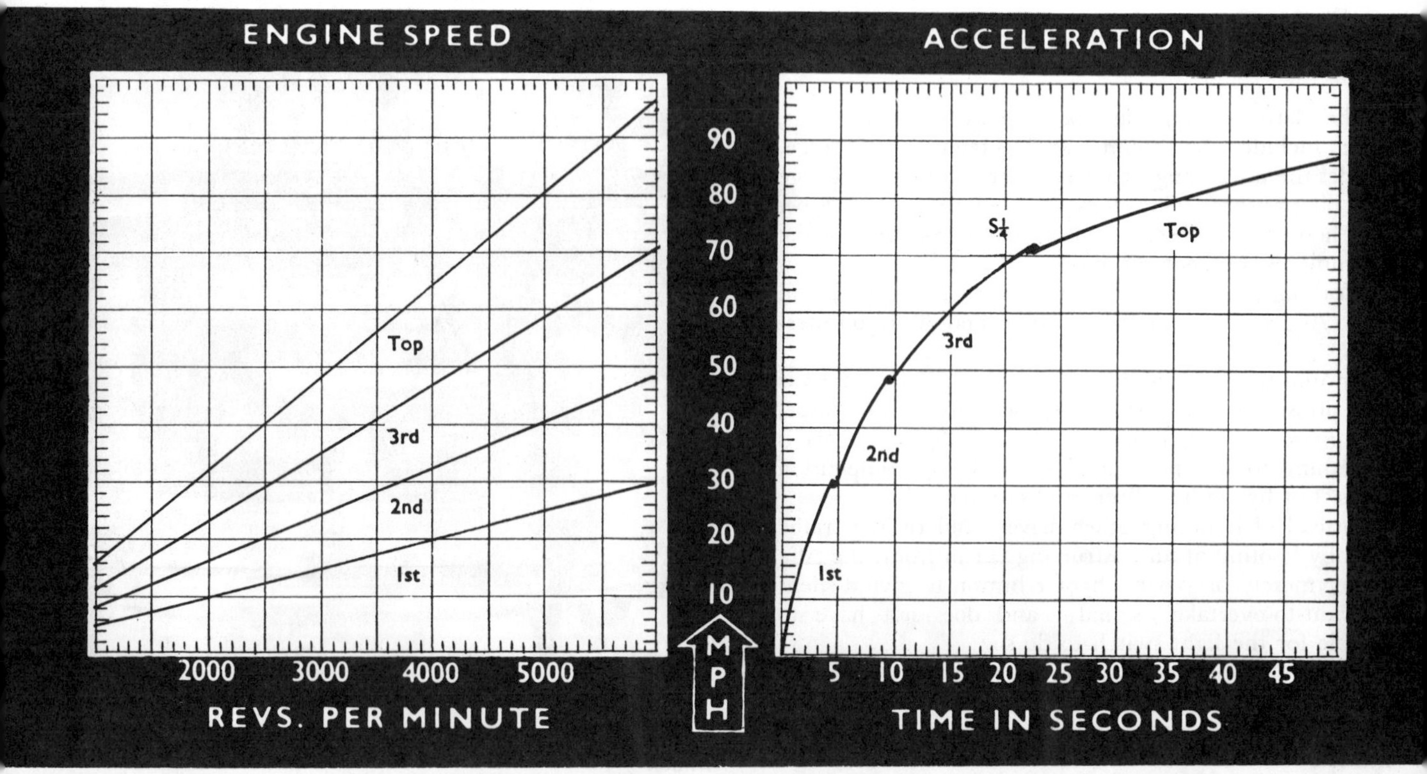

The engine is not big, but largely through the abundantly efficient breathing provided by a twin-choke Solex carburettor and high-lifting twin overhead camshafts, it develops 84 brake horse-power to provide for rapid motoring.

The factory specifications do not give the maximum torque level, but it seems to come in fairly high up — about 3,500 r.p.m. — so that the gearbox is there to be used and enjoyed. The safe rev. limit is 6,100, but the car exceeds 90 m.p.h. before reaching 6,000 and cruises very happily and smoothly (and economically) at 80 (5,000 r.p.m.).

A study of the performance figures compiled on test will show how efficient and capable the motor is. When pushed it is a bundle of restless energy, yet it is well suited to docile top-gear motoring about town with a fair amount of flexibility.

When pushed it is a noisty engine, with plenty of valve gear clatter and a tearing exhaust note above 5,000 r.p.m. in the gears, but in top, apart from a fair noise level in the region of 70 m.p.h., it does not make itself felt so noticeably.

At 80 — which seems to be the natural cruising speed — the motor settles down to a steady and pleasant hum.

The bulkhead is well sound-proofed, and to further minimize interior noise level the models now coming off the assembly line have a sound-absorbing roof trim.

The Alfa — as one would expect from a competition-bred car — behaves itself well at speed. The steering is firm and accurate. Right through the speed range there is an early oversteer characteristic which tends to cause the car to sway. This is particularly noticeable at high speeds, but is controllable. In cornering, the car starts to oversteer, and then converts to a comfortable understeer.

"Once she is in to a fast corner and the oversteer corrected, she sits beautifully," in the words of the driver on this test.

Among the virtues the T1 showed at speed are efficient windscreen wipers (they continue to function perfectly at 90), good aerodynamics and cool running. Even when worked very hard on test, on a mild, sunny day, water temperature never rose above 180 deg. F. and oil temperature above 200 deg. F. — both well within safe limits.

The instruments also proved to be exceptionally accurate, both speedometer and rev. counter showing only a fractional error at top speed.

The seats provide comfortable posture for long trips without any fatigue, and the headlights provide sharp, clear beams which dip low.

Braking, of course, is of a very high standard, with no fade whatsoever. In stops 1, 2, 6, 7 and 10 during the test pedal pressure was eased to prevent the rear wheels from locking, with a consequent effect on stopping time. With rear wheels locking, the car stopped consistently from 50 m.p.h. within 3 sec., and five stops each with and without locking provide a fair and commendable average stopping time of 3·17 sec.

The car's handbrake was out of adjustment (it was just due for its 1,500-mile service) so that a fair handbrake emergency stop could not be recorded.

The twin-o.h.c. engine is a tidy and well-finished unit with lots of power.

The TI's raised suspension causes some lean and early oversteer in the tight 30-m.p.h. corner, but the car handles very well.

Fuel economy at steady speeds is unusually good for a five-seater sports saloon. An owner can expect from 25 to 30 m.p.g. at high speed on the open road (without being heavy-footed), and in city family motoring over a year the average consumption might be as high as 35 m.p.g.

Beyond doubt the Alfa T1 is a reasonably priced, well-built, safe and economical car with a wide appeal. Its manoeuvrability and superb gearbox make it very manageable. It is a car out of the ordinary, which looks good and is great fun to drive. ●

Ruddspeeding to Scotland

by J. W. Anstice Brown

From coast to coast in a right-hand drive Alfa Romeo

AS one who specializes in being badly travelled, and can think of nothing better in life than sitting in the garden taking sun and iced lager in equal quantities, I must admit that Fate has been extremely kind on the rare occasions when she has made me stir my stumps. I was sitting in my office thinking of what not to do next when Fate, in the shape of an editor, summoned me the other day. Could I go to Munich at once overnight and return with a story the next day? This is a difficult shock tactic to counter and I was at a loss for rapid excuses. Ten minutes later I was summoned again. Would I instead kindly go to Scotland for a week and report the Scottish Rally? By this time I was past even thinking of weak excuses. There is a definite and quite low rev. limit for Brown's brain and it had already been reached.

I had thought of making the journey in a Citroen ID which would have let me call this article "To Scotland by Air," but things did not work out that way. I had nearly settled for "Rootes to Scotland" when a chance phone call to Ken Rudd found me asking him if he had anything interesting to lend me for a week. Apologetically, he replied that the best he could do at such short notice was an Alfa Romeo Giulietta Sprint with slight engine tuning and, of course, Rudd's conversion to right-hand drive which has made these cars so much more attractive to British buyers. I said I thought I might suffer this vehicle, rapidly abandoned the telephone and took train to Worthing before he changed his mind.

There is always a selection of mouthwatering cars at Rudd's and it is easy to understand how people with enough money succumb to this temptation when in Worthing. A sports racing Lagonda, P1800 Volvo, various Porsches and assorted Alfas were rapidly bringing out kleptomaniac tendencies in me when I was shown the machine shop where the engines are breathed upon. The most interesting one there at the time came from a Maserati which was raced by Raymond Sommer in the mid-forties. A 1½-litre supercharged unit (but not a 4CLT), it has only two valves per cylinder in a monobloc cylinder casting. Rudd's are completely rebuilding this after a blow-up that partially wrecked the block and called for the making of new pistons and crankcase halves from scratch. The car should be worth watching in historic racing-car events.

Derby week is not the best time to drive back to London through Epsom, but KR3 and I finally got home and prepared for the Great North Road. Until now A1 has been a mystery to me beyond Biggleswade, and I was pleasantly surprised to find that England has a road of such length which, even if not quite a motorway, is over much of its length not unlike one, and almost as fast for cruising.

A1 suits the Alfa ideally. The car's 1,290 c.c. twin-overhead-camshaft engine develops maximum power at around 6,000 r.p.m., and torque at low revs is not its strong point. Furthermore, this one was fitted with Veloce camshafts and a modified cylinder head with raised compression. Twin S.U.

The inscription on the John Cobb memorial showed that this almost storybook Englishman had greatly impressed the locals as a man, as well as a sportsman, during his tragic stay at Loch Ness.

Home by day for a week. The right-hand-drive Alfa differs little inside from that with left-hand drive. The instruments are neatly grouped but the handbrake is inaccessible and there is a shortage of cubbyholes. Note the clever use of leather and cloth on the seats and the leather pad for the driver's clutch foot, a Rudd refinement. There is a horn ring, the central button flashing the headlights.

The "Scottish" is worth watching, as this view of a competitor climbing Inverfarigaig, above Loch Ness, shows.

carburetters had previously been fitted experimentally, but these had been temporarily removed in favour of the original single Solex. All this meant that some rather surprising cars out-accelerated us from roundabouts, particularly if I had forgotten to get into second gear for the exit, but a few seconds later, snicking from third to top gear at a little over 80 m.p.h., we were well past everyone and away. Everyone? Well, almost.

We came upon a TR travelling quite slowly and, as we passed, I overheard a sound that, even from inside my car, was quite obviously "big-ends." Surprise, then, when he pulled out and gave chase. Once when I slowed for other traffic on a dual carriageway he came level. By now the noise was unbearable. What should I do? If I slowed he would slow—and I had a long way to go before dinner. Yet surely this could not last? It didn't. The 2-litre Triumph engine is one of the toughest, but when all's said and done it is made of metal. Once more we motored alone.

Like Snow Driving

The outward journey was made easy by spending the night at Barnard Castle, a short distance from Scotch Corner. Next day Glasgow. Very few people watch the Scottish Rally. This is a pity as the special sections are many, varied and quite spectacular. The sight of a car ascending Rest and Be Thankful with hardly a spectator in sight has quite a period air about it. The special sections through Forestry Commission woods were interesting, too, and a change both for watchers and crews. The technique is like driving on snow; pluck up courage, travel flat-out and you can't go wrong. Perhaps.

It is astonishing what a glutton one becomes for motoring punishment when the car is an Alfa. One evening, having done some 150 miles or more following the rally, we decided to drive round Loch Ness as added entertainment. No, we didn't see it (the locals told us the wind was in the wrong quarter), but we did stop to have a look at the John Cobb memorial at the side of the road that runs beside the loch. We learned something as well, namely, that in this part of Scotland at any rate it is virtually impossible to get a meal after about 8 o'clock, an old-world custom that does not appeal to the weary and ravenous traveller. However, when you do get one even the *table d'hôte* menu offers five or six courses at a very reasonable price, so one can make up for mistiming deficiencies.

One could go on about the idiosyncrasies of this or any other country, but the end of the rally and the journey home loomed up. On the last day the local police held a special test of their own just before the final control and awarded tickets to several competitors and at least one official. Fortunately, my navigation is not of the best. I approached the point from the opposite direction.

We turned the Alfa back towards the smoke. This time the run was done in a single day and provided one of those escapist periods when it is possible to drive for six or eight hours with hardly a stop. For a while the car becomes one's entire world and its qualities and deficiencies really stand out. A following wind and the excellent road allowed the speedometer needle to remain on the 100 m.p.h. mark for long periods, and this did not worry the machinery or the Pirelli Cintura tyres, although the latter were prone to squeal on the slower corners. Fast bends were particularly easy in this car and even when taken at around maximum speed did not worry the driver or passenger in the least. The rather heavy steering and poor lock did not make it so pleasant to manœuvre at less exotic velocities. Although there is a typical overhead camshaft noise from the engine, particularly on tick-over, the noise is such when motoring fast that it does not tire the travellers. There is little or no drumming or vibration except for a resonance, seemingly in the exhaust system, at about 80 m.p.h. on the overrun. Another item that could be appreciated on a long run was the cloth that was used on the vital parts of the seats. Not only did it grip the driver, but it also helped to keep him cool.

The drum brakes on this car are another feature that give confidence when cramming the miles into the hours, and demonstrate that, even if discs are generally superb, drums, too, can stop a fast machine. They felt the same at the end of the journey as at the beginning, like the rest of the car. In fact, the only trouble in 2,000 miles of motoring was a fuse that shook loose over one of the forestry sections and a petrol tank that leaked very slightly, I fancy from the start. Unfortunately, this leak made an accurate check on petrol consumption impossible but, even so, at between 25 and 26 m.p.g. overall it compared well with the TI saloon which we road tested.

Home again after a week with KR3, I next drove my old Commer. Its comparatively light steering, torque at low revs, accessible hand brake and high-geared window winders demonstrated that the Alfa Romeo does not have everything, but when I obey the sign at Gretna that says "Haste ye back to Scotland" I know what car I should like.

Note: Conversion of the Alfa Romeo Giulietta Sprint to right-hand drive by Rudds of Worthing, High Street, Worthing, costs £185, and other floor-change Giulietta models can also be modified at prices varying according to exact details.

CARS ON TEST

ALFA ROMEO GIULIA SPYDER

ELSEWHERE IN THIS issue you will read that over a period of thirty years, up to 1940, only about ten thousand Alfa Romeo cars were built. Since the last war, of course, production has increased enormously, and nowadays the cars are manufactured in series and in much larger numbers. This is true of several manufacturers who have, in the past, been associated with cars of character: with the introduction of more modern production methods, some of that character has, in many cases, been all too sadly lost. Nothing, however, could be further from the truth in the case of Alfa Romeo. The 1,600 c.c. Giulia Spyder, which recently underwent a full road test in our hands, is probably one of the most delightful small sports which will ever be produced. Faultless road manners, quiet, refined tractability, brisk acceleration, high standards of comfort and a maximum speed of well over 100 m.p.h. make it the sort of car with which one seeks out long journeys: at the end of them, the driver and passenger arrive refreshed and even exhilarated.

The mechanical specification is similar to that of the Giulia Sprint model in that the engine power output is the same and much of the car's equipment is common to both models. The Spyder, however, has a slightly shorter wheelbase and a correspondingly smaller turning circle: overall width and height (with the hood erected) are slightly greater than those of the Sprint. Both cars, of course, have the all-synchromesh, five-speed gearbox which is one of the features which makes the car such a pleasure to drive.

A high-revving four-cylinder engine has bore and stroke measurement of 78 mm. × 82 mm., giving a total capacity of 1,570 c.c. With a compression ratio of 9:1 and a single twin-choke Weber carburettor, the net power output is 106 b.h.p. at 6,200 r.p.m.; the cylinder head is light alloy, valves are operated by twin overhead camshafts and are sodium-cooled, and the crankshaft runs in five main bearings. As installed, the engine was always an instantaneous starter, whether hot or cold, and idled well. The warm-up period is commendably short, and the power unit pulls well from cold, while once warm it is extremely smooth, right up to the limit of the safe crankshaft speed, and mechanical noise is at extremely low level. The power is developed through a usefully wide range, and low-speed pulling in the upper ratios is satisfactory, although naturally one is encouraged by the engine characteristics to make full use of the excellent gearbox. Clearly, the designers have expected the car's performance to be used to the full, and sustained running at high r.p.m. has no adverse affect on engine oil pressure or temperature.

The engine transmits its power through a single dry-plate clutch and a five-speed and reverse gearbox, with synchromesh on all five forward ratios. It sounds trite to say that the gearchange is like a knife through butter, but, truthfully there is no better way of describing it. The lever is spring-loaded in the third and fourth gear plane, and pressure must be overcome to engage the two lowest ratios or fifth gear. When one first drives the car, some care is necessary when making fast down-changes to avoid going from fifth to second, but a little practice overcomes this danger, and gears can be changed as fast as the rigid, central floor-mounted lever can be moved. The clutch is light to operate and engages precisely and smoothly, and at no time during the test was there the slightest tendency towards slip, even during the racing changes employed during the performance testing. The ratios themselves are particularly well-spaced in a way which can only be possible with five forward speeds: fifth is far from being an over-drive gear, and can be engaged at speeds as low as 35 m.p.h. without snatch or "fussiness". The synchromesh is powerful and virtually unbeatable.

The Alfa Romeo Giulia does not feature independent all-round suspension, and if all rigid axle systems were as good as this, there would have been no need for independence to have been invented in the first place. On the Giulia, radius rods and an "A" bracket provide positive location of the axle, with coil springs forming the suspension medium; at the front, inclined wishbones and coil springs provide conventional independence. Telescopic dampers are fitted all round. The ride is soft but extremely well-controlled, and the car really does combine sports car road-holding with saloon car comfort. If maximum acceleration is employed, there is some wheelspin on take-off, while the rear tyres will give a pronounced "yelp" as second gear is engaged, but in normal fast driving there is no trace of axle tramp and all four wheels remain firmly anchored to the ground. Major bumps are well absorbed, but an undulating surface taken at speed induces a vertical movement: hard cornering brings to light a noticeable amount of body roll.

The interior is well-appointed, and features two of the best bucket seats for driver and passenger we have ever encountered. More than adequate support for the thighs and shoulders is provided, and despite some fairly hard driving at no time did passengers feel the need to hang on to a grab handle to prevent themselves from sliding about. It is, of course, a two-seater car, and no attempt has been made to accommodate additional passengers in the hideous discomfort of what is sometimes described as an "occasional seat".

The driving position is excellent. The steering wheel—that fitted to the test car was a non-standard wood-rimmed wheel—is set at precisely the right angle, and the steering is light and delightfully precise. It is sufficiently high-geared to make it unnecessary to move the hands on any but the tightest hairpin bends. Visibility, with the hood lowered, could scarcely be bettered: with the top erected, it remains good with the exception of an awkward blind spot formed by the hood's quarter panels.

The hood itself is a substantial device which proved itself to be completely weatherproof and draught-free when erected, although the process of raising and lowering it is one which needs considerable practice. When lowered, it stows neatly out of sight behind a trim panel behind the seats. Wind-up windows fit snugly into the raised hood, in bad weather conditions.

Instrumentation is complete, as might be expected, and is arranged in three dials set immediately in front of the driver. Sensibly, the rev.-counter and oil pressure gauge are incorporated in the central dial, with the speedometer to the right and, to the left, a dial incorporating oil and water temperature gauges and a fuel gauge. Hand controls are grouped along the centre of the dashboard, and comprise choke, hand-throttle, lighting switch, and cigar lighter; beneath them are switches for panel lights, heater fan and the self-parking windscreen wipers. A foot-pedal operates the windscreen washers, and also brings the wipers into operation. The inner of two concentric buttons on the steering wheel boss sounds the strident horn, while the outer "ring" flashes dipped headlight beams. Warning lights on the left-hand instrument dial relate to side-lights, low fuel level, and flashing direction indicators, while the speedometer dial incorporates heater fan operation and headlamp main beam warning lights. The heater fan is not controlled through the ignition switch, and if due regard is not paid to the warning light it can inadvertently be left on.

The particularly elegant body is a Pininfarina design and is in fact constructed at the Pininfarina workshops; the beautiful lines are immediately apparent, whether the hood is raised or lowered, and there can be few prettier small sports cars. Nevertheless, we were disappointed to find, in view of the coachbuilt nature of the body, that the boot lid allowed rain-water to leak in, while the offside door had dropped appreciably.

The Giulia Spyder is available with servo-assisted disc brakes as an optional extra, but the car tested was fitted with the standard drum brakes, with no power assistance. We found them faultless; generous size and efficient cooling leads to a complete absence of fade or uneven pulling, and pedal pressure was not unduly high.

However, the feature most associated with Alfas is the performance, and this is supplied in full measure by the Giulia. Acceleration is brisk and satisfying, and if it is not what can honestly be described as electrifying one does well to remember that there is only 1,600 c.c. to propel a comfortable, well-appointed car weighing almost 1,900 lb. The maximum speed is comfortably over 100 m.p.h., and in one direction of our two-way test we achieved a maximum of over 108 m.p.h. Fourth gear will encompass more than 80 m.p.h., making it a most useful gear for rapid overtaking; roughly 20 m.p.h. separates the maxima in each gear, so that a little brisk gearchanging has the Alfa off the mark in double-quick time. From a standstill, 90 m.p.h. can be reached in little over thirty seconds, while only 12.5 seconds is needed to reach 60 m.p.h. The fuel consumption is moderate and, indeed, an improvement on the figures returned by 1½-litre cars with less performance. Driven hard throughout the test, and adopting a cruising speed wherever possible of around 100 m.p.h., the car returned an overall figure of 26 m.p.g., and this improved to better than 28 m.p.g. with a cruising speed in the nineties and with less use of the gearbox. Allied to a tank which contains 11½ gallons, this provides a cruising range of barely less than 300 miles at the worst figure, a point to be appreciated to the full on long fast runs.

The road-holding is all that might be expected of the car. When cornered on full throttle the tail slides out, to be corrected instantly by the precise steering. Under less urgent conditions the car goes round on rails, and while there is sufficient power to enable the tail to be slid deliberately, the little jewel of a car has no apparent vices whatsoever, and remains completely controllable all the time whether the road is wet or dry. Directional stability is good, and cross winds, even on the exposed roads of East Anglia, have little effect.

Recent reductions in purchase tax and by Alfa Romeo in the basic price have reduced the cost of this machine to less than £1,400, which must make it among motoring's best value. It is a car of immense personality, and one which fully lives up to Alfa traditions.

Cars on Test

ALFA SPYDER

Engine: Four-cylinder, 78 mm × 82 mm. (1,570 c.c.); twin overhead camshafts; compression ratio 9 : 1; single twin-choke Weber carburetter; 106 b.h.p. at 6,200 r.p.m.
Transmission: Single dry-plate clutch; five-speed and reverse gearbox with synchromesh on all five forward gears. Central, floor-mounted remote control gearlever.
Suspension: Front, independent with wishbones and coil springs: rear, rigid axle with radius rods, "A"—bracket and coil springs. Tyres: 155 × 15.
Brakes: Drums front and rear.
Dimensions: Overall length 13 ft. 5 in.; overall width 5 ft. 3 in.; overall height, 4 ft. 4 in. (hood erect); turning circle 32 ft.; dry weight, 2,130 lb.

PERFORMANCE

		m.p.h.			secs.
MAXIMUM SPEED		108.1	ACCELERATION	0–30	3.5
(mean over two ways)		106.5		0–40	6.1
				0–50	9.4
SPEEDS IN GEARS	First	27.5		0–60	12.5
	Second	43.0		0–70	17.0
	Third	64.2		0–80	23.8
	Fourth	81.6		0–90	32.4
				0–100	40.5
FUEL CONSUMPTION		27 m.p.g.	Standing quarter-mile		22.4

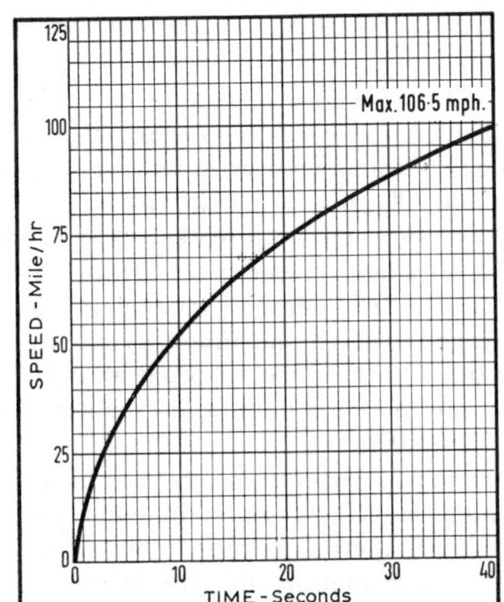

Concessionaires: Alfa Romeo (G.B.) Ltd., Sloane Street, London, S.W.1.
Price: £1,155, plus £241 3s. 9d. purchase tax. Total price: £1,396 3s. 9d.

A CAR OWNERS' REPORT

ALFA-ROMEO Giulietta TI

SPORTS saloons are built for motorists who want cars just a shade more exhilarating than the average family saloon, and they are strongly in vogue in South Africa.

Four or five different makes of sports saloons are among the most popular cars in their price grouping in the Republic. In some measure, this modern trend began with the Alfa-Romeo Giulietta TI — though other makes were equally in the field.

But the Giulietta, in spite of its modest 1·3 litres of engine capacity, deserves some special mention for its achievement in motor sport in this country. It has featured in some of our finest saloon car racing, and was the rock upon which was built one of the strongest one-make car clubs in South Africa.

As many owners remark in their reports, it is a car of strong character, and there have been countless arguments about whether it is a genuine "tonner" in standard form.

The heavy folder of completed questionnaires going to make up this report makes fascinating reading, coming as they do from a wide cross-section of owners with more than 900,000 miles of TI driving behind them. Individual average ownership works out to 10 months.

The first fact to emerge clearly from the averages is that this is no weekend motoring car: Alfa owners really get through the miles, averaging nearly 1,400 a month. Quite a few reported that they cover 2,000 miles and more a month, and exceptionally few claimed less than 1,000.

One man — who obviously uses his car professionally — had done 18,000 miles in eight months! On the other hand, there were quite a few whose cars were only two or three months old, the averages varying widely enough to give a good picture of general use.

Fuel consumption worked out to 27·7 m.p.g. in town driving, and 32·5 on the open road, which is a very fair reflection of the car's economy. More than a quarter of owners gave this as one of the "best-liked features" about the TI.

This list of best features shows up the sports saloon aspect strongly. Roadholding heads the list with specific mention by two-thirds of owners, followed by brakes, gearbox and performance (all acclaimed by about half of owners).

The car's high cruising speed capability, high-speed handling and engine smoothness get the votes of one owner in five as most popular features, with fair mention for the high standard of finish, accurate steering, body styling, full instrumentation and inherent safety.

This is a wide range of special points, and few people are likely to quarrel with them.

The many complimentary remarks made in reports would require more than a page of space to themselves but we can give a fair selection:

"I have never been so happy with a car," reports a man from Kimberley.

"My 12th car — and at last I've found one that thrives on hard driving," comes from Durban.

"Lovely, lively and altogether satisfactory . . ."

"Outstanding for its size."

"Effortless high-speed cruising is a grand feature . . ."

"At its best on the open road."

". . . Offers so much for so little."

"The perfect car for my tastes and conditions of use."

"Its faults are far outweighed by its good points."

"An absolute dream to drive!"

"Impossible not to be enthusiastic about it."

"A car with character . . ." (several made this point).

"Driver and car become very good friends."

"For a sporting man . . . absolute safety. Not a car for teenagers and drivers over 50!"

Such phrases as "fabulous", "a pleasure", "outstanding" and "best car I ever owned" crop up repeatedly, but this selection is enough to give an idea of the enthusiastic ownership.

And in fairness, owners do not let their enthusiasm run away with them when it comes to the reverse side of the coin. They give a clinical and critical assessment of the TI as a car.

The inside dimensions of the car come in for fairly stiff criticism: the rear leg space and the headroom at front are singled out in particular, together accounting for 48 per cent of votes against.

The umbrella-handle handbrake, suspension noise and high service costs are complaints from one owner in five, and others mention front seat discomfort and engine noise at speed as jarring notes.

There is also a minor criticism of 2nd gear synchromesh, door rattles, the ribbon speedometer, the battery position in the boot with poor access, and rough road riding.

The adverse comments are comparatively few in number, and here are representative ones for readers to judge by:

". . . Not worth the price."

"The Alfa is useless in the country: too low . . . "

". . . Definitely not built for bad roads."

". . . Too much expected, too little received, from such a boosted car."

"Expensive to keep on the road, but otherwise wonderful . . . "

"A much-overrated car."

Nearly two-thirds of owners report having no trouble at all with their Alfas, and it is significant that the troubles reported are almost entirely confined to comparatively minor and easily replaceable features (many owners mentioning that these were covered by warranty).

For instance, the biggest offender was shock absorbers — but only 14 per cent of owners had trouble with them. Electrical faults listed concerned the generator (11 per cent) and starter motor (an insignificant proportion). One owner in 10 had to replace a tachometer cable or have the toe-in adjusted in the first 10 months.

A similar number reported clutch failures of various kinds, specifying either thrust mechanisms or linings.

On the whole, and considering the volume of components going into a modern performance car, the troubles reported in this case are reasonably low both in frequency and in terms of cost.

Replies on dust-proofing and weather-proofing yielded high proportions of "yes's", and again this is something covered by guarantee, as owners pointed out.

Nearly three-quarters of owners reported that servicing was attended to by Alfa dealership, and the quality of dealer service averages out as "good, tending to excellent". Most of the "excellents" went to the dealerships in Pretoria, Johannesburg, Maritzburg, Cape Town and Durban respectively.

This again is a case of a fairly specialized car in a country of low population and great distances, so that a round 15 per cent of owners said they were forced to go elsewhere for service for geographical reasons.

The other 13 per cent were divided almost equally between those who preferred to do their own servicing at home, and those who found service facilities unsatisfactory.

Two representative comments on the cost of service are: "Primarily a young man's car, and as such the service is too expensive" (a Pretoria owner).

And: "Servicing as per the book is too expensive" (a Cape owner).

Cars considered before buying T1s reflects a substantial proportion of sports-type saloons, with the Volvo and Fiat 1500 as the stiffest competition.

But after owning the T1, most owners specify "Alfa-Romeo" for next time. As one would expect, the new Giulia comes in for the lion's share, followed by the Giulietta and the Sprint models.

No other car gets a 10 per cent mention, though such worthy contenders as the Mercedes 220, Jaguars, Zephyr 6 and Volvo, in that order, are desired by the few, and one owner wants a G.S.M. Flamingo G.T. next.

On the whole, though, the attitude of T1 owners is summed up by a Vanderbijlpark man, who mentions the "pride and charm of Alfa ownership", and his Johannesburg counterpart, who says: "I shall never buy another car, other than Alfa-Romeo!" ●

SUMMARY OF OWNERS' REPORTS

Make and Model: Alfa-Romeo Giuletta T1 (1960-64).

Total Distance Covered: 904,335 miles.

Average Period of Ownership: 10·0 months.

Average Monthly Mileage: 1,384·9.

Average M.P.G. in Town: 27·7.

Average M.P.G. on Long Trips: 32·5.

Most-liked Features: Roadholding (66 per cent), brakes (51 per cent), gearbox and shift (43 per cent), performance (42 per cent), fuel economy (26 per cent), high cruising speed (23 per cent), high-speed handling (21 per cent), engine smoothness (20 per cent), standard of finish (18 per cent), steering (17 per cent), styling and instrumentation (both 14 per cent), safety (10 per cent).

Least-Liked Features: Rear seat legroom (34 per cent), soft suspension and body roll (28 per cent), handbrake position (22 per cent), suspension noise and drumming (20 per cent), high service costs (19 per cent), front seat discomfort (15 per cent), front headroom (14 per cent), engine noise at speed (11 per cent).

No Trouble Encountered: 63 per cent.

Most Troublesome Components: Shock absorbers (14 per cent), clutch and generator (both 11 per cent), toe-in and tachometer cable (both 10 per cent).

Dustproof: 72 per cent.

Weatherproof: 85 per cent.

Serviced by Alfa Dealer: 72 per cent.

Quality of Dealer Service: Good, tending to excellent.

Reasons for Not Using Dealer Service: Geographical (15 per cent), do own service (7 per cent), poor quality and expense of dealer service (6 per cent).

Cars Considered before Buying: Volvo (55 per cent), Fiat 1500/2300 (32 per cent), Sunbeam Rapier (21 per cent), Peugeot 404 (12 per cent), Ford Zephyr 6 and Austin A60 (both 10 per cent).

Next Car: Alfa Giulia 1·6 T1 (43 per cent), Alfa Giulietta T1 (23 per cent), Alfa Sprint models (12 per cent), don't know (5 per cent).

ALFA GIULIA SPIDER VELOCE

An excellent sports car by any standards

BEYOND QUESTION, lust for a particular automotive vehicle goes well beyond the processes of reason. Mere logic cannot explain the blind certainty, the grand sublimation, the quasi-religion felt by a devotee of one marque. Another mental process, however, can explain it—pure and simple romance.

All cars have the ability to engage man's emotions, in some way and to some degree, from the backyard cobble-up to the stately land-yacht. But none can provide its owner with more mystique, image and personal identification than an Alfa Romeo. We have shown Alfa permutations regularly, often with enthusiasm, in R&T, and have thereby earned the ire of a few readers. Yet few would deny our obligation to issue deserved praise as well as criticism.

The Alfa's charm is both as individual and as varied as human nature, including overtones of nostalgia for the Great Days of the Thirties when Nuvolari and Alfa beat bigger and better machines through sheer brilliance and nerve; the Great Names of Alfa *piloti* like Varzi, Farina and Fangio; as well as the basic fact that it is Italian and, therefore, undeniably the product of a poetic and romantic race.

On the other hand, with all nostalgia and past associations set aside, it takes no more than a few miles of open (preferably convoluted) road for any driver worthy of the name to feel growing appreciation and start to form his own personal attachment to the responsiveness, control and engineering of the firm's latter-day designs. Notable among these are the Alfa Super Spider 1300 (Road Test, October 1959) and the Alfa Giulia 1600 Spider (June 1963), both contributing antecedents to the 1600 Veloce Spider.

These three models are markedly similar to the eye. In fact, one observer asked, "When are they going to change that design? I thought it was a '59 with a new paint job." This canard will be instantly rejected by *Alfisti* as both irrelevant and irreverent, but it is nevertheless true that Bertone's original Sprint lines (The Giulietta Sprint, R&T December 1964), modified by Pininfarina into the svelte Spider version, have been with us without significant change for nearly a decade. And, just as true, the lines are as clean, honest and pleasing to the eye today as they were in the prototype.

Under its skin, the 1600 Veloce is also similar to the recently tested 1600 in suspension (independent front, live-axle rear), basic engine (1570 cc, 4 cyl, dohc), clutch (single plate, dry) and electrical system (Bosch), so we will make our primary concern the differences between the three models.

The most obvious of these differences is additional power: the 1300 Super Spider's 103 bhp (risen to 108 in the 1570-cc 1600) has reached 129 in the 1600 Veloce, bestowing a comparable and thoroughly welcome rise in torque from 86.8 @ 4500 rpm to an estimated 105 @ 4000.

ROAD TEST

ALFA GIULIA VELOCE

PRICE

List price.....$3595
Price as tested.....$3670

ENGINE

No. cylinders & type...4 cyl, dohc
Bore x stroke, in.....3.07 x 3.29
Displacement, cc.....1570
Equivalent cu in.....95.8
Compression ratio.....9.7:1
Bhp @ rpm.....129 @ 6500
Equivalent mph.....132
Torque @ rpm, lb-ft...96 @ 4500
Equivalent mph.....91
Carburetors, no. & make..2 Weber
No. barrels & dia.....2-40 mm
Type fuel required.....premium

DRIVE TRAIN

Clutch type.....single plate, dry
Diameter, in.....7.9
Gear ratios: 5th (0.79).....3.60:1
4th (1.00).....4.55:1
3rd (1.36).....6.17:1
2nd (1.99).....9.06:1
1st (3.04).....15.05:1
Synchromesh.....on all 5
Differential type.....hypoid
Ratio.....4.55:1

CHASSIS & SUSPENSION

Frame type.....unit with body
Brake type.....disc/drum
Swept area, sq in.....345
Tire size.....155-15
Make & model.....Michelin X
Steering type.....worm & roller
Turns, lock to lock.....2.8
Turning circle, ft.....32.8
Front suspension: unequal length A-arms, coil springs, tube shocks, anti-roll bar.
Rear suspension: live axle, lower trailing arms, upper triangulated link, coil springs, tube shocks.

ACCOMMODATION

Normal capacity, persons.....2
Seat width.....2 x 20.5
Head room.....38.5
Seat back adjustment, deg.....0
Entrance height, in.....45.0
Step-over height.....16.2
Door width.....36.0
Driver comfort rating:
For driver 69-in. tall.....90
For driver 72-in. tall.....80
For driver 75-in. tall.....70
(85–100, good; 70–85, fair; under 70, poor)

GENERAL

Curb weight, lb.....2150
Test weight.....2510
Weight distribution (with driver), front/rear, %.....54/46
Wheelbase, in.....88.6
Track, front/rear.....50.8–50.0
Overall length, in.....153.6
Width.....62.0
Height.....52.6
Frontal area, sq ft.....18.1
Ground clearance, in.....5.8
Overhang, front/rear.....29.5/35
Departure angle (no load), deg..19
Usable trunk space, cu ft.....5.7
Fuel tank capacity, gal.....13.8

ACCESSORIES

All equipment on this model comes as standard equipment. Seat belts, side mirrors, radio, etc., are available at extra cost.

MISCELLANEOUS

Body styles available: roadster (as tested).

INSTRUMENTATION

Instruments: 8000-rpm tachometer, 140-mph speedometer, oil pressure, oil temperature, water temperature, fuel.
Warning lights: lights, high beam, fuel level, ignition, heater, turn signal.

CALCULATED DATA

Lb/hp (test wt).....19.5
Mph/1000 rpm (5th gear).....20.2
Engine revs/mi.....2970
Piston travel, ft/mi.....1626
Rpm @ 2500 ft/min.....4560
Equivalent mph.....92
Cu ft/ton mi.....61.1
R&T wear index.....48.5

MAINTENANCE

Crankcase capacity, qt.....6.8
Change interval, mi.....2500
Oil filter type.....full flow
Change interval, mi.....2500
Chassis lube interval, mi.....2500

ROAD TEST RESULTS

ACCELERATION

0–30 mph, sec.....4.6
0–40 mph.....6.2
0–50 mph.....8.1
0–60 mph.....10.5
0–70 mph.....13.3
0–80 mph.....17.4
0–100 mph.....32.5
50–70 mph (3rd & 4th gears)...5.0
Standing ¼-mi, sec.....17.4
Speed at end, mph.....80

TOP SPEEDS

High gear (5400), mph.....109
4th (6500).....104
3rd (6500).....76
2nd (6500).....52
1st (6500).....29

GRADE CLIMBING
(Tapley data)

5th gear, max gradient, %.....8
4th.....12
3rd.....17
2nd.....25
1st.....off scale
Total drag at 60 mph, lb.....130

SPEEDOMETER ERROR

30 mph indicated.....actual 28.7
40 mph.....38.2
60 mph.....57.2
80 mph.....76.2
100 mph.....95.2

FUEL CONSUMPTION

Normal driving, mpg.....19–23
Cruising range, mi.....250–315

ACCELERATION & COASTING

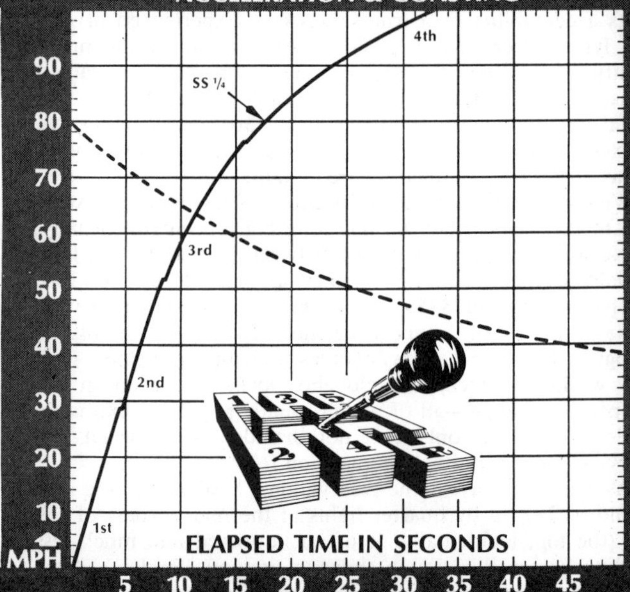

Owners of 1300s have all experienced stop-light-despair syndrome as they delicately fiddled their little jewels off the line, trying to maintain sporting aplomb as the hulks roared by. The 1600 Veloce has no such problem: its Weber-fed engine leaps eagerly up into the meaty high-revolution range, and even pulls well farther down the tachometer. This extra strength, controlled by the superb 5-speed all-synchromesh transmission, makes the car much better suited than before to increasingly fierce U.S. traffic conditions.

Alfa's transmission always has been one of the prime keys to its acceptability as a true sports car. Experienced drivers wax lyrical over its fast, positive operation, the near-impossibility of faulting the synchromesh, and their sustained pleasure in its day-to-day use. We can't argue. After a few initial contretemps in trying to locate 3rd gear quickly, we relaxed, let the effective spring-loading take over, and found it was just as they claimed—pure delight.

Wind roar, often the bane of convertibles, was not excessive in the Veloce. During one long high-speed run (legally, in Nevada), the noise increased between 60 and 80, but then actually seemed to diminish somewhat from 80 to over 100.

The list price of $3595 seems high when compared with those of the rest of the sports-car field. However, it should be recognized that this price includes tachometer, heater, window washers, gas cap and glovebox locks, and even an oil-temperature gauge—all of which usually appear as extras on dealers' lists. The only cost not included is the usual $75 "servicing and handling" preparation fee.

As usual, we have some reservations. Visibility to the rear would be helped by quarter lights at the rear corners of the top (the top, incidentally, is of an improved and much less baffling convertible design than on previous Spiders). Also, though most of the staff found the seats fully supporting and comfortable, our 99th man stated that the side pads were fundamentally deadening over a long trip, and definitely were not designed for the broad of beam. The spark plugs loaded and fouled a little after a few days of pottering about town, but cleared nicely for the timed tests after a short run up to 6000 rpm in 3rd gear. The brakes, discs in front and drums in the rear, were very good, and gave us no difficulty at all beyond announcing their presence with a slight squeal during the first couple of stops every morning.

All things considered, the Alfa Romeo 1600 Veloce is an excellent sports car by any standards: its responsiveness, accurate handling and ease of operation make it a continuing pleasure for either the skilled driver or the novice, and it undoubtedly has all those intangible personality factors which have always enabled Alfas to form close, rewarding relationships with their owners.

ALFA GIULIA SPIDER VELOCE

AT A GLANCE...

Price as tested $3670
Engine .... 4 cyl, dohc, 1570 cc, 129 bhp
Curb weight, lb 2150
Top speed, mph 109
Acceleration, 0–60 mph, sec 10.5
50–70 mph (3rd & 4th gears), sec 5.0
Average fuel consumption, mpg 21

GENE GARFINKLE

30,000 Mile Report
ALFA ROMEO GIULIA SPRINT

BY JAMES T. CROW

IN OUR REPORT on the 60,000-mi Peugeot 403 in the June issue of *Road & Track* we came up with some interesting figures about an imported sedan that proved to be low in operating cost but relatively high in overall cost per mile because of a higher-than-average depreciation rate. This month we have a sports car to examine, an Alfa Romeo Giulia Sprint, and the results should be no less illuminating.

In general, the Alfa Giulia Sprint is a highly regarded car. It has a fine light-alloy, double-overhead-cam engine, a displacement of 1600 cc and a power output of just over 100 bhp. A 5-speed full-synchro gearbox is standard, the brakes are excellent and its appearance is pleasing to almost everyone. It is not a low-priced car by U.S. standards, the list price being $3595 and the delivered price averaging $3740 once state and local taxes have been included. The long suit of the Sprint, in common with all Alfas, is the pleasure that can be had from driving it. It is responsive, has superior handling and offers brisk performance. Alfa owners are a prideful lot and take a back seat to no one in being car-proud.

The owner of the Sprint on which we are reporting is Evan Wilson, a student at the Johns Hopkins University in Baltimore. Evan, who is 25 and single, bought the car in France while in military service. Because of his status at that time, some of the costs associated with his Sprint are not typical. He paid $2880 for the car, for example, and even including the cost of shipping it back to the U.S. ($160), he still saved something like $700 over what the same car would have cost him in the U.S. Also, as a member of the U.S. military service, he was able to buy gasoline tax-free so that the 1200 gal (approx) required for the 30,000 mi cost him only $240 while the same quantity would have cost the average driver about $420 in the U.S.

Of greatest interest to the prospective buyer of any car, judging by the very large response we have had from earlier articles on similar subjects, is the cost of routine maintenance and unscheduled repairs and replacements. Mr. Wilson has kept a complete record of these and they are shown in chronological order in Fig. 3. Though these were performed in France, we have made no attempt to adjust these costs to what they might have been had they been done in the U.S. Concerning the comparative price of such work in France and the U.S., Evan says, "Labor is cheaper in Europe but taxes on parts are higher. In general, I'd say the total cost of labor plus parts is about the same."

Almost all of the 30,000 mi on the Sprint have been touring miles, not city traffic miles. While in Europe, Evan used the car only for trips. Now, going to Johns Hopkins, he uses it only on weekends as he has no need for everyday transportation by auto. The car has had regular maintenance in accordance with the factory recommendations and Mr. Wilson, appreciating a fine piece of machinery, has kept it in excellent condition at all times.

Operating cost

As shown in Fig. 1, Col. A, Mr. Wilson's cost of operation, including everything except the cost of the car and transporting it back to the U.S., has been exactly 5¢ per mi. This is not typical of what the average U.S. driver could anticipate because it is distorted by the comparatively low-cost tax-free gasoline the owner used in Europe and also by the very high cost of insurance in France. If the gasoline had been purchased at U.S. prices and the insurance based on U.S. rates for a driver over 25, the operating cost would have been 4.71¢ per mi (Col. B, Fig. 1).

Even at this lower figure, the Sprint is a comparatively expensive car to operate, as can be seen from the following:

Operating Cost per Mile of Typical Cars

Volkswagen 1200	2.13¢
Ford Falcon 6	2.32¢
Mercury Marauder V-8 w/automatic	4.00¢
Cadillac De Ville w/automatic	4.50¢
Alfa Romeo Giulia Sprint	**4.71¢**

Volkswagen, Ford, Mercury and Cadillac costs based on "The Real Costs," R&T, Feb. 1964. Alfa costs based on information in this article.

The comparatively high operating cost of the Alfa results

directly from the expenditures made for maintenance and for repairs and replacements. These two items cost Mr. Wilson a total of $655 during the 30,000-mi period. The Peugeot on which we recently reported required only $230 worth of comparable service and repair during the first 36,000 mi. That the Alfa was considerably more expensive in this respect does not surprise us. It is a far more sophisticated piece of machinery than the Peugeot and it is to be expected that it would not only require more attention to keep it running properly but would also cost more when repairs and replacements were required.

Overall cost

Using the amounts actually spent by the owner and taking $2250 as the present value of the car (the average retail price for this model according to NADA, the National Auto Dealers Used Car Guide), the overall cost has been 7.65¢ per mile as shown in Fig. 2, Col. A. Again, however, this figure is not typical of what the average American buyer could expect because of the special circumstances under which Mr. Wilson bought the car in France. If we use typical U.S. figures (Col. B, Fig. 2) for the original cost of the car and for gasoline (and, in all fairness, a typical U.S. insurance cost of $340 instead of the $605 actually paid by the owner), and use the actual cost for maintenance and for repairs and replacements, we find that the overall cost is 9.68¢ per mile.

Overall Cost per Mile of Typical Cars

Volkswagen 1200	5.13¢
Ford Falcon 6	6.67¢
Alfa Romeo Giulia Sprint	**9.68¢**
Mercury Marauder V-8 w/automatic	10.00¢
Cadillac De Ville w/automatic	14.90¢

Volkswagen, Ford, Mercury and Cadillac costs based on "The Real Costs," R&T, Feb. 1964. Alfa costs based on information in this article.

Because we do not have comparable figures for other sports cars, it is difficult to say whether the Alfa's overall cost per mile of 9.68¢ is high or low for this type of car.

Depreciation

As we demonstrated in our article "The Real Costs" in February 1964 *Road & Track*, depreciation alone accounts for approximately 45% of the total cost of the average new car over a 3-year period. This earlier article did not include sports cars, however, so we have no direct comparison to make with such a car as the Alfa Sprint. We can compare the depreciation rate of the Alfa with that of other marques, however.

Depreciation of Typical Sports Cars (Original price vs. average retail value)

	Percent of Original Value lost after 1 yr	2 yrs	3 yrs
Alfa Romeo	30%	37%	51%
Austin-Healey Sprite	20%	36%	46%
Austin-Healey 3000	30%	40%	47%
MG-B	23%	40%	52%
Porsche	14%	27%	39%
Sunbeam Alpine	23%	37%	51%
Triumph TR-4	27%	42%	50%

Based on average retail values of typical 1, 2, and 3-year-old cars in NADA, the National Automobile Dealers Used Car Guide.

If these figures seem rather high at first glance, they can be put into perspective by comparing them to typical cars in other categories. For example, in a 3-year period, the average small imported sedan depreciates by 56%, the average American compact by 52%, the medium-range American sedan by 54% and the typical American luxury car by 50%. In other words, the average sports car (including the Alfa) depreciates at about the same rate as a typical American sedan. The lowest depreciation among our examples is that of the Porsche (39%), which is well under the average sports car but does not approach the car with the all-time low depreciation, the Volkswagen 1200 (32%).

Reliability & Service

Regarding the reliability and service of his Sprint, Mr. Wilson has been disappointed in two regards. First, he was surprised that of the five Pirelli Cinturatos that came with the car, one needed replacement after 18,500 mi and the other four at 21,800. This comparatively high rate of wear may have resulted in part from an alignment problem that was delivered to him with the car and was not completely corrected until it had been realigned three times. As the car was still under warranty at that time there was no charge for the alignment correction. But there was no allowance made for damage to the tires either. This high rate of wear does not seem to be typical of Alfas and other readers have reported getting up to 35,000 mi on the original set of Cinturatos.

The most expensive repairs on Mr. Wilson's Sprint resulted from the necessity of replacing the synchromesh mechanisms in the gearbox after 24,250 mi. Although he was disappointed that his Sprint should have this trouble, Mr. Wilson does not believe this difficulty to be typical. "The mechanics who fixed it said that improper lever adjustment caused the failure of the synchros. Because of this improper adjustment, the gears were not completely engaged and failure resulted long before it would normally have been expected."

Although the high rate of tire wear and the gearbox difficulty experienced by Mr. Wilson may not be typical, new Alfa owners might be a bit cautious in the same areas. We add this note because another Alfa owner from whom we've heard has had trouble with the 5-speed in his Alfa 2600. This owner, a doctor in Tucson, has 22,000 mi on his car at present and is certain that the transmission, which is howling and clattering, is ready to fail at any time. Coincidentally, this doctor also experienced a high rate of tire wear, as the front end of his car was sufficiently out of alignment on delivery to ruin one tire in less than 10,000 mi and to be a constant source of irritation for several thousand more before it was finally corrected.

Owner evaluation

Although Mr. Wilson likes his Alfa Sprint very much, he has several suggestions for improvement. He does not like the seats, for example, and puts this at the head of the list. He also believes that a wider range of rear end ratios should be offered so the owner could select that which was best

Fig. 1: Alfa Romeo Giulia Sprint Operating Costs for 30,000 miles

	A	B
Gasoline	$ 240	$ 420
Routine maintenance	210	210
Repairs and replacements	445	445
Insurance	605	340
Total operating cost for 30,000 mi	$1500	$1415
Operating cost per mi	5.00¢	4.71¢

Col. A shows actual cost to owner Evan Wilson. Col. B shows typical figures for U.S. buyer. (See text.)

ALFA ROMEO GIULIA SPRINT

Fig. 2: Alfa Romeo Giulia Sprint
Overall Cost per Mile for 30,000 miles

	A	B
Delivered price	$2880	$3740
Transportation to U.S.	160	n.a.
Gasoline	240	420
Routine maintenance	210	210
Repairs and replacements	445	445
Insurance	605	340
Total investment in 30,000 mi.	$4540	$5155
Retail value at end of period	2250	2250
Cost of driving 30,000 mi.	$2290	$2905
Overall cost per mile	7.63¢	9.68¢

Col. A shows actual cost to owner Evan Wilson. Col. B shows typical figures for U.S. buyer. (See text.)

Fig. 3: Alfa Romeo Giulia Sprint
Repairs & Replacements in 30,000 miles

Front end realigned 3 times	no charge
Fuel gauge defective, replaced	no charge
Sump drain cock replaced	$ 2.06
Tire repaired	0.90
Cracked distributor cap replaced	6.70
1 new Pirelli Cinturato tire at 18,500 mi.	18.82
4 new Pirelli Cinturato tires at 21,800 mi.	85.30
Dented bumper brace trim replaced	5.30
Hubcap and license light replaced	9.40
Gearbox rebuilt & brakes relined at 24,250 mi	302.32
Headlamp replaced	1.80
Dipstick lost, replaced	1.05
Sparkplugs required before time for routine replacement	4.05
Brakes bled and adjusted before time for routine servicing	3.00
Dented trunk positioning arm replaced	2.06
Total	$442.76

"No charge" items repaired under warranty.

suited to his kind of driving rather than accept the "compromise" necessary for the average driver. He thinks too that the quality of the carpeting, the exterior trim and the paint should be improved. He would also like to see stronger bumpers, longer maintenance intervals and, understandably, more careful attention paid to proper gearbox adjustment and front end alignment before delivery.

All things considered, however, Mr. Wilson has been pleased with his Alfa Sprint. He ruefully admits that it is somewhat staggering to look at the list of repairs and replacements that have been necessary (Fig. 3) but is quick to point out that these were spread out over more than two years.

"Since I bought the car for pleasure, it was a successful purchase," he says. "It has been a lot of fun to use."

He sums up his feelings about the car this way: "I have never been fooled by those who contend they are looking for an economical car as an investment. A car is never an investment and is always a potential liability. They should admit that they, like everyone, are buying a car for social status, as a virility demonstrator, to beat the neighbors, to satisfy insufficiency feelings, or to enjoy the exhilaration of speed. . . Therefore, much as I am conscious of the comparatively high cost of owning the Alfa, I rationalize by saying that because the car was bought for pleasure, which it has given me, my 'investment' has been an exceptionally prudent one."

ROAD TEST
ALFA ROMEO
GIULIETTA SPRINT

Continued from page **123**

fast car, and a bore for the passenger. Alfa has been accused of having a weak-kneed electrical system. By fitting an ammeter to keep a running check on the sparks department, and running above 3000 revs at night (which is the way the car should be driven anyway) this problem disappears. Early Alfas suffered a spate of sticky gearboxes. Introduction of the Porsche-type box eliminated many of these complaints. Use of the proper gear lube (Shell Dentax 90—a mineral-based transmission oil without additives) will keep the new Alfa cogs happy and smooth. These few little nuggets were garnered from Tom O'Brien, one of the better Alfa tuners in the country, who spends his time running Competition Services at Paterson, New Jersey.

Alfas are fascinating automobiles. They are almost feminine in their appeal, being delicate to look at but horribly strong when put to the test. If you are a car buff you'll eventually own one —and then you will know what we mean. *C/D*

Alfa Romeo GIULIETTA TI

ENGINE CAPACITY 78.90 cu in, 1,290 cu cm
FUEL CONSUMPTION 33.2 m/imp gal, 27.7 m/US gal, 8.5 l × 100 km
SEATS 4 - 5 **MAX SPEED** 96.9 mph, 156 km/h
PRICE list £ 1,046, total £ 1,264

ENGINE front, 4 stroke; cylinders: 4, vertical, in line; bore and stroke: 2.91 × 2.95 in, 74 × 75 mm; engine capacity: 78.90 cu in, 1,290 cu cm; compression ratio: 8.5; max power (SAE): 84 hp at 6,200 rpm; max torque (SAE): 77 lb /ft, 10.6 kg/m at 3,500 rpm; max number of engine rpm: 6,200; specific power: 65.1 hp/l; cylinder block: light alloy, wet liners; cylinder head: light alloy, hemispherical combustion chambers; crankshaft bearings: 5; valves: 2 per cylinder, overhead, Vee-slanted, thimble tappets; camshafts: 2, overhead; lubrication: gear pump, full flow filter; lubricating system capacity: 10.91 imp pt, 13.11 US pt, 6.2 l; carburation: 1 Solex 35 APAI-G downdraught twin barrel carburettor; fuel feed: mechanical pump; cooling system: water; cooling system capacity: 13.20 imp pt, 15.85 US pt, 7.5 l.

TRANSMISSION driving wheels: rear; clutch: single dry plate; gearbox: mechanical; gears: 4 + reverse; synchromesh gears: I, II, III, IV; gearbox ratios: I 3.258, II 1.985, III 1.357, IV 1, rev 3.252; gear lever: steering column; final drive: hypoid bevel; axle ratio: 4.55.

CHASSIS integral; front suspension: independent, wishbones, coil springs, anti-roll bar, telescopic dampers; rear suspension: rigid axle, trailing lower radius arms, upper A-bracket, coil springs, telescopic dampers.

STEERING worm and roller; turns of steering wheel lock to lock: 2.75.

BRAKES drum, 2 front leading shoes; braking surface: front 73.18 sq in, 472 sq cm, rear 66.05 sq in, 426 sq cm, total 139.23 sq in, 898 sq cm.

ELECTRICAL EQUIPMENT voltage: 12 V; battery: 38 Ah; dynamo: 240 W; ignition distributor: Marelli or Lucas; headlights: 2.

DIMENSIONS AND WEIGHT wheel base: 93.70 in, 2,380 mm; front track: 50.87 in, 1,292 mm; rear track: 50 in, 1,270 mm; overall length: 161.65 in, 4,106 mm; overall width: 61.22 in, 1,555 mm; overall height: 59.05 in, 1,500 mm; ground clearance: 4.13 in, 105 mm; dry weight: 2,029 lb, 920 kg; distribution of weight: 52.3% front axle, 47.7% rear axle; turning circle (between walls): 34.1 ft, 10.4 m; tyres: 155 × 15; fuel tank capacity: 8.8 imp gal, 10.6 US gal, 40 l.

BODY saloon; doors: 4; seats: 4 - 5; front seats: separate, adjustable backrests.

PERFORMANCE max speeds: 29.2 mph, 47 km/h in 1st gear; 48.4 mph, 78 km/h in 2nd gear; 70.8 mph, 114 km/h in 3rd gear; 96.9 mph, 156 km/h in 4th gear; power-weight ratio: 24 lb/hp, 10.9 kg/hp; carrying capacity: 1,014 lb, 460 kg; acceleration: 0 — 50 mph (0 — 80 km/h) 12.1 sec; speed in direct drive at 1,000 rpm: 15.6 mph, 25.2 km/h.

PRACTICAL INSTRUCTIONS fuel: 95-100 oct petrol; engine sump oil: 9.68 imp pt, 11.63 US pt, 5.5 l, SAE 20-30 (winter) 30-40 (summer), change every 2,500 miles, 4,000 km; gearbox oil: 2.64 imp pt, 3.17 US pt, 1.5 l, change every 5,000 miles, 8,000 km; final drive oil: 2.11 imp pt, 2.54 US pt, 1.2 l, change every 5,000 miles, 8,000 km; steering box oil: 0.35 imp pt, 0.42 US pt, 0.2 l; tappet clearances: inlet 0.018 in, 0.47 mm, exhaust 0.020 in, 0.52 mm; valve timing: inlet opens 25° 20' before tdc and closes 68° after bdc, exhaust opens 61° 20' before bdc and closes 18° 40' after tdc; tyre pressure (medium load): front 21 psi, 1.5 atm, rear 23 psi, 1.6 atm.

VARIATIONS AND OPTIONAL ACCESSORIES 4.10 5.12 axle ratios; central gear lever; 300 W dynamo; fuel tank capacity 14.3 imp gal, 17.2 US gal, 65 l.

ROAD & TRACK ROAD TEST

ALFA ROMEO SPRINT SPECIALE

A fleeting touch of high fashion, Italian style

THE ALFA ROMEO Giulia Sprint Speciale is a highly individual sports car, designed for a highly individual market. Almost everything about it is unusual to some degree, including its price, accommodations, interior and exterior appointments, and especially its body configuration. Thus, it can be expected to attract a rather sharply delineated stratum of buyers, with distinctly *outré* tastes.

The arrival of the Sprint Speciale set off a styling controversy at our offices that continued throughout the car's stay. Every staff member formed a definite opinion regarding its appearance, ranging all the way from warm empathy to downright antipathy. Admirers praised the sleek, low litheness of Bertone's design and cited the obvious air-penetration advantages of its diminutive front end and flowing lines. The critics, however, maintained that the lines are too feminine for the masculine behavior of the car, one even descending to the unforgivable, "The rear end should be more truncated—Kammed down."

This, of course, is not a new design. The Bertone-bodied SS was introduced late in 1958 and the 1290-cc Giulietta version was tested in July 1961 R&T. At that time we found the construction and handling excellent, but the performance rather disappointing.

Then strength, vigor and a new name came with the advent of the 1570-cc engine, the designation "Giulia" being chosen to indicate its presence in various Alfa models. For some reason the Giulia Sprint Speciale was not originally slated for export to the U.S., though the Giulietta had been sold

ALFA ROMEO SPRINT SPECIALE

AT A GLANCE...

Price as tested	$4961
Engine	4 cyl, dohc, 1570 cc, 129 bhp
Curb weight, lb	2273
Top speed, mph	112
Acceleration, 0–60 mph, sec	12.0
50–70 mph (3rd gear)	6.6
Average fuel consumption, mpg	20

ALFA ROMEO SPRINT SPECIALE

here since its introduction. This decision has been reversed, however, and the higher powered model is now available.

The additional torque of the 1570-cc engine is at once delightfully apparent to those who have driven the 1290-cc version. The improvement is, in fact, much more noticeable when accelerating in any given gear than the zero-to-speed times would indicate. One can keep abreast of stop-light surges easily or execute a neat avoidance under power when necessary. This engine's liveliness and flexibility unquestionably adds greatly to the driver's pleasure.

Four-cyl, dohc engines have certain pronounced operational characteristics, some praiseworthy and others somewhat less appealing. The Giulia engine is designed for high-rpm running, the red-line appearing between 6200 and 8000. We used a shift point of 7000 throughout our acceleration runs and got off the line in very good order, as indicated by the data panel figures. The car felt strong and eager when driven hard, always seeming to invite action and adventure.

Many sports car drivers enjoy the whirring, clicking and general mechanical activity of a busy engine. Though the Sprint Speciale seemed quieter than some, the sounds are there for the connoisseur's delectation. There were a few unwelcome noises, however, including an engine vibration period at around 70 mph, a muffler baffle resonance at 2300 rpm, and a startling and annoying throttle linkage rattle.

Starting up the engine when cold, a rite formerly approached with dread and occult supplication by Alfa owners, proved child's play after mastery of the "Start" fuel-mixture control. However, we had a little more difficulty in finding accurate low-speed settings for the manual throttle control.

Our test car was served up with a set of Sprint GT camshafts, rather than the long-duration Veloce cams originally installed in this engine, for "smoother running during the break-in period" and, though it was prone to buck a little when cold, a trait forgivable in any high-bred steed, it quickly settled down to run easily and dependably throughout its range, including a quiet, even idle at 500 rpm.

With discs in front and big, finned drums in the rear, the brakes are powerful and dependable, among the best we have tested. In an era of over-weight, under-braked cars, really authoritative stopping power is heady wine indeed.

A driver's primary means of identifying with an automobile must be through the engine's response, and the feel of the steering and the transmission. The latter always have been areas of particular Alfa glory, and both are superb in the Sprint Speciale. The steering is as quick, positive and rock-steady as ever, and manipulation of the gearbox is utter, continuing delight. The throws between gear selections are short and quick and the action is firm; and spring loading draws the lever toward its two center positions (3rd and 4th) with just the right degree of guidance. Finding the "inner" gears in a 5-forward-speed gearbox is often a tense, hesitant probing action, particularly when downshifting in a hurry. There need be no doubt or hesitation in the Sprint Speciale.

Handling, too, is excellent. At low speed, there is a good deal of road noise and vibration from small bumps, but the suspension smothers large dips remarkably well. And, one of the hallmarks of the true sports car, as speed increases, the handling gets better and better. One half-serious comment was, "Alfas make you realize what a good driver you really are . . ."

With a 21-gallon fuel capacity and 22-mpg consumption figure while cruising, the Sprint Speciale must be appraised as

a long-distance touring car. From this standpoint, the strong virtues of its performance, handling and brakes must be weighed against one peculiar fault: There simply is not enough headroom for drivers even slightly taller than average. The combination of Bertone's extremely low roofline and the 4-in.-thick seats put a slightly out-sized driver's scalp in unwanted intimacy with the headlining. Also, the driver's eyes are at such a high level that lowering the sun visor leaves only a narrow slit for forward vision. One is forced to adopt an extreme version of the classic Italian driving stance—arms straight and head well back. It is no small testimonial to the accessibility of the Speciale's controls that it actually can be driven with confidence from this exaggerated position. The installation of low bucket seats would add at least 2 in. to the headroom, and would make this car much better adapted to a large portion of the U.S. market.

The interior appointments are tasteful and of a quality commensurate with the car's substantial price, except for the surprising placement of a yellow window-washer bag close by the driver's left foot. There are no frills—everything inside the SS has a definite purpose, and serves it with both utility and understated luxury. It should be mentioned that, as in all extremely low sports cars, entering and alighting with speed, grace and aplomb require practice and dexterity of a high degree, even for males. The combination of a short skirt and a Sprint Speciale would be both a show-stopper and a boon to oglers, oculists and body repair shops, Ah, the Italians. . .

Instruments and controls are both in the Alfa tradition, and thus are among the very best in legibility and accessibility. When the driver grips the substantial steering wheel, all necessary knobs, levers and switches are within a few inches of his hands, and Alfa's concept of a proper instrument panel should be required reading for all automobile designers. A highly satisfying amount of information regarding the car's functions is available within a minimum amount of space, yet all is discrete, uncrowded and presented in lucid black and white. Colors are reserved for a multitude of supplementary warning lights—red, blue, green, orange and yellow indicators which signal the operation of everything from ignition to the heater blower. All are useful and well placed, except that the high-beam indicator was hidden by the key fob, and we would have preferred an ammeter to the existing red light.

Both the glove box and the trunk were too small to carry more than a minimum of goods, the trunk already brimming with spare tire, battery and tool kit. Long-distance travelers will find soft luggage *de rigeur*. However, an upholstered area behind the front seats can be used for luggage, and even has provision for the attachment of securing straps. This area also can carry an equable child or two for moderate distances.

In utter defiance of Alfa tradition, the heater worked quickly and well on cool mornings, undoubtedly aided by the small passenger compartment and the car's good weather sealing. We had no opportunity to try it in really frigid weather. The extremely low-angle rear window accumulated its expected layer of silt, as all such windows must, but a small bug-deflector strip on the hood effectively boosted insects and detritus over the windshield.

The Alfa Romeo Sprint Speciale does not lend itself to a brief and tidy summing up. It is a unique, controversial machine which can be regarded invidiously, as a victory of Bertone over Alfa Romeo, or with rapture—one driver said, ". . . it has beauty, performance, handling, mechanical interest, personality, class . . . it has everything." Our conclusion, then, is that the Sprint Speciale will not attract everyman, but that its owners will regard it as an exquisite jewel beyond comparison or price.

R&T

ROAD TEST

ALFA ROMEO SPRINT SPECIALE

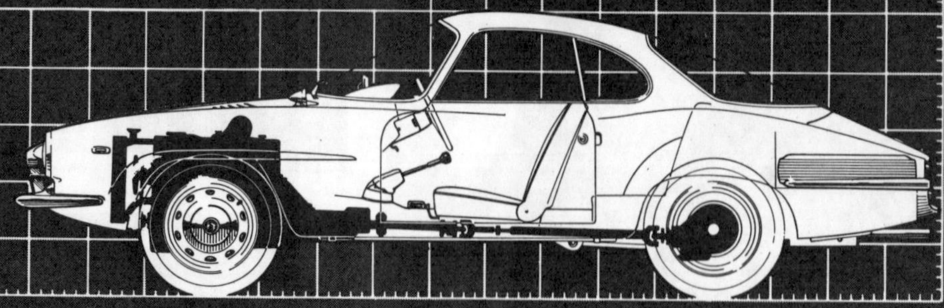

PRICE

Basic list	$4961
As tested	$4961

ENGINE

No. cyl & type	4 inline, dohc
Bore x stroke, mm	78 x 82
In	3.06 x 3.23
Displacement, cc/cu in	1570/95.6
Compression ratio	9.7:1
Bhp @ rpm	129 @ 6500
Equivalent mph	125
Torque @ rpm, lb-ft	96 @ 4500
Equivalent mph	85
Carburetors	Weber 40 DCOE
No. barrels, dia	2 x 1.58
Type fuel required	premium

DRIVE TRAIN

Clutch type	sdp
Diameter, in	8.0
Gear ratios: 5th (0.79)	3.60:1
4th (1.00)	4.56:1
3rd (1.36)	6.20:1
2nd (1.99)	9.06:1
1st (3.31)	15.09:1
Synchromesh	on all 5
Differential type	hypoid
Ratio	4.56:1

CHASSIS & SUSPENSION

Frame type	unit with body
Brake type	disc/drum
Swept area, sq in	345
Tire size	155-15
Make	Pirelli Cinturato HS
Steering type	worm & roller
Turns, lock-to-lock	2.5
Turning circle, ft	33.0

Front suspension: independent with unequal A-arms, coil springs, tube shocks, anti-roll bar.

Rear suspension: live axle on trailing arms and upper triangle link, coil springs, tube shocks.

ACCOMMODATION

Normal capacity, persons	2
Seat width, in	2 x 22.0
Head room	33.2
Seat back adjustment, deg	30
Entrance height, in	46.0
Step-over height	15.0
Door width, front	37.0
Driver comfort rating:	
Driver 69 in. tall	80
Driver 72 in. tall	65
Driver 75 in. tall	55

(85–100, good, 70–85, fair; under 70, poor)

GENERAL

Curb weight, lb	2273
Test weight	2638
Weight distribution (with driver), front/rear, %	54.3/45.7
Wheelbase, in	88.6
Track, front/rear	50.8/49.9
Overall length	162.2
Width	61.2
Height	50.2
Frontal area, sq ft	17.1
Ground clearance, in	5.9
Overhang, front/rear	34.3/39.3
Departure angle, deg	21
Usable trunk space, cu ft	3.5
Fuel tank capacity, gal	21.1

INSTRUMENTATION

Instruments: 8000-rpm tachometer, 140-mph speedometer, oil pressure & temperature, water temperature, fuel level.

Warning lights: generator, heater, high beam, choke, directional signals.

MISCELLANEOUS

Body styles available: coupe only.

Warranty period: 6 mo./unlimited mileage.

CALCULATED DATA

Lb/hp (test wt)	20.4
Mph/1000 rpm (5th gear)	18.8
Engine revs/mi (60 mph)	3200
Piston travel, ft/mi	1722
Rpm @ 2500 ft/min	4640
Equivalent mph	88
Cu ft/ton mi	67.0
R&T wear index	55.2

EXTRA COST OPTIONS

Blaupunkt AM/FM radio, exterior mirror, chrome wheels, seat belts.

MAINTENANCE

Crankcase capacity, qt	7.4
Change interval, mi	2500
Oil filter type	full flow
Change interval, mi	2500
Chassis lube interval, mi	2500

FUEL CONSUMPTION

Normal driving, mpg	18–22
Cruising range, mi	380–463

ROAD TEST RESULTS

ACCELERATION

Time to speed, sec:

0–30 mph	3.9
0–40 mph	6.1
0–50 mph	8.5
0–60 mph	12.0
0–70 mph	15.4
0–80 mph	19.7
0–90 mph	26.0
50–70 mph (3rd gear)	6.6

Time to distance, sec:

0–100 ft	3.6
0–500 ft	10.0
¼-mile	18.0
Speed at end, mph	77

Passing exposure time, sec:

Car ahead going 50 mph	6.3

SPEEDS IN GEARS

5th gear (5800), mph	112
4th (7000)	106
3rd (7000)	79
2nd (7000)	54
1st (7000)	32

BRAKES

Panic stop from 80 mph:	
Deceleration, % G	84
Control	good
Parking: hold 30% grade	no
Overall brake rating	very good

SPEEDOMETER ERROR

30 mph indicated	actual 27.0
40 mph	36.3
60 mph	55.3
80 mph	74.3
100 mph	93.6
Odometer correction factor	0.961

ACCELERATION & COASTING

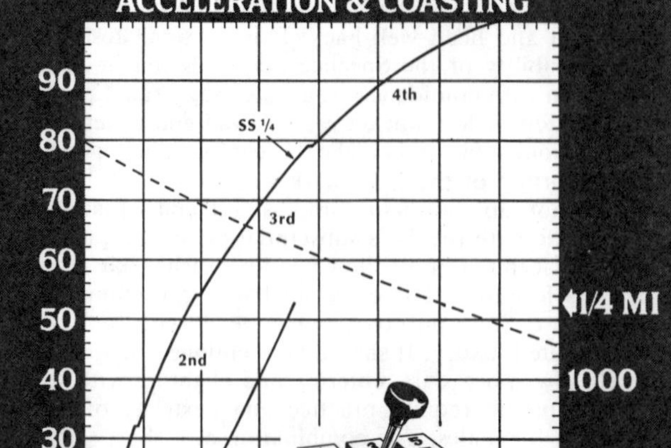

1950s

ALFA ROMEO GIULIETTA SPIDER VELOCE

IN GERMANY in 1956 a man named Porsche was showing just how far the Volkswagen concept could be taken. In America Zora Arkus Duntov was demonstrating his camshaft design that gave Chevrolet's new small-block V8 the power to take the still-poorly-regarded Corvette to 150 mph. In Britain MG was raking in the US dollars with a design that owed far more to the 1930s than the '50s, while at Jaguar the XK series was cruising, still some years away from the E-type and a second high water mark to match that of the original XK120 when launched in 1948. In Italy, meanwhile, the sports car of the future was already on the road.

While Mr Ferrari was rewriting the rules for cars at the top end of the market Alfa Romeo was tentatively, almost accidentally, working toward a car, to be called Giulietta, that would not only provide a styling/performance/handling benchmark for mass produced 'affordable' sports cars but whose engine would go on to become a landmark in automotive design.

The Giulietta Sprint was launched in 1954. The Sprint, quickly followed by a multitude of derivatives, was the first Alfa to be called Giulietta and the first to use the newly-designed (by Orazio Satta) all-alloy 1290 cm^3 twin-cam engine. The Giulietta series was Alfa's first big post-war push with an all-new car, and it marked a considerable change of direction for the company, into an area of mass produced cheaper-costing cars which, after the more exclusive nature of prewar Alfas, understandably caused problems both at the factory and at chief bodybuilders, Bertone, then a small *carrozeria*/racing company hardly capable of meeting Alfa's first order for 1000 Sprints. That these cars were for Alfa to give away as prizes in a lottery, held to raise money to get the car into production, says a lot about just how close to the wind Alfa was sailing. It's easy to say today that if the Giulietta had failed, if its remarkable engine had failed, it's likely there would be no Alfa today.

Fail it did not; the Giulietta Sprint, the first popular car to offer all the advantages of a twin-cam engine, was readily accepted, its Bertone styling acclaimed. The car was good-looking, practical, light of weight, had spirited performance and fine handling; a live axle, drum brakes and a four-speed, all-synchro column gearchange had never been made to work as well in this class of car, though even in the '50s these features were hardly in the vanguard of sports car innovation. All except the live axle, which is a feature of the Spider, sole surviving relative of that first Giulietta series, still made today by Pininfarina, were to be changed in time.

The car we've picked for the Top 10 is the 1956-introduced Giulietta Spider Veloce, the model that combined a pretty Farina-designed open body with what was then the hottest version of the twin-cam, twin Weber carburettors boosting power to 90 bhp, 67 kW, at 6500 rpm. It was enough to send the 865 kg convertible from 0 to 100 km/h in a bit over 11 seconds and on to a top speed of respectably past the old ton. And this model had a floor gearchange, though still with four ratios.

Ninety horses from 1290 cm^3 wasn't bad; the Fiat X1/9, also with 1290 cm^3 of four-cylinder engine, could boast only 75 bhp two decades later, while even today such a high specific output is not readily beaten in a non-turbo engine. There was a lot about the mid '50s Veloce that was common to racing — apart from the alloy engine, aluminium was used for gearbox housing, bonnet, doors and boot lid. Super-hot versions gave the Porsche Spyders of the time a run for their money, but all too often braking and handling showed up in competition less than well; there was no doubt the Giulietta series first and foremost was for the road, however hot the engine.

"If the Giulietta had failed, if its remarkable engine had failed, it's likely there would be no Alfa Romeo today"

For three decades the engine has literally been the backbone of Alfa Romeo cars, and it says a huge amount for its design that this charismatic unit, with the so-familiar crisp response and wide-spread torque, was far from disgraced when pitted against the competition even as recently as two or three years ago. A Giulietta Veloce story in SCW almost 30 years ago got it right. The last par reads: "The longer you consider it the more obvious it becomes that this car, for all its smallness, is a true Alfa and will make its mark on the automotive world as surely as did the 2.9." Today, that mark has indeed been made.

TOP TEN

CLASSIC BY CHANCE

When Alfa approached Bertone in his modest workshop and asked him to design a four-seater GT, nobody imagined that more than 500 would be built, much less almost 40,000. Griffith Borgeson relates the hitherto untold story of the car that became a classic, merely by chance

Few classic cars of the past were a preconceived masterpiece but each acquired that status due to immediate and spontaneous public response. None has been more successful nor was conceived in a more accidental manner than the Alfa Romeo Giulietta Sprint. The story behind this eternally satisfying design, this pioneer among small-displacement GT cars, is almost totally unknown, as the historic records have been kept locked in Bertone's office safe.

Long before its huge and diversified factories had been smashed by Allied bombs in World War II Alfa Romeo had been a ward of the Italian state. During the 1930s much of its production had been directed to the practical demands of belated empire-building in Africa and, contrary to general opinion, its production of passenger cars was a minor sideline. But through a subsidiary called Alfa Corse and the campaigning activities of Scuderia Ferrari it managed to keep its name and the Italian colours very much alive in international racing.

After World War II Alfa rebuilt its production facilities slowly. A token automobile output was based on the old six-cylinder 2500 chassis of the late Thirties, and Alfa Corse continued to race the pre-war Grand Prix Type 158/159 Alfettas. In late 1950 the more modern, more sophisticated, four-cylinder 1900 began to replace the old six but production figures remained microscopic. (The 1900, incidentally, was the creation of Alfa's new chief engineer, Orazio Satta, who masterminded the technical side of the marque's conversion to large-scale automobile production. All subsequent Alfa Romeo cars bore his signature—including the Giulietta—up to the time of his death in March 1974.) The Grand Prix cars, although they were developing absolutely unprecedented specific horsepower, were ancient and at the very end of their tether. Italian economic recovery was beginning to boom, change and growth were in the air, and it obviously was time for Alfa to come to grips with the future.

Alfa was and still is under the wing of the IRI—the industrial reconstruction agency of the Italian state—and, more specifically, under the administration of Finmeccanica, the sub-wing with which IRI embraces mechanical industries. And Finmeccanica's automotive consultant attached to Alfa Romeo was former Porsche engineer Hruschka who, with Carlo Abarth and Eberan von Eberhorst, had worked on the Cisitalia GT and racing-car projects of the late Forties.

In the early Fifties Alfa Romeo and Alfa Corse co-ordinated a plan for future racing development. It was decided to retire the Type 158/159 GP machines at the end of the 1951 season and to concentrate upon the newly booming field of sports-car competition. Working with the components of Satta's quite brilliant Type 1900 touring car, Ing. Gioacchino Colombo developed the mechanics for the *Disco Volante*, a two-litre, 130 bhp sports-competition machine for which the racing world waited breathlessly. *Barchetta* and coupe bodies were designed on Alfa's own premises and prototypes were built by Carrozzeria Touring. It was at this point that Ing. Quarino, Alfa's managing director at the time, first contacted Carrozzeria Bertone. His interest and respect had been aroused by Bertone's brilliantly conceived coachwork for the Arnolt MG and then for the Arnolt-Bristol. He wanted to find out what it would cost to have Bertone manufacture a small run of bodies for the *Dischi Volani.*

A few years earlier, when Nuccio Bertone had had little work to do and nothing but time for the indulgence of his love for racing, he used to find himself competing every weekend against masses of Cisitalia *monoposti.* The Cisitalia *equipe* would consist of from 20 to 30 identical cars, all based on Fiat 1100 components and all prepared and managed under Carlo Abarth's thorough and secretive Teutonic authority. Bertone used to show up at each track with, at that time, a Stanguellini 1100 on a small trailer. He would roll it off, take a practice lap or two, and then garage his car until the start of the race. With himself and one mechanic against Abarth's 50 or 60 team members Bertone used to finish consistently in second or third place and, when everything was right, he occasionally came in first. This did not amuse the grimly determined Abarth but Hruschka, who was always on the scene, thought it was hilarious. Out of mutual respect a strong friendship grew between

The lower photographs are of one of two pre-production prototypes fitted with a lift-up tailgate. Actual production followed the same basic shape with only minor changes to cosmetics and in 1962 the 1300 cc model was replaced (above and below) by the 1600

the ex-Porsche engineer and the young Torinese coachbuilder.

So, when Ing. Quaroni said to his government consultant, "I know of a coachbuilder who not only knows his job but who knows racing too," Hruschka recognised the description of his friend and immediately approved giving Bertone a crack at the *Disco Volante* project. They went to Turin, examined Bertone's facilities and ingeneous methods and asked him for a cost estimate. His advanced techniques enabled him to quote a price that was almost half that of Carrozzeria Touring, and the job was as good as his.

However, the exciting *Disco Volante* project was abandoned, to the eternal loss of the sport. All the powers involved in Alfa's administration finally had chosen a new course for the company.

The most promising future, they decided, lay in the mass production of moderately priced cars of modest size. This meant total reorganisation of the firm's automotive division, which would leave no time, money, nor manpower for sports-car production on an artisan scale.

The new car was to be the basic, four-passenger, 1300 cc Giulietta, austere and utilitarian in spite of its beautiful little dohc engine. The tooling for its production demanded a huge amount of new financing. To raise this capital Alfa issued and sold bonds, which paid a normal rate of interest. Then, to give extra incentive to investors, Alfa announced a sort of lottery: the holders of bonds with the lucky serial numbers would receive a new Giulietta free. The rush for a couple of hundred free cars brought in the needed funds.

Since this was Alfa's first venture in large-scale production many things had to be learned the hard way, and progress dropped ever farther behind the projected schedule. Finally the time came when the lucky winners of the free cars had to be named. But there were no cars to give them and many of them became quite understandably disturbed. Alfa found itself in an intolerable situation. The Giulietta's mechanical components were beginning to come off the production lines but it would be perhaps a year before the body plant was in operation. Something had to be done, and done without delay.

For testing the new engine and running gear Alfa technicians had built a pair of *muletti*—"mules"—whose design had been hastily roughed out by the same internal talent that had drawn up the *Disco Volante* and many other "house" designs. The workmanship of these mules also was rough since they never were intended to be seen by the public. But in them and in their inherently good form Quaroni and Hruschka saw the way out of

their dilemma: build a small run of *fuori serie* special-bodied cars, using the already available *Giulietta* parts. Build enough to calm the irate winners of free vehicles. Beyond that, use them as show cars, as a glamour version heralding the introduction of the standard Giulietta. And beyond that, sell the remainder of the run to a few hundred distinction-minded motorists and get them circulating on the public roads, creating further interest in the long-awaited mass-production car.

This was already the autumn of 1953. Investors and press were on the warpath and the Turin Salon was coming up in the spring. Quaroni grabbed a phone and asked Bertone to rush to Milan, then repeated the same command invitation to Boano of Carrozzeria Ghia.

The resulting session was hectic. First, a prototype had to be ready for the Salon. Second, a production run had to be completed in the shortest possible time. Third, the size of the run was up in the air. Hruschka wanted a thousand units. Quaroni said that it would be hard to unload a maximum of five hundred.

There was no talk of making a *gran turismo* car of it; no one dreamed of that then. The objectives were, simply, distinctiveness of conception and appearance and, hopefully, silghtly heightened performance due to a little streamlining. Boano and Bertone were shown the rough *muletti,* each was given a plaster scale-model copy and was sent packing back to Turin. The assignment was to refine the model into something attractive to the public and to return with the results in ten days.

Boano's well-known love for elegance led him to work up a very different design, much longer than the mule, with flatter surfaces and less of a frankly sporting look. Bertone recognised the basic grace and harmony of the rough mule's lines and contented himself with refining them and, as an innovation, treating the entire rear panel as one large door, giving access to the rear interior of the body. It was Bertone's version which Alfa chose.

Still, there was a key role in the project for the reliable Boano. Hruschka's insistence upon a production run of a thousand units prevailed. Bertone could responsibly promise to deliver four bodies per day, in the form of unfinished sheet-metal shells. Boano, having much larger facilities, could deliver the same number of bodies, complete with interiors, paint, glass, and trim. A verbal pact was concluded without any discussion of money. It was more than an urgent situation; it was a desperate one. And all parties concerned enjoyed each other's fullest confidence.

It was well into 1954 when all these arrangements were concluded and Bertone attacked the construction of the first Giulietta Sprint prototype.

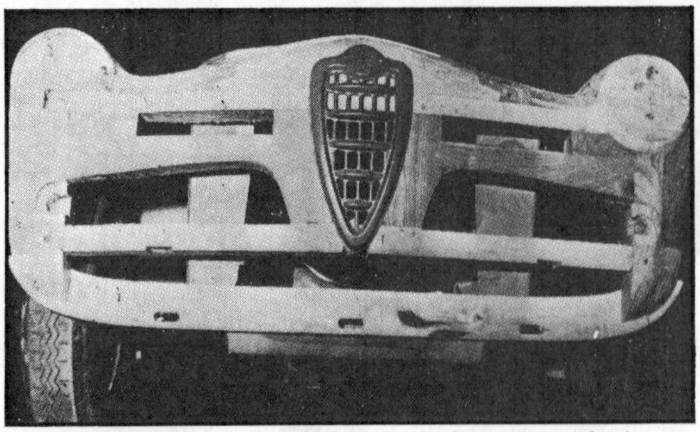

This wooden mock-up was the pattern for the first series production of Sprints. Demand was so great that the mock-up soon had to be replaced by steel dies

Twenty days before the April 21 opening of the Turin Salon he trucked the first shell over to Ghia for finishing. And then all hell broke loose.

Of all the controversial personalities that have lent colour to the pages of Italian automotive history Gigi Segre is far from the least notable. He was a brilliant salesman who arrived at a position of power within the Ghia organisation. Although they were friends, he and Boano had frequent strong differences and Segre chose the moment of the arrival of the Sprint prototype to physically throw the distinguished Sig. Boano out the door of his own establishment. Boano called his lawyer, instructed him to liquidate his interests in the firm, and never went near Ghia again.

This sensational squabble filled the front pages of Italy's newspapers and Bertone was still reeling from the shock when Hruschka called him from Milan. What was to be done now, he wanted to know? How were the deadlines to be met? Segre, excellent salesman though he was, was an unknown quantity as a coachbuilder in his own right. Of course only a verbal commitment was involved but in a country which still held the word of honour in high respect, the situation was more than delicate. Finally, Hruschka spent many hours with a truculent Segre in order to convince him that the interests of all concerned would best be served if the whole job would just be turned over to Bertone. And it was.

The prototype was received with lively public acclaim at the Turin Salon. Alfa's press release for the show read, "This is not yet the car expected by Alfa Romeo enthusiasts, the four-seater Giulietta produced in quantity. But it means the positive announcement that this car will, within a short time, be revealed to the public." Bertone kept cranking out bodies and the winners of free cars were finally calmed. They should have been, since the cars they received surpassed anything that they had had reason to expect. Still, the normal Giulietta did not appear and at the London Show that winter Alfa was constrained to fall back on the same, procrastinating press release.

Bertone is convinced that the Giulietta Sprint phenomenon never could have taken place without Hruschka's insight and foresight. It was he who insisted upon its high stage of engine tune, upon its Type 1900 Ti brakes, upon its whole *gran turismo* personality. Without performance to match its appearance it would have been just a handsome mediocrity.

Instead of the anticipated one thousand units, Bertone was called upon to build six thousand copies of the first-series Sprint. Two months after its introduction its powerful impact upon the market had been sensed and, this time, Alfa's top management came to see Bertone, in the modest shop which his father had built decades before. They came to tell him that the anticipated production never would meet the demand and that it would have to be doubled. At this time the car was made entirely by hand and there was not a single stamped piece except for the radiator grille. It had been lifted from the Romeo *Autotutto* van.

Bertone sagged, having already given more than his all.

"Gentlemen, how much time can you give me for this?", he forced himself to ask.

"Just do it as soon as you can," was the reply. "We know you well enough. You'll find a way."

Bertone recalled later that if Alfa had, at that time, made a flat demand for 20 cars a day, he would have given up the project as being impossible.

But the men of Alfa were prepared to give him a chance to increase his output by degrees. He worked at it furiously, inventing new processes as he went along. To his clients' amazement, he shared with them the benefits of his increased production. The bodies which he had had to domand $1120 for in the beginning he eventually was able to deliver for almost half that cost. This involved the building of an immense new factory and the transformation of Carrozzeria Bertone into one of the world's giants among specialist coachbuilders . . . later to become *the* greatest.

The demand for the beautiful little Giulietta Sprint *berlina* grew steadily until, in 1959, they were coming off Bertone's production lines at the rate of 34 per day. After that peak, sales remained strong for years, diminishing at a very gradual rate. When, in the late 1960s, the painful decision finally was made to retire the model, 39,654 units had been made and sold. And in the beginning, five hundred units had seemed the optimistic maximum that the market could absorb.

Early in the car's career Hruschka was haunted by the likelihood of a sudden loss in its popularity. So, every couple of years, Bertone would develop a new, carefully studied design to replace it. Each was more modern, more sophisticated than the last; each was an improvement over the original, while preserving its essential qualities of charm and grace. Each of these proposed replacements was approved and purchased by Alfa Romeo . . . only to be reconsidered at the eleventh hour. There was always a doubt: Will this really be an improvement over what we have right now? Hruschka had a charming way of rationalising these decisions:

"This girl is not pretty," he would say, "but she is pleasing. That one is pretty but she does not please. The first one is better: she is *simpatica.* We have a *simpatica* car and we'd better be content with her."

"Her" success helped Hruschka on his way to being placed in charged of the huge Alfa Sud development in southern Italy. Her success was the key to launching Bertone on the road to world leadership in his field and to the highest honours at home, including knighthood. In her practical beauty and spirited performance she was a fitting bridge between the marque's artisan past and technocratic future. She was small, but as significant as she was *sympa.*

READERS' REBUILDS

Rover 105 : Alfa Guilietta Sprint

Robert McManus tells how he rescued an Auntie Rover 105 from the scrap heap and turned it into a stylish – and cheap-method of transport. His story proves that, with the right materials, a home restoration can be worthwhile

: all began when a friend, who owns a Rover 105, ought another to use for spares. However, he ran ıort of money and decided to either sell the latest cquisition or scrap it.

Fortunately I heard about it before it was too late nd before long became the owner of a delightful old eap of a 1959 Rover 105 for the princely sum of 150. Character costs money these days.

ncredibly sound chassis

On getting it home my long-suffering wife nearly emanded a quickie divorce, the third party being a elapidated Rover. Still, having once been a alesman, I managed to persuade her, just, that one ay she would be proud to be conveyed in style.

I started work on the interior as the mechanics were passable but the cosmetics were terrible. The leather upholstery was rotten and had split, well beyond redemption. A scrap yard provided a complete interior but the seats, though perfect, had no colour left on them. The interior, a spare wheel and bumper cost £45.

Re-Connollising is very expensive but a trip to Connolly Bros brought forth two tins of the original colour paint for £14 and I did them myself . . . a lovely result even if I do say so myself.

Next stop was the body and chassis. A thorough check of the latter showed that though covered in rust it was incredibly sound and, other than a good coating with old engine oil, needed no other major attention.

The superstructure was a different matter. The bottoms of the door pillars had rusted away to such an advanced state that they were no longer connected to the floor. The boot floor was partially missing and the metal moths had long been and gone as far as the front wings and rear wheel arches were concerned. The cost of professional welding on so many different odds and ends would have bankrupted me, so I invested in a small arc welder and carbon arc brazing torch. It was fine in theory until I tried to strike an arc and then weld sheet metal. So, when at first I didn't succeed, I tried and tried again.

I found the best way is to cut out the plates to fit the holes, pop rivet them into place and then weld or braze. By this method I can honestly say it is as strong as when it first came off the 'line', although the finished result may not have been as pretty.

Now I came to painting what was now very much like a patchwork quilt as you can imagine, with plates and filler here, there and everywhere. The most reasonable quote for a good quality respray was £300, still too much for my meagre funds, so in for a penny in for a pound, I purchased a reconditioned Sherry spray plant – £150 with hose and gun. It was one of the best investments I have made and it has now sprayed several cars and more than paid for itself.

I had never sprayed before but a bit of practice made perfect and now I can do it as well as the next man. Joining the Rover Drivers Guild helped no end and their technical department gave me the part numbers of the two-tone grey paints used. Essex Cellulose mixed them while I waited and what a perfect match they turned out to be, the best I have ever seen.

Gearbox needed work

Next came the gearbox. It was a bit 'clonky' in first so I thought I'd have a look inside to see what was going on. I found that when trying to drain the gearbox/overdrive unit no oil would come out, so I poked my finger up the hole, cleared the debris of what once had been a bearing and out came the oil, needle rollers and all.

A new bearing for £4 cured the trouble, but where first gear (the one in the bottom of the box) had ground the bearing bits to pieces the teeth were in a bit of a sorry state but they still soldiered on. Now, nine months and a couple of thousand miles later, I've put a 'new' 'box with overdrive in . . . it cost a whopping £12.

The car, as you can see from the photo, looks pretty good and I must say a few heads turn when she passes their way. She returns 26mpg on a run and round town about 22.5mpg so it is a pretty economical car for the size and weight. And, boy, do I enjoy driving her!

Don Pither in full flight in his rebuilt Alfa Romeo Guilietta Sprint at Curborough a couple of years ago. He calls it 'an outstanding example of automobile design'

The word Alfa is music to the ears of enthusiast Don Pither. He regretted selling his Guilietta Sprint so much that he bought it back, and started on a long rebuild

t was back in 1967 when the name Anonima ombardo Fabbrica Automobili first became music) my ears . . . I was, at last, the owner of a long-dmired car, an eight-year-old Alfa Romeo Giulietta print. Sadly its delightfully sensuous bodywork was lready suffering from deep seated rust, but since I ad exchanged it for a Mini which had spent a hard fe rallying, I was not too dismayed.

asting love affair

So started a love affair that has hardly abated to ıis day, despite a period of non-ownership lasting ome 18 months. This refers to the occasion when I ad to sell the car in order to get married due to its xtraordinary success as a courting car! Luckily I appened to spot an advert in *Motor Sport* a couple of ears later for a similar car when fortunes were ecovered only to discover that it was my own iiulietta still owned by the person to whom I had old it.

Little discussion was necessary before it was back ı the fold again despite even deeper seated rust and ss the original valve radio. Two more years of timulating motoring were enjoyed before both

Classic twin cam engine in 1300cc form on its home-made moveable test 'bench'. Pither would love a 2-litre power plant

This 1965 Alfa Romeo Giulia Sprint Speciale is one of 1399 that were built between 1963 and 1966. It is owned by Charles Weber of Atlanta, Geo

Handsome, purposeful, and *all* Alfa.

Alfa emblem, unchanged for over 50 years.

Rear e

1965 Alfa Romeo Giulia Sprint Speciale

by Paul Spruell
photos by Roy Query

Alfa Romeo has been in the business of producing exciting automobiles for some eighty years now. Each succeeding generation of motoring enthusiast has had the thrill of seeing yet another sensational series of Alfas become a part of automotive history. A few far-sighted individuals have selected their favorite car of an era and squirreled it away to savor on special occasions. Charles Weber's 1965 Giulia Sprint Speciale is one of these coveted selections.

The Sprint Speciale is a member of a family of Alfa Romeo cars produced between 1954 and 1966. After World War II, Alfa painfully resurrected itself from almost total destruction and production figures remained microscopic. The Milano firm decided that the most promising future lay in the mass production of moderately priced cars of modest size. Chief engineer Satta masterminded the marque's conversion to large-scale automobile production and a new line of 1290cc Giulietta cars was selected to be the first small-displacement Alfas ever built. In 1963 the Giulietta series of cars was fitted with a more powerful 1570cc engine, plus updated mechanical goodies, and renamed the Giulia series. Today, these cars have largely disappeared from the street and have begun the long process of becoming collector items. Because of its classic beauty and limited production, the Sprint Speciale reigns supreme as the queen of the Giulietta/Giulia series.

Late in 1953, Alfa officials, in desperation, commissioned Nuccio Bertone to design and produce a handful of two-seater coupes to satisfy the demands of enraged investors who for quite some time had been waiting for cars promised to them in return for their investment. In an incredibly short length of time, Bertone built a prototype from scratch and began cranking out finished Giulietta Sprint bodies. This superhuman effort staved off disaster for Alfa Romeo, and Bertone was subsequently awarded a contract to mass-produce his hastily-conceived creation.

In 1955, the first Giulietta series cars rolled out at the Portello factory near Milano. These cars were a mixture of sports cars and sedans that shared a common unitized chassis layout, raoe inspired running gear, and a snappy little 1290cc twin-cam engine.

The Bertone Giulietta Sprint was an overnight success, and the transformation of Carrozzeria Bertone into one of the giants among specialist coach-builders had begun. Nuccio Bertone, on several previous occasions, produced prototype bodies for the engineers at Alfa. These projects were for exotic 1900cc Gran Touring machines and all were unceremoniously shelved when Alfa, for economic reasons, elected to concentrate production on the Giulietta. With the success of the Sprint assured, Bertone was again able to turn his attention to producing a very special two-seater coupe for the company.

Long-standing cooperation between Alfa Romeo and Carrozzeria Bertone dates back to 1932 when Nuccio's father, Giovanni Bertone, began manufacturing bodies on unfinished rolling chassis delivered to his workshop by Alfa. During the difficult war years, the younger Bertone kept the family business going by selling his brilliantly-executed auto designs. After World War II, Nuccio spent much of his time racing sports cars. There he came to know everyone in the world of racing. In the competitive field of experimental racing machines, he acquired the knowledge that would serve him so well in the coming years as Italy's premier coachbuilder and designer. He acquired a hands-on experience in aerodynamics, chassis rigidity, weight distribution, and countless other skills that would eventually be embodied in the Sprint Speciale. Most important of all, in those years Nuccio Bertone single-handedly

shows Bertone's fine hand.

Only minor trim changes differentiate the Giulia Sprint Speciale from the Giulietta.

humbled the mighty Carlo Abarth's legions of Cisitalia race cars and won the respect of Rudolph Hruschka. A strong friendship developed between the ex-Porsche engineer and the young coachbuilder from Torino. Hruschka became the chief government consultant for Alfa and their friendship resulted in Bertone getting a crack at the Disco Volante project in 1953.

The principle feature of the Sprint Speciale is its aerodynamic design. To truly appreciate the flowing lines of this car, one must examine the five years of development that preceded Bertone's final statement of aerodynamic design in the Sprint Speciale. Bertone was never allowed to finish his work on the Disco Volante. These exciting Alfa Romeo C-52 racing sports cars were abandoned, to the eternal loss of the sport, but the knowledge gained by Bertone was soon put to use. His brief participation on this project had thrust him to the forefront of aerodynamic automotive design.

The smooth flowing lines of the early fifties' cars were intended to improve performance by reducing the horsepower required just to move a vehicle against a wall of air. As styling changes smoothed out the old-fashioned radiator grille work and reduced frontal area, the air stream began to follow the continuous curves of the body envelope. The resulting aerodynamic lift caused havoc with rear end stability at high speeds. Bertone's idea was to cleverly shape the car so that the air flow would exert its pressure toward the tail of the car, and help keep the rear tires solidly on the ground.

In 1953, Alfa commissioned Carrozzeria Bertone to explore performance gains that might be realized through streamlining. Nuccio Bertone stunned the auto design world when he revealed the first of his BAT (Berlina Aerodinamica Tecnica) series prototype cars built on Alfa Romeo standard 1900 SS chassis. These cars had swooping bodywork and tail fins that would have delighted the Caped Crusader himself. The experience gained on the BAT prototypes provided the final exercise needed for Bertone to create the Sprint Speciale. Many basic elements of the BAT-9, less tailfins of course, are quite evident in the Sprint Speciale. The aborted Disco Volante, the BAT prototype, and finally the crash program to produce Giulietta Sprints had provided the homework that gave the Sprint Speciale the look of a finely-polished product with no traces of prototype grotesqueness. The combined genius of Alfa Romeo and Carrozzeria Bertone had at last produced a classic to enjoy for all time. Aerodynamic design would soon degenerate to more functional wedge and Kamm-inspired configurations. Government safety regulations created further restraints. Never again would automobiles be produced in the grand tradition of pure automotive aesthetics that motivated the Sprint Speciale.

Bertone's long-awaited aerodynamic Alfa coupe was finally introduced at the Torino auto show in March 1957. That year, Bertone produced the first five Giulietta Sprint Speciales and introduced the motoring world to the most beautiful Alfa since the 1750 Zagato of the late twenties and early thirties.

For Bertone's new Giulia Sprint Speciale, Alfa Romeo went all-out and furnished state-of-the-art mechanical bits and pieces from its engineering department. The lid was still clamped down on the racing desires of the former giant of the world's racing circuits and with financial prosperity at hand, the die-hard racers at Portello were eagerly awaiting just such a project as the Sprint Speciale. The Giulietta Sprint Zagato project had whetted their appetite for a hot street machine. Now they put themselves to the task of developing the engine for the Speciale that would

The view from the driver's seat; gearbox is a five-speed.

Four-cylinder engine displaced 1570cc, developed 129hp.

later be the beginning of their re-entry into racing with the GTZ's, GTA's and street hot-rod Giulia T.I. Super. The finished product was Alfa's most powerful four-cylinder engine to date. This enlarged version of the basic Giulietta 1290cc engine featured a much longer stroke on the crankshaft and a modest 4mm bore enlargement to increase the capacity to 1570cc. Valve sizes were larger, and new Sprint Speciale camshafts and exhaust headers were designed to increase mid-range torque lacking in the Giulietta engine. The added capacity and a higher 9.7 to 1 compression ratio turned out 112bhp per 6500rpm. Our American horsepower rating would put this figure at 129(SAE). By the time the air inlet box and headers were enlarged, the new SS engine would no longer fit into the chassis. To solve this, Nuccio Bertone himself revised the chassis to accept the larger Giulia 1600cc engine.

Many other things were done to upgrade the performance. The gearbox was already a five-speed, so the ratios and final drive were slightly changed. Girling front disc brakes replaced the three-shoe drum types in the front and large-finned aluminum drums were added to the rear. Externally, the only changes were the addition of side winking lights, and the new designation on the side and tail.

The Giulia Sprint Speciale first appeared in 1963, and a total of 1399 were produced before the last one was completed in 1966.

Charles Weber literally grew up with sports cars. As a lad in Florida, his father had sparked his interest with visits to Daytona and Sebring to see exotic foreign machinery first-hand. Seeing the racing Alfa Zagatos of the early sixties had left Charles with an unfulfilled desire to own an Alfa, and in 1979 he saw his first Sprint Speciale in the flesh at a local Alfa dealer. He was overcome with desire that only a true Alfisti can understand. The next day he became the owner of one of the best-preserved Sprint Speciales in existence.

Weber's Sprint Speciale is one of 103 units produced in the 1965 calendar year. The following year it was delivered to its first owner in Zanesville, Ohio, who kept it only a year and sold it because the locks froze in cold weather. The name of the second private owner and the next three years of its existence are not recorded. In 1969, Anatoly Arutunoff found it languishing on a used car lot in Nashville, Tennessee, and purchased it. "Toly" was attending Vanderbilt University and had long since been bitten by the sports car racing bug. He decided to enter a car in the Daytona 24-hour race and his exotic Alfa Romeo seemed to be the way to go. From this point on, the story of the Weber Sprint Speciale becomes reminiscent of the fairy tale where Prince Charming (Toly) discovers Sleeping Beauty, revives the fair lady and rescues her from an ignoble fate.

Toly enlisted the services of Joe Marina and Bill Pryor to co-drive and supervise the preparation for this great endurance race. Joe was experienced with Alfas, since he was at the time racng a G-Production Alfa Spider and working as service manager for the Nashville Alfa dealer, Madison Smith. Race preparation, in those days, was a relatively simple matter. The sanctioning body for the Daytona 24-Hour, the Federation Internationale de l'Automobile (FIA), allowed an absolute minimum of modifications on production category automobiles. This meant that the SS could run in practically original condition without undergoing serious modifications that would later render it useless as a collector's car. Under the FIA rules, only factory-homologated shock absorbers, springs, and swaybars could be substituted on the car. Also allowed was a factory side pipe exhaust system. The addition of the roll bar, competition seat belts, and fire extinguisher completed the safety requirements. The next stop was the body shop, where the minor dents were removed, the white paint was touched-up, and a set of green fuzzy contact paper numbers affixed to the hood, doors, and decklid. The lonely little Alfa had been rescued, revived, and was ready for the adventure of a lifetime.

It was entered in the under two-liter Gran Touring class. This class was the exclusive domain of the Porsche 911 and few would have given the little four-cylinder 1600cc Alfa a chance against these more powerful 2000cc, six-cylinder machines. The method behind Toly's madness was the aerodynamic shape that would give the Sprint Speciale better gas mileage and the stock 22-gallon fuel capacity that would necessitate fewer pit stops than the Porsches. He also figured that Alfa engineering excellence was as good or better than the Porsche and would therefore place them higher because of attrition when the gruelling twenty-four hour race was done.

The story of the race is now history, and the record shows that the little Sprint Speciale did indeed outdistance and outlast the bigger and more elaborately-prepared racing machines. Sleeping Beauty finished a respectable third in its class and twenty-fifth overall out of the fifty-nine cars that started the race. It should be noted that it finished only one place down from the Ferrari GTB co-driven by Sam Posey.

After the race, the car was sold to Bill Hutton of Clarksville, Tennessee, who undertook the complete mechanical rebuild needed to put the Alfa back on the street for daily transportation. Subsequent owner Donald E. Hughes of Nashville painstakingly relocated some hard-to-find bits and pieces, removed and misplaced during the rush to prepare the car for Daytona.

For several years, the SS was lost in the hinterlands of Alabama and then re-surfaced in the hands of an Atlanta, Georgia, Alfa dealer. Since that day in 1979, when Weber claimed the car for his own, it has taken a well-deserved stand against the ravages of time. When he goes to the SCCA sports car races, his SS draws more attention in the paddock than the race cars. The car has been an instant winner in every concours d'elegance it has entered. The greatest single thrill in owning it came last April when Weber entered the SS in the Walter Mitty Challenge at Road Atlanta. For almost two glorious hours that weekend, he was able to put his foot to the floor and let the Alfa negotiate the eleven wicked turns and relive its moments of glory from the past.

Now that his car has proven itself still raceworthy after all these years, Weber intends to include more vintage-type speed events in his plans. It is refreshing to know that this spring Charles Weber will be ready and waiting to roll out his Alfa Sprint Speciale and allow less fortunate admirers another chance to see it in action.

Paul Spruell, a long-time Alfa Romeo enthusiast, has owned and raced Alfas since 1965. He captured the SCCA National Championship for G-Production in 1969, driving a Giulietta Spider. The same year, he was awarded the President's Cup as the most outstanding amateur driver in the nation.

Today, he is still campaigning an Alfa in national SCCA races and owns Paul Spruell Alfa, Inc., in Atlanta. He also serves as a National Director of the Alfa Romeo Owners Club which can be contacted at 2304 San Pasqual Valley Road, Escondido, California 92027.

LOVE STORY

History in the making – three generations of Alfa 1.3 saloons. Left, the 1955-1962 Giulietta Berlina, centre, the 1970-1973 Giulia Super and, right, the 'new' Giulietta

Jon Dooley looks back at Alfa Romeo's charismatic Giulietta family, which in its day set standards few manufacturers could equal

It is refreshing to know that Alfa's latest Giulietta has much in common with its forbear, at least in saloon form, both in concept and in some actual components. Both old and new are solutions in their time to the problem of conveying four people far and fast within a compact exterior. Both offer considerable rewards to the keen driver with a quality of feel and communication way above their price, and a level of overall safety outstanding in their time.

The original Giulietta was a rarity in the UK. Offered first as the two-door Sprint coupé from 1954, then the four-door Berlina (saloon) and two seat Spider (convertible) from 1955, the Giulietta was an expensive luxury by the time British duties and taxes had been added – and to add to the grief it was only available with left-hand-drive. Almost Aston Martin money for a 1300cc car with no right hand drive, ridiculous! With only some 40 to 80 cars being sold each year during the late fifties, there was no need for an owners club – a trip to Thomson and Taylor (Brooklands) Ltd at Cobham, the sole importers, sufficed with either a chat to Ken Taylor or Jack Fifer and if you were really lucky a drive round the block in the latest variant with test driver Frank Cox. And George in the stores knew not only every part back to the vintage days but your car as well. As an alternative, if you happened to be in town shopping in Bond Street, S. Morris and Co would sell you an Alfa from Conduit Street having conducted you rapidly, and sideways, around one of the Royal Parks just to convince you.

Crossing the channel was quite an experience. On the battleroads of Europe the stately British trogmobile, occasionally exercised to 65mph on the Kingston bypass, got repeatedly zapped by the dreaded foreigners who all appeared to have those incredibly sporting Simca Montlhérys, Fiat 1100s and big Citroëns. And there was always the appalling hazard of being hustled by one of those horribly quick, downhill Deux Chevauxs. A kingdom for a Jaguar! In this company an Alfa Giulietta, even in four-door saloon trim, was like holding the Ace of Spades. Quite simply, for a run across Europe, down the *Route Nationale,* (yes the dangerous one), over an Alp or two and then an *Autostrada* trip to the sun, there was nothing with anything like its practicality to beat it. Sorry, the Porsche 356 just needed too much concentration. And when you got there, the suspicion that had grown through France from seeing so many that you were not alone in this opinion was confirmed, but in Italy, the Giulietta was just another ordinary car.

Fast small car phenomenon

Looking back on those years I now see that we were experiencing a first glimpse of the fast small car phenomenon with which we are now surrounded, be it a Golf GTI, Escort XR3, Astra GTE or of course Alfa's own Sud TI, as well as the first glimpse of the elegant, small and practical, sports coupé, in the Giulietta Sprint. That trail blazing products should have emerged from a small company that only four years previously had won two Grand Prix World Championships on the trot, was all the more remarkable.

Production cars with light alloy twin overhead camshaft engines were a real rarity in those days, yet Alfa proceeded to offer a light van with one fitted. The initial, single carburettor variants, were followed in 1956 by twin carburettor Veloce versions of the Sprint and Spider placing twin double choke Webers within reach of the man in the street. Soon the journals were marvelling at this little gem which could be wound up to 7700rpm in the gears while giving 90bhp (DIN) from its 1300cc, and this in 1956. 0-100 test figures of under half a minute were common but the best was from an Italian test at Monza where a good Sprint Veloce pulled a full 125mph around the banking and reached the ton in 26 secs. Those early Sprint Veloces were quite exceptional with lots of lightweight trim, alloy doors, bonnet, boot and bumpers, and perspex side and rear windows. Built until early 1958 they were the 1300 GT competition car to have; drivers like Richard Shepherd-Barron, who were served up with the later heavyweight version, were to find it not the same beast at all for racing. Naturally enough the really keen types spoilt it by taking their shiny new Sprint Veloces to Zagato, who hacked off the Bertone shell and fitted a lightweight, compact alloy shell of their own, causing all sorts of confusion for historians with a non-existent model, the Sprint Veloce Zagato, but all perfectly correct under the rules at that time.

At the Frankfurt Show in 1957, a TI version of the four door saloon was announced, essentially with an engine in Sprint form, and this soon became a favourite. Some startling competition appearances, especially in France, really put the saloon on the map. The rules permitted free carburation at that time and a trip to Conrero in Turin soon put a full-house, Veloce spec 115bhp engine into the lowly-looking saloon shell with surprising results. Bernard Consten, who won the Tour de France touring Category in a TI, is said to have embarrassed Gawaine Baillie's 3.4 Jaguar by chasing it the full length of the Mulsanne straight in the little Giulietta.

and then beat him. According to Consten the car would pull 192kph and would run a 7800 rev limit with total reliability. Consten went on to win the German Rally, Marathon de la Route and Coupe des Alpes in 1958 with a Sprint Veloce Zagato. Altogether Alfa had discovered that with a product ahead of its time, it sold itself, and eager sportsmen all over Europe had to have it to win. No fewer than 54 Giuliettas turned up for the 1957 Mille Miglia and similar quantities would appear for events like the Monte.

To keep the impetus going, Alfa showed a Giulietta Sprint Speciale prototype at the '57 Turin Show. Incredibly long, low and sleek, the body by Bertone owed much to that coachbuilder's series of BAT aerodynamic studies on 1900 chassis, as well as his experience with Abarth record breakers. Coachwork was alloy and, looking at the car, it is not hard to appreciate that winning at Monza was the important objective. With Cd figures all the rage these days, it is interesting to note the Speciale's Cd of better than 0.30 all those years ago, with a small frontal area as well. As a competition car though, the SS project was to be embarrassed by the more practical lightweight Zagato solution, so that although Alfa eventually presented the SS at Monza in July 1959 as a competition car, its admitted 110kg weight penalty relative to the Zagato made it a rather half hearted affair.

Nevertheless 1959 was an important year for Alfa and the Giulietta and these Speciales were to prove effective marketing flagships. Briefly through 1958 and 1959, Alfa had taken the Giulietta through a round of production improvements and cost reduction exercises. During 1958 the early gearboxes had given way to more robust and delightful-to-use affairs with Porsche ring type synchromesh, while the bodies across the range had been subtly altered to make them more susceptible to production techniques. By 1959 the engines too had been redesigned to take advantage of a switch to diecasting from sandcasting, and to prepare the way to allow subsequent enlargement to 1600cc for the forthcoming Giulia which was to phase out the Giulietta from 1962. Production volumes were geared up through 1959 and prices were lowered substantially, much to the chagrin of those who had bought early. Giuliettas prior to the change were 750 series, those after were 101 series, though with the "smudged" changes there were some overlaps.

80 Sprint Speciales at Monza

The 80 odd SSs that lined up at Monza for the launch were in fact 750 series, as far as the engines were concerned, coded Tipo 750 SS, and although the suspension was to 101 spec, the bodies were seamless, gutterless and hand-built. Like the early lightweight Sprint Veloces, they had alloy doors, bonnet and boot, alloy grilles, trim and rear bumper (no front) and perspex rear window. Those subsequent to this early series had the taller 101 series engine, raised headlights to meet US regulations, gutters and front bumpers. As time went on, most of the lightweight trim was deleted from production in favour of steel items.

DNW 4 which you see featured here was the car that captivated me at Earls Court in 1959 with its fabulous dramatic lines and which Gregor Grant road tested for *Autosport* in 1960. To me, it represented the pinnacle of the classic Giulietta era, and if it does not quite have the competition credentials of the Zagatos, it is a far more practical grand tourer and cross Europe express. She was actually raced briefly by Peter Bolton in the *Autosport* Cup series in 1960 but, knowing Lotus Elites and the emphasis on British circuits of light weight, it is not surprising she showed poorly.

It was all the more remarkable some ten years later to be presented with the opportunity to buy the very same car I had fancied all those years before and I still feel a debt of gratitude to the former owner, Peter Piper, so that of all the Alfas I have she would be the last to go. Now 24 years young, she really needs a spot of yoga or aerobics that could be provided by the incentive of another trip around the circuits. But alas, as time goes on, you begin to know too much . . .

Driving the SS you receive a number of reminders of the progress made in the past 20 years. Improvements in tyre and brake technology have been significant. The brakes when warm do work pretty well and the handling, on the tyres used, is beautifully neutral but would probably only be ruined by more grip.

Perhaps the most outstanding feature of the SS, and any Giulietta come to that, is the steering quality. It is quick at only 2½ turns lock to lock, communicates well, and allows most changes of direction on the road to be made without conscious effort. That, combined with the classical Alfa five-speed box and total lack of wind noise at speed, makes her a restful partner on a journey with the bonus of being able to enjoy her good looks on arrival. Good looker, good performer, this old girl has what it takes.

Giulietta Berlina – a wolf in sheep's clothing

Giulietta Sprint Veloce – truly, sweetness itself

Giulietta Spider – Alfa's delectable open-top

Giulietta Sprint Zagato – short and stubby

Giulietta SS – Jon Dooley at Donington's chicane

DNW4 as she is today. Few would disagree that this beautiful car represents the pinnacle of the classic Giulietta era

Bob Swann in the Giulietta Spider which he has progressively restored since buying it 18 years ago

ROMEO AND GIULIETTA

Alfa Romeos seem to arouse more passion in their owners than most cars, and Bob Swann is no exception in his love for his Giulietta Spider. Mike Walsh discovers why

Just to sit in a classic Alfa Romeo conjures up the company's motoring heritage — that great lifelong, sporting pedigree. Emblazoned before you on the steering boss is the green serpent and red cross of Milan, immediately recalling heroes of the Alfa Romeo's illustrious racing career — names like Giuseppe Campari, Achille Varzi, the 'flying Mantuan' Tazio Nuvolari, and Juan Manuel Fangio.

This pretty Giulietta Spider should have had the same effect, but instead of elation one feels a little hesitant and nervous. A previous owner dominates the car's aura, a character closely involved with a 'great' crime of the sixties, which British Rail would prefer to forget. He had a discerning taste, or perhaps a weakness, for 'choice motors' and even dabbled in motor racing when sufficiently flush. Rumour has it that Mr X had strong affections for this Italian beauty, and it is perhaps telling that its present owner, Bob Swann, had a great deal of trouble acquiring the log book when he purchased the Spider from a Greenwich scrapyard back in 1967. Even to this day Bob jokes of 'de boys' recognising a certain 'guvnor's moter', and it is hardly surprising that the Spider's distinctive metallic blue was quickly resprayed white just as a precautionary measure.

In fact, little else is known of 147BPK's history before this. It is chassis no 001495 00425, which indicates the 1956 production run, and is reckoned to be the first Giulietta to be imported to Britain by Thomson & Taylor of Chobham. Ken Taylor and Jack Fifer of that famous Alfa Romeo agent must have had trouble convincing buyers of this Italian jewel's merits. In 1957 the Giulietta Spider had a total list price of £2349, a sum which could have bought you both an MGA *and* a basic Jaguar 2.4 Mk I. Such was the price of Italian flair should your personal style require it, but you certainly were guaranteed exclusivity . . . until, of course, you ventured over the Alps to the homeland where even Alfa vans were graced with that light alloy twin overhead camshaft engine.

Great refinement

It was the lucrative American market, and US agent Max Hoffman in particular, that we must ultimately thank for such an automotive gem. The new Giulietta Spider set new standards of refinement matched to excellent performance, a package unheard of on the mass sports car market. British designers had insisted that sports cars must be draughty, uncomfortable and impractical to justify such a title. Here at last was an open car with wind-up windows, a spacious cabin, and a simple hood that didn't take till the end of the storm to put up. All this in a car with superb handling and performance. Only British taxes and import duties prevented greater sales among loyal British sports car drivers.

Records show that 6381 standard 750D Spiders were built and a further 1253 Spider Veloces until the Duetto sister replaced it in the mid-sixties. Production was a drop in the ocean compared to Abingdon's MGA at 58,750, but the Twin Cam was more comparable at 2111.

The later Spider Veloce was the most desirable of the open Giuliettas, with twin Webers and a floor-mounted change as opposed to the column selector of the early standard models. It produced a healthy 90bhp, took 11secs to race to 60mph and, if you could tolerate the internal noise with the hood up, would produce 110mph. And all that from a mere 1300cc!

Although the performance was very tempting, it was Pininfarina's styling that made the Giulietta so desirable. After the near perfection of the Sprint, Bertone's proposals for the Spider were a little too adventurous for the Alfa management, his prototype hinting at the Duetto shape 10 years later. Battista Farina and his son Sergio produced the perfect solution, an ideal combination of refined elegance and assertive sporty cheek with lines owed a little to the pretty Lancia Aurelia Spider. Only the bumpers caused a minor styling headache, and thankfully the 'Cadillac' influence was dropped for the production model.

Like all great works of elegance, it is simplicity that is the key to perfect form. The Giulietta Spider comes close to the ideal, particularly when one looks at more recent solutions to the same problem such as the TR7 drop-head and the new Scimitar SS. From the 'tin toy' pressed grille, across the modest bonnet to the exacting rake of the shapely pillarless screen, over the harmonious hard-top to the subtle, perfect tail, it was love at first sight for any Romeo seduced by open cars. Only the Italians are capable of such automotive temptation. Somehow the original is always the most desirable, and the later Giulia was less harmonious with its fake bonnet power bulge.

A Spider in competition was very much a rarity, although many saw extensive racing in American

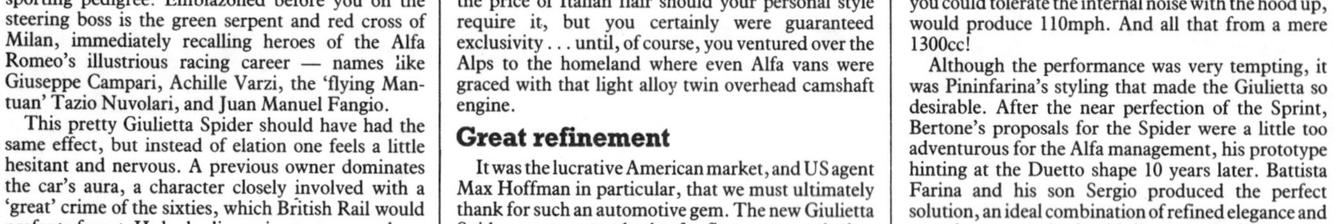

The purity of the Giulietta Spider's shape is beautifully displayed in these original factory drawings of the 750 series car, with measurements for the uninitiated

Thomson & Taylor 1959 line-up includes five Giulietta Sprints, one Spider, a 2600 Spider, an SS and a Berlina

national championships, leading the way even as late as the seventies. Denis Jenkinson led many of the competitors unofficially during the 1958 Coupe des Alpes at the helm of the Spider camera car. The model's competition pedigree was earned by Sprint Veloces and Zagatos, while the Spider was more at home on a Hollywood movie set in California or parked on the quay in St Tropez. It was just too pretty to race.

Bob Swann has always had a taste for carefree open top motoring, an essential ingredient of his life-style back in the swinging sixties. A previous Austin Healey 100/4 (arguably the prettiest British sports car) of the fifties had seen extensive competition in sprints and hillclimbs, not to mention European touring holidays through the Alps. An aviation engineer by profession, he is naturally very discerning about the mechanical quality of his cars. After several years stationed in the Middle East with the RAF with little but sand and motoring magazine classified columns to contemplate, he returned to 'Blighty' in 1967 with a yearning for the model of the moment — not the seductive curves of a blonde or brunette (he was already hooked on his present wife), but a Lotus Elan. His financial assets and romantic commitment, however, would not permit such an indulgence.

He continued to look for something with the same attributes of performance and style. He had always been impressed by the quality of design and engineering of Alfa Romeos, not to mention their charisma, so when a friend tipped him off about the sale of a Giulietta Spider with many performance modifications he took more than a passing interest. On viewing the car in Greenwich, he took little convincing to part with £180. The car was fitted with a four-branch exhaust manifold, high lift cams, twin Webers, a Veloce 17 gallon tank, the later four-speed floor change 'box and ZF back axle and steering box. In fact, someone had gone to great lengths and expense to convert the car to Veloce specification, indicating that competitive plans may have been cut short by an unexpected change of address.

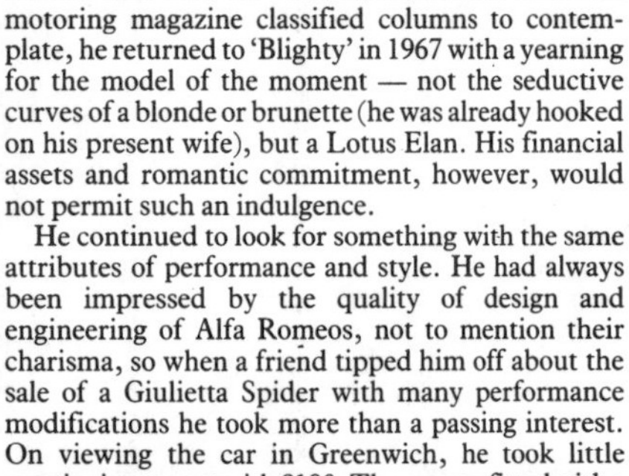

Simple cockpit with lhd and conversion to floor change

Bob and his wife became very attached to the car, and its fond romantic associations help to explain his reluctance to part with it. His wife, in fact, has continually deterred him from selling. The car became such a part of their life that his father's wedding present was not useful household gifts but a new hood for the car, while his wife's first salary went towards recovering the seats! Certainly the Swanns had the right priorities in their wedded bliss.

Spares found abroad

Bob was stationed abroad many times during the sixties, so the Spider was stored with his tolerant parents in Bromley. While with the RAF in Belgium, he acquired a rare and desirable Giulietta Sprint Veloce in Brussels. The intention was to break it for spares — it was later sold still complete, which he now deeply regrets. Relocation in Sardinia gave ample opportunity to search Italian scrapyards for vital replacement parts, then so expensive back in the homeland. All were shipped back courtesy of the RAF, while on further bombing raids with Buccaneers he discovered the spacious spares carrying facility of drop tanks. Even 'Red Flag' exercises in the Nevada desert in 1977 produced exciting discoveries in breakers yards in Las Vegas! By now Bob was involved with Tony Stevens, the leading light in the Giulietta Register who became a crucial link for spares for club members.

All through his ownership, the body was regularly protected with a mixture of engine oil and petrol to deter the horrors of Italian rust, and generally the car has stood up well. The heavy gauge of the steel panels compared with its stablemate, a modern Giulietta saloon, is quite a revelation: the saloon's bodywork has become ventilated and lightened over the years, although it must be said its present mileage is nearly twice that of the Spider.

The restoration of the Spider is very much a progressive project stretching over the last 10 years. Initially an incident on the M4 prompted an engine rebuild. While travelling at about 80mph, the rubber doughnut from the first half of the propshaft disintegrated, catapulting it dramatically through the floorpan, wrecking the gearbox and fracturing the crankcase. The car locked up solid and neatly spun in front of an on-coming truck. Fortunately contact was avoided and the Spider was towed back home. With the help of that devoted Alfisti, Dave Butcher, now of EB Spares, a new casing was found, and the salvageable mechanics were transferred. The gearbox was stripped and carefully rebuilt.

Since leaving the RAF and joining Hants and Sussex Aviation, Bob has settled in England and has more time to devote to his 'toy', as his wife affectionately dubs the Spider. With the mechanics now well sorted, he decided to strip the body down to bare metal. All the areas of localised rust were treated

Bertone's prototype shape for Giulietta Spider was rejected

and leaded up, and the body carefully prepared for a new coat of deep Alfa 'Monza' red. The task of respraying was entrusted to his brother, and the result can clearly be seen on the colour page. The next project is to re-chrome all the brightwork.

The Giulietta is a delight to work on. Everything is simply yet precisely engineered, making for easy access and maintenance. One could imagine nightmare problems locating the simplest spares for a fifties Italian classic, but in reality nothing could be further from the truth. The brakes and clutch are Girling, and most of the electrics (including the starter and distributor) are Lucas. It has generally been a question of careful cross-checking of catalogues to find parts. The main problem is those infuriating window and trim seals. Inside, the cockpit carpets and panelling originate from simple patterns which could easily be copied by a loyal wife, but so far the pretty works hard-top has generally protected the interior from the rigours of the English climate.

Bob's attitude of progressive restoration has meant that the car has a strongly original character, and over-restoration would destroy any memories even worn details rekindle. The car has a continuous development, particularly with its tuning modifications, but somehow they are all in keeping with its character. Like old houses, antique furniture and favourite leather jackets, they lose much by a complete facelift or total restoration. As long as the car is mechanically perfect, performs as it should, looks smart and tidy, yet maintains that used aura, it will always have more charm than 'pristine' untouchable show cars. Bob's Giulietta has become part of the family, is regarded with great affection, and will be enjoyed for many summers to come.

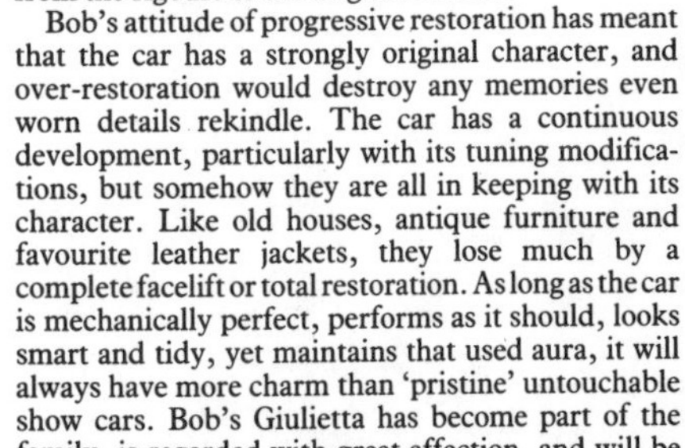

Twin-cam engine produces around 90bhp from just 1300cc

SPECIALE

Patrick O'Brien traces the history of the Alfa Giulietta Sprint Speciale and tells us how he restored one

ANNOUNCED at the 1957 Turin Motor Show, and based on the Giulietta Sprint structure and running gear, Alfa Romeo's Speciale was the culmination of five years' intense aerodynamic studies by Bertone's chief designer Franco Scaglione. Made in small numbers, the car was really a 'special', providing performance out of all proportion to its 1.3-litre engine capacity — a 125mph top speed. Built entirely at the Bertone works rather than by Alfa Romeo Milano, the Speciale was expensive and very well finished. Consequently production figures

were never great with only 742 cars made in the peak year, 1961. Nuccio Bertone professed himself "profoundly satisfied with the car from every standpoint. Its aerodynamic qualities are truly excellent. The structural integration of the design is such that there is not a body being made today that has greater torsional rigidity."

In all, 1,366 Giulietta Sprint Speciales were made between 1959 and 1962. There were two series, the first 100 units becoming known as the 'low-nose' model. They were devoid of bumpers and had 'plexiglas' side and rear windows, and weighed 1,968lb. In 1960 Alfa decided to expand production and revised the Speciale. It now featured curvaceous bumpers, a longer boot lid aperture and, to comply with American requirements, the headlights were raised 2.36ins, as was the nose. Extra headroom was facilitated by lifting the roofline by 1.5ins. This second series model was consequently 143lb heavier.

A grand touring car *par excellence*, contemporary reports praised the Speciale highly. *Autosport* reckoned it was "a real jewel of a motor car", while *Motor Sport* said "The car is a revelation. . ."

The Speciale's forte was long distance touring. At 100mph, the absence of wind noise was, and still is, uncanny and restful. Strangely, engine noise too, is quieter than in the standard Sprint. Particularly advantageous are the five-speed gearbox and the 17.5-gallon fuel tank. With fuel consumption between 28 and 35mpg, this provides a range of 500-600 miles! The boot is virtually consumed by spare wheel, battery and huge tank, but luggage accommodation on the carpeted shelf behind the seats is adequate. The all-round visibility is superb, with so much glass and such slim roof pillars. Build quality is excellent. Made before the age of plastics, everything is of steel or aluminium, welded or fixed with fine-thread metric bolts and machine screws. According to *Road and Track* in 1961: "As a road car, the Speciale is ideal: fast, comfortable, safe and economical."

As this Alfa Romeo was so rare and little-known, it helps to put it in its contemporary market context by comparing it with other cars. Because of the flowing style and proportions, the Speciale looks very long. Yet it is no longer than a VW Karmann Ghia or Renault Floride. At 49ins to the top of its domed roof, it is as low as the Austin-Healey 100/6. The kerb weight of 2,128lb is within a few pounds of the Triumph TR3, MGA, Austin-Healey 100/4 and its principal rival, the Porsche 356.

Despite its Mini Cooper-sized engine the Speciale's 200km/h puts it ahead of the contemporary 4-litre Jensen 541R, 3.4-litre Jaguar XK150, 3-litre Aston Martin DBIII, 1.5-litre Porsche Carrera, 2-litre Aceca Bristol and 5.2-litre Ford Thunderbird. In 1957 only the Ferraris, Mercedes 300SL, Pegaso, Facel Vega, fuel-injected Corvette and Maserati 3500 GT were faster. In acceleration up to 62mph, the Speciale's 11 seconds is quicker than the Austin-Healey 100/6, Porsche 1600, Allard Palm Beach 2.6, BMW 503, Jensen 541, Lancia Aurelia GT and Peerless GT — all with much larger engines. To 100mph it will beat the Porsche, Healey, Aceca Bristol, Lancia Aurelia and Jensen.

Aerodynamics

No wind tunnel figures are available for the Speciale, but by comparing frontal area, horsepower, gear ratios and maximum engine revolutions with the best of the opposition, the 1972 Lotus Europa Special and Lotus Elan +2, whose

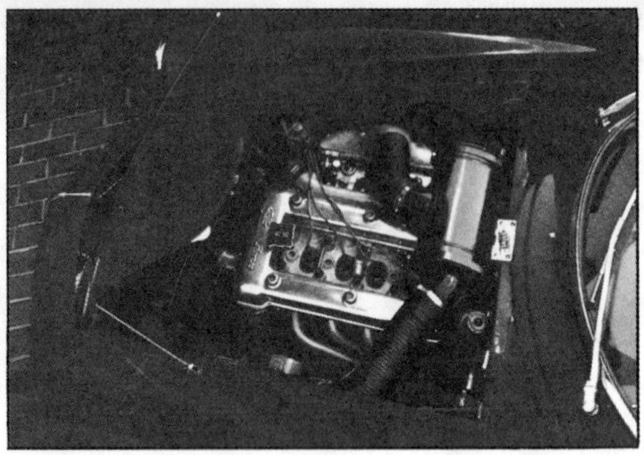

Above, fully stripped out and waiting for paint

Far left, the home-restored interior looks excellent

Left, classic Alfa twin-cam in 1300cc form

values are 0.29 and 0.30 respectively, one can extrapolate that the value for the Speciale must lie between 0.27 and 0.28. To put this remarkable figure in perspective, the Jaguar E-type and Porsche 928 have figures of 0.44, the Ferrari 308, 0.40, Mazda RX7, 0.38 and Porsche 924 Turbo, 0.34. The Speciale's 125mph from only 1,290cc — representing 96mph per litre of engine capacity — must rank it as one of the most efficient road cars ever made.

In its major overseas markets — the USA and UK — the Speciale was expensive. Its £3,100 (or so) price tag put it in the same bracket as the Porsche Carrera, Lancias Aurelia and Flaminia, Mercedes 190 SL, Jensen 541R, Aston Martin DB III and Corvette. In South Africa at R3,300 it compared with the Porsche 356 and Daimler SP 250, but was over 50% more than the Triumph TR3, MGA or Sunbeam Alpine and over 30% up on the Austin-Healey 3000. No wonder then that only 1,366 rolled off Bertone's line, making the Speciale as rare as the Mercedes 300 SL Gullwing, the Aston Martin DB4 and Facel Vega HK500/2. By comparison, during this period its most prolific competitors were VW Karmann Ghia at 200,000, MGA 92,000, Triumph TR3 75,000, Sunbeam Alpine 70,000, and 356A Porsche at 21,000.

The exceptional performance of the Speciale is really highlighted when realising that in its class — 2-seater coupés up to 2-litres, very few competitors have surpassed its top speed even now — 30 years later!

Local history

Of the handful of Speciales which came to South Africa and Rhodesia between 1959 and 1965 only the first two became well known. Imported by the Pieterse brothers of Continental Motors in 1959, the first (chassis no 78) cleaned up the sports car races at its debut at Grand Central. The field included three-litre Austin-Healeys, two-litre Triumphs, Porsches and Twin Cam MGAs. Greatest success was achieved by their second Speciale, one of only four Allegerita models with all aluminium bodywork (chassis no 5) — outright victories in the 1960 Maritzburg Six Hour Race in November, and in the Lourenço Marques Three Hour Race. During the Rand Daily Mail Nine Hour Race at Grand Central, the car was running third overall — all 1290cc of it — to a Porsche Spyder and a Lotus XV when a stone shattered the windscreen. After removing the rear screen, the intrepid drivers battled on, dropping to fifth at the end!

The car illustrated (chassis no 491) is a 'second series' with bumpers and raised headlights. It was produced in April 1961, and imported by Continental Motors for Philo Pieterse. Nick Kingwill later took it over, selling to Koos Roets in the mid-sixties. It then passed through four more hands before Johnny Lauryssen of Venturi Motors, Pretoria, acquired it in 1974. Dawie Breed of Alfa Boksburg took over in 1978. It was in Dawie's showroom that I found the car in December 1980 — the first Speciale I'd seen since Ernest Pieterse's at Grand Central twenty years ago!

Alfa fan

My first Alfa Romeo was purchased in September 1979, a four-year-old one-owner, immaculate 1600 GT Junior with only 35,000km on the clock. Very impressed with the dynamics and engineering, and having just joined the Alfa Romeo Club of South Africa, I was casually asking how many Sprints, Spiders and TIs had survived, and if they knew of any Speciales? One languishing in a Boksburg showroom was casually mentioned. . . It had been looked at by some club members but they steered clear from the car which had been badly and incorrectly repaired after a front end crash, lacked a grille and numerous other items and had bonnet scoops, two radio aerials and disc brakes added. Having been impressed by the car's sheer beauty in 1960, I was hooked. Within a week my brother and I drove the faded, patchy, tomato-red car home — slowly and listing to starboard. It was December 7, 1980: price paid £1,750. (All prices quoted here were for the period 1980-1984 when S.A. R2.00 = UK £1.00.)

So much was missing or incorrect on my car that the first six months were spent examining it and looking at Sprints, Spiders and the only other complete Speciale I knew of — in Pretoria, 40 miles away.

With so many owners, the inevitable deterioration had set in. Sills and door skins were badly corroded; the seats, upholstery and carpets were all incorrect; disc brakes had been fitted at the front and narrow Berlina drums at the rear. Worst damage of all was to the nose, which had suffered a front end crash. The grille was missing altogether, the once-shapely front bumper damaged and so badly repaired that it was scrap; the bonnet was too flat, fitted 1½ins too low and equipped with forward-facing scoops. The nose panels above and below the mis-shapen grille opening were built up with filler up to ½in thick. The original engine was replaced with a later Giulia 1300 unit. The Veloce-type air cleaner, aluminium fan, the ventilation ducts, one chassis cross-member, electric fuel pump and filter, hooters, toggle switches, spare wheel, tools, jack, rear view mirror, visors, badges and script, all were gone.

The rebuild

The daunting task of searching for parts was made interesting and enjoyable by the generosity and enthusiasm of the many Alfa Romeo Club members and enthusiasts throughout the country. Leslie Hegedus of the Sprint Speciale Register in Toronto was a great help with his vast knowledge of the model. Most critical were the grille shield and side bars which were non-existent on my car.

Within two weeks I had located a grille after telephoning many Alfa agencies. Unbelievably a Pretoria dealer had imported one in 1962, but the owner had never collected! Price charged £2.50! A side bar was given by the owner of a superb white Giulia SS, also in Pretoria. This was later copied by an ex-Loganda sheet metal worker, who refused payment. Body parts and trim for this rare model are virtually non-existent. Four approaches to Alfa Romeo Milano and Arese were unsuccessful, but a fifth, obviously landing on the right desk, yielded the last bonnet skin and the upper and lower nose panels! They kindly let me have them for the 1960 sticker prices: £140 for the lot. A separate request resulted in a new windscreen at £180 — complete with good quality packing case. The screen on the car was OK, but scratched by a bladeless wiper. The gold 'Giulietta Sprint Speciale' script and Bertone badges for the flanks came from AFRA — a new-and-secondhand Alfa parts specialist in Milano. They also supplied one sun visor, rear tail light lens and a new tool bag.

When originally looking to purchase a Classic sports-tourer, I went for Alfa, not only for their inherent qualities but because many components were shared with the TIs, Sprints and Spiders. In addition, almost 90,000 Alfas were made up from CKD in South Africa between 1958 and 1985, so mechanical parts are readily available. Suspension components, bolts, bushes, brackets, plates were taken from abandoned or scrapyard TIs and members' Sprint or Spider parts' bins. Similarly interior door handles, window winders, toggle switches, rear numberplate lights — even a chassis crossmember and fuel filter — all of which were missing! Most of these were generously supplied free of charge; most notably the Bendix electric fuel pump which an enthusiastic member whipped off his running 2600 Sprint saying: "I just want to see your SS going"! No charge.

Meanwhile progress on the body took place — slowly and intermittently, a common problem here with restorations. They always seem to be treated as fill-in jobs, despite offers/bribes of monthly or progress payments. Two-and-a-half years later the shell was complete, beautifully sprayed in the original paint colour formula Alfa Rosso — that deep blood red used on the early Speciales, not the more common light orange tone. Some phone calls to club members and eight turned up, one with trailer. They carried the shell into my garage on December 7, 1984.

> ***"Everything went together easily — Alfa's engineering is really superb"***

Assembly

Once each part, nut, bolt and washer had been cleaned then assembly began, working nights and weekends. The first job was the wiring, which was a mess. Having waited for three weekends in a row for an auto electrician who didn't turn up, I climbed in myself. Fortunately Alfa were possibly the earliest users of vinyl for cable sleeving; the main loom behind the dashboard was filthy but otherwise OK. All the ancillary wiring was cleaned off with thinners, wire wool and rags — new connectors/ends soldered on, new sleeving slid on with the aid of talcum powder where long lengths were required. Very satisfying: the job was completed in three weeks. The next job was to replace all the brake pipes, unions and connectors, again DIY with the aid of a small flaring tool-kit and careful hand bending. This took about a week. Assembly was delayed as more missing bits were discovered. Whenever there was a problem, knowledgeable club members came to the rescue. One of the items of most concern was the brakes. The original huge finned aluminium drums had long since been removed; narrow Berlina units being substituted at the rear, ATE discs at the front. The correct items were gathered over a period of four years — rear backing plates from AFRA, rear drums individually from Port Elizabeth and Pretoria, the huge three-shoe fronts from Johannesburg. Bleeding and adjusting the latter were quite tricky, but help from my brother and a friend had it done in a weekend.

The 1300cc engine was totally rebuilt at home. All the pistons, sleeves, bearings, gaskets, etc, were bought direct from a Johannesburg Alfa dealer for £325 in 1982. Everything went together easily — Alfa's engineering is really superb. A new clutch was put in and the differential serviced. A full set of the original-type brand new Pirelli Cinturato tyres were obtained, still with the stickers on — one coming from a club member who had bought it recently, the other three from another source.

In October of 1984 we went for the first test drive — the car still incomplete. Over the next year it was driven on temporary permits to have the exhaust system and alignment done and then to the upholsterer, who worked from my full-size drawings of each carpet and panel. An authentic PVC black and grey scheme was chosen. In between, missing bits such as the horns, regulator, correct handbrake system and other minutæ were collected, cleaned and fitted. The last item still missing, and definitely unobtainable worldwide, was the lovely curvaceous front bumper. As a last resort I had it made up by David Cohen's restoration shop. This work of craftsman's art was completed in one week, copied from another, and even better than the factory job. The first attempt at obtaining a roadworthy certificate failed — but only on details. A split-pin was required for the handbrake compensator, the foot-brakes were pulling to one side and unbelievably — the chassis number stamped on the firewall was "too difficult to read. You must scratch the paint off"! A week later and after obtaining a police clearance on the chassis number, the car sailed through. It was October 1986, five and three-quarter years after initial purchase. Total cost, excluding labour, was about £10,000, despite all except the bodywork and upholstery being done at home.

Over the next year Speciale no 491 managed a third place in the 1986 National Classic Concours d'Elegance and third in the Classic Section of the 1987 Alfa Romeo Club of SA Annual Concours. Far better than I had expected. Without the help and encouragement of my family, friends and Alfa Club members, I don't think I'd have managed. Our high altitude dry sunny climate is also an advantage. The car goes very well: smooth, softly suspended and quiet, totally untemperamental — and it's very fast. Having just driven a Triumph TR6, Austin-Healey 3000 MK III and DB4 and DB6 Aston Martins, I'm amazed at how much more modern and refined the Alfa is. After falling for the beauty of Ernest Pieterse's Speciale twenty-six years ago at Grand Central and now realizing the dream, it would be difficult to part with this Classic coupe.

BACK TO BACK

ELITE GROUP

Alfa Romeo Giulietta Sprint, Porsche 356 or Lotus Elite? Mick Walsh drives all three small, fast coupés to find out which one he prefers

Every era of motoring evokes certain styles of car. The Edwardian age: long, grand tourers. The twenties: rakish two-seater, road-equipped Grand Prix cars. The thirties: elegant, supercharged sports cars. The fifties: mean, sculpted competition prototypes. But the sixties for me is the age of the Gran Tourismo, the coupé, the fastback in the true sense of the term.

It was an era before the exclusive specialised prototype returned to sports car racing, a time when the grid at Le Mans could be easily translated to true production cars. It was an era that dawned with the trendsetting 1946 Cisitalia 202 by Pinin Farina, blossoming out with other masterpieces in styling like the Lancia Aurelia B20, the Ferrari 250MM, Mercedes-Benz 'Gullwing' and Aston Martin DB4GT, to name just a few pinnacles of the art. All were exclusive performance toys for the wealthy, but slowly a new generation of junior GTs became available: some, like the Alfa Romeo Giulietta Sprint, arrived by accident, others, like the Porsche 356, were born out of mass production saloon design (albeit rather distantly), and others again, like the Lotus Elite, were calculated marketing decisions with small-scale production disciplines.

Here, in fact, were three cars with strong nationalistic styles, almost identical in concept and price, but with very different characters.

The saga of the pretty Giulietta Sprint could not have developed in a more accidental manner. Alfa Romeo's diversified state-owned factories had been smashed by allied bombers in the Second World War, but slowly the empire was rebuilt around pre-war designs, while Alfa Corse dominated the Grand Prix scene with the pre-war Alfetta, bringing much prestige to the battered company's image. The new four-cylinder 1900 model arrived in 1950, but its engineering was already dated and production remained tiny. Change was in the air, and Alfa needed a new model to match the new post-war outlook.

Under the wing of the IRI (an industrial reconstruction agency), the administration appointed a new consultant to supervise development. He was the Porsche engineer Rudolf Hruschka, an Austrian who was one of the co-ordinators of the Volkswagen Beetle, and worked on the Cisitalia project in the forties.

Suddenly Alfa Romeo had new priorities: the Type 159 GP cars were retired and there was a new focus on sports cars. The first new project was the Disco Volante, and it was at this point that Alfa approached Carrozzeria Bertone to build bodies for the new design. The project was sadly short-lived, as Alfa's administration decreed that the future of the company lay with the mass production of moderately priced models of very modest size. This meant a serious reorganisation of the car division which naturally demanded huge capital investment. Tooling began on a rather staid – in spite of its wonderful dohc heart – 1300cc saloon Giulietta, and to raise the much needed capital Alfa sold securities at a good rate of interest. As an extra incentive to stockholders, they announced that a draw would be made, and that the lucky number shares would win a brand new Giulietta.

This was Alfa's first venture into mass production, and the project was way behind schedule

But even when the lucky shareholders were named months later, there were still no cars available. This was Alfa's first venture into mass production, and the project was way behind schedule. Alfa found itself in a very embarrassing situation, with the press screaming 'fraud' and 'scandal'. The Giulietta's mechanicals were coming off the production line, but the body plant was almost a year behind schedule. Something had to be done without delay!

For testing the new chassis, two *muletti* had been built, with bodies hastily roughed out by the internal talent that had conceived the Disco Volante. Strangely these two prototypes, never intended to be seen by the public, bare a remarkable similarity to the Giulietta coupé, and *Ing* Quarino, Alfa's managing director, and Hruschka saw a way out of the dilemma. The grand plan was to build a small run of *fuori serie* custom bodied cars, to appease the irate winners of the lottery draw, and use them at motor shows to preview the introduction of the standard saloon model.

It was already autumn 1953, and investors and press were on the warpath, so the Turin Salon in the spring of '54 became the target launch for the new prototype. The original plans were very much in the air when Quaroni summoned Bertone and Boano of Ghia to Milan, but their assignment was clear – to refine the *muletti* into something acceptable to the public within 10 days! Boano's concept was ultimately far too removed from the original prototype, but Bertone recognised the basic harmony of the rough *muletti*, and simply refined the shape with an innovative fastback door. Naturally, Alfa chose Bertone's simpler solution, and Hruschka's insistance on at least 1000 units was finally accepted by the management.

The original plan was for Bertone to build the primered shells and for these to be delivered to Ghia for interiors, paint and glass. But days before the Turin Salon deadline, Boano left Ghia, liquidating all his interests in the company, and so the whole task of production was given over to Bertone.

Thankfully, the story had a happy ending after so many upsets. The prototype was enthusiastically received at the show, and the lottery winners found themselves getting a much more exciting car than they expected. Great credit is due to Hruschka for his insight throughout the project: it was he, according to Bertone, who insisted that the new model had a sporting character, demanding a higher state of engine tune and 1900 brakes.

Instead of 1000 units, Bertone was called upon to build 6000 of the first series Sprint. This was quite amazing for a design that was almost entirely built by hand, and the only single stamped body component was the grille that was pinched from an Autotutto van. Alfa management were patient with Bertone's production methods, and by 1959 he was producing 34 a day. As the model was phased out in the early sixties, the schedule declined to the original production level of four a day by 1965.

According to Bertone, Hruschka insisted that the new model had a sporting character

Unlike the Pininfarina Spider, the Giulietta Sprint originally was fitted with a steering column gearchange, an incredible decision for such a sporty model. In 1956, the Sprint Veloce was introduced, and by employing twin sidedraught Webers, 90bhp and a potential 111mph were available, compared to the standard Sprint's 80bhp and maximum speed of just over the ton, and thankfully a floor gearchange was introduced. In 1962, the Giulia Sprint arrived with a bored out 1570cc producing Veloce performance, and equipped with vastly improved electrics, always a criticism of the Giuliettas. The bigger engine may not have produced a dramatic change in performance, but in terms of torque and flexibility it was much improved. In 1964, the 1300 Giulietta returned to production with disc brakes now on the front, and finally in 1966 the Giulia Sprint was blessed with discs all round on the Veloce model. But by then Bertone's new coupé had superseded the original Sprint concept.

Unlike the development of the Giulietta, there was nothing accidental about the conception of the Lotus Elite. Colin Chapman was fully aware of the market potential of a small capacity coupé, a sales area already dominated by the Alfa Romeo and Porsche products featured in this test. Large capacity sports racing prototypes were losing favour after the Le Mans disaster, and new classes were being instigated for sports GT cars with a strong production car emphasis. Of particular interest to Chapman was the roadgoing GT class up to 1300cc, which he felt represented a very lucrative market.

Any form of metal unitary construction bodywork was out of the question for such a relatively low-budget project, and Chapman had for some time been examining the potential of a glass-fibre design. Jensen and Turner (and GM in the USA) had tried the moulding process for the same reasons as Chapman: glass-fibre did not rust, it was far lighter than steel, and required considerably cheaper tooling. But no-one had yet had the idea of creating a complete unitary construction bodyshell from this new material. All Chapman's background as a qualified stress engineer (which included bridge design) and his expertise with performance chassis development proved invaluable when he and accountant Peter Kirwan-Taylor started designing their revolutionary new coupé.

The basis for the new model was the Eleven, and when the two men finally settled on a shape, Frank Costin's aerodynamic wizardry made final modifications to the body's aerodynamics. He was responsible for the Kamm tail and minor changes to the nose, but essentially Kirwan-Taylor's beautiful profile remained unchanged. Progress was slow on the Elite project at the secret workshop in Edmonton, as one man, John Flayling, single handedly turned the drawings into a full size mock-up. Money and staff were in short supply, and the planned debut at Le Mans in 1957 was missed, the Elite finally making its bow at Earls Court in October of that year. The result was little short of sensational. Not only had the talented team produced one of the cleanest, most well proportioned coupé shapes of all time, but, by no accident, it was one of the most slippery, with an incredible drag coefficient of 0.29.

The suspension had impeccable racing pedigree, owing much to the Lotus 12 Formula 2 car. There were coil springs and double wishbones at the front, and an independent rear by Chapman struts. The top ends of these struts were anchored high in the tail of the cockpit, and the bottoms, carrying alloy hubs, were located by double jointed driveshafts, forward radius arms and triangulated brackets. It was simple, cheap, ingenious and very effective. The brakes were no less advanced than the rest of the chassis, with discs all round, outboard at the front and inboard- to cut down the unsprung weight – at the rear.

Because of production headaches, a year elapsed before the Elite's first customer, jazzman Chris Barber, collected his car

The light alloy overhead cam Coventry Climax powerhouse had begun service as a high performance drive for portable fire pumps before it was adopted and developed for a long line of Chapman racers. The Elite version was the FWE (Feather Weight Elite) with 10:1 compression ratio, and even with a single SU14 carburettor it developed 75bhp. Naturally, more power came over the years with the twin carb S2 giving 90bhp, with potential for even more if you desired ...

Because of production headaches, more than a year elapsed before the Elite's first customer, jazzman Chris Barber, collected his new Lotus, and the new model went on to achieve the most comprehensive competition pedigree. Too fragile for rallying, the Elite dominated the 1300cc GT class in racing just about everywhere, and among other things launched the career of a young Scot, Jimmy Clark. While on a touring holiday with rallyman Peter Riley, John Lumsden won the 1300cc class in the Nürburgring 1000kms ahead of 16 Alfas ...

But it wasn't all roses for the pretty Elite. It was neither the most refined nor the most dependable of designs. Like all Lotuses, bits had a habit of falling off, or needed constant lubrication to avoid failure. And the Coventry Climax engine was a real oil guzzler with an inherent design feature of wide clearances, as it was originally designed to rev immediately at 5000rpm from a cold start for emergency pump work. Quality control was never a strong point of Lotus production either, but it greatly improved for the Elite when production moved to the Bristol factory.

Brilliant though it was, the Elite was a commercial failure, Chapman claiming that he lost £100 with every car sold. It was exclusively priced at £1950 (rising to £2500 by 1962), and with a negligible service back-up abroad most foreign buyers were frightened off. As the acknowledged Lotus historian Dennis Ortenburger stated in his excellent history, 'maintaining an Elite in the early sixties was like surgery without anaesthetic – very painful'. Production finally ceased in September 1963 after only 988 Elites had been completed, but now they are commanding quite incredible prices all over the world, particularly in Japan.

No cars – not even Bugattis, pre-war Bentleys or Ferraris – inspire such 'white-gloved, blind devotion' as Porsches. There is no such thing as a mere Porsche owner – you either worship the cars or you hate them. To me their styling is functional at its best, and no amount of sporting camouflage can hide the fact that the basis for the 356 family is the Volkswagen Beetle. But like a Braun shaver by Dieter Rams, or a Leica camera, the first true Porsche represents the rational style of German design; simple shapes, undecorated surfaces, purity of detail and colour, all combining to produce an expensive, rugged and beautifully made character. The brilliant shape, full of wide radius curves, has a functional charm perfect for a Le Corbusier cityscape or parked in a precious Boilerhouse exhibition in the V&A, but high on an Alpine pass or charging through the *Pflanzgarten* at the 'Ring in the mist it would look ridiculous. It lacks the red-blooded automotive flair which both the Lotus and Alfa ooze just standing still.

The car was designated 356 simply because it was the 356th project to come from the Porsche Bureau since 1930. It is the only Porsche in which Ferdinand had a hand, albeit an indirect one. The first car, completed in 1948, used VW running gear, VW torsion bar suspension all round, and a modified VW 1131cc engine positioned ahead of the rear wheels in a tubular frame. For practical reasons, the first 356s built at Gmünd utilised a pressed steel chassis frame and aluminium bodywork, and the engine was positioned behind the rear axle. The air-cooled – or should it be oil cooled? – flat-four produced 40bhp at 4000rpm, which pushed the smooth coupé to a maximum speed of 85mph.

Throughout its 76,303 production run, the 356 remained basically unchanged in its general conception. A wide range of engines was offered over the years, from the pushrod overhead valve 1100cc unit to the advanced four camshaft 2-litre, but the basic layout of four air-cooled cylinders, horizontally opposed in pairs, remained unchanged.

In 1951, 1.3 and 1.5-litre engines were introduced, and a one-piece 'screen and synchromesh gearbox came in 1952. The 356A had unitary construction with five ohv engine options in various states of tune. Coupé and open bodies were offered, including the desirable Speedster, which was later superseded by the Carrera as the top model in 1958 with its 115bhp engine and all-round disc brakes.

Through the 1959-63 life of the 356B, the body was revised and gained other designations related to engine power. The 356C featured in our test was the last of the line, and essentially a stop-gap product before the new 911 was introduced. Steering was a refined ZF design instead of Volkswagen componentry, better rear suspension using a compensating spring (this acted in the opposite way to an anti-roll bar to make the rear end do more work) was developed, and disc brakes were finally adopted all round. The range was by now simplified to three engine options – the 1600S (75bhp), the 1600SC (95bhp) and the hot Carrera 2 (105bhp). The final 356 rolled off the line in September 1965.

Out on the road, these three coupés could not have more diverse characters. Like its external shape, the interior of the Elite is elegantly simple. There is an aura of purpose, just like the cockpit of any other great GT racer such as the Ferrari GTO. Lotus fanatics will argue that the interior is snug, but for me it was more claustrophobic with strong aromas from glass-fibre, oil, petrol, leather and damp carpets. In every aspect of its road manners, the Elite is alive. Very little has been done to dilute the driving sensations, and it demands to be driven quickly. Its ergonomics are perfect, even with its classic stretched back driving position.

The Climax engine has a raucous bark, and with the ZF gearbox this S2 model lugs away fiercely from low revs, coughing and spluttering until it can breathe cleanly from about 3000rpm upwards. The boomy Climax is never happy until you are revving close to the red line, if, of course, you can stand the rorty resonance throughout the car. It's a road racer (with the emphasis very much on the 'racer'), and above the whine of the gearbox and the final drive your mind is on greater things. The steering is beautifully precise and responsive, the gearchange is slick with perfectly chosen close ratios, and the acceleration eager and exciting.

Above all, it is the smooth ride that is the real surprise, proving how advanced Chapman's concept of long suspension travel was. And of course there is advanced braking power too – talk about being last of the late brakers! Compared to the Giulietta and the 356, the handling is dynamic and at least a decade ahead of its time. But underneath all this advanced thinking is a very unrefined car. With longer acquaintance, its lack of tractability in traffic became annoying, the interior just too intimate, and above all I couldn't live with the Climax's raucous boom unless I was wearing a crash helmet.

After the intensity of the Elite, driving the Alfa is pure pleasure. It may be nowhere near as dramatic, but its more relaxed character increases the enjoyment. Again, it is very much a driver's car with a low seating position, the offset engine and transmission allowing more room for the driver's feet. The typical hooded group of Alfa instruments can all be read at a glance, and the forward vision is greatly helped by the thin corner posts.

The sturdy gearlever with its large knob fits perfectly into your palm and connects to one of the finest gearboxes ever designed. Behind an engine so flexible that it cleanly revs to 7000rpm, and produces those wonderful sounds that only Alfa twin-cams can make, you have a wonderful driving experience. The handling is nowhere near as capable as the Elite's, producing far more roll, but it is neutral, very forgiving and an absolute pleasure to conduct through your favourite twists and turns. As the road tests always concluded, it's 'a joy to drive'. That all sounds like a salesman's advertising copy, but every time I experience a classic Alfa I feel that my motoring requirements match perfectly their chemistry and charisma. The most complicated and advanced music is certainly not the most enjoyable, and the rhythms of the Climax are no match for the Alfa twin-cam.

I'm sure the Porsche 356 evokes similar responses from its fans, Denis Jenkinson among them. But that Teutonic spirit is not for me. I hated its floppy gearchange, even though it may allow the owner to perform party tricks by changing gear with his foot (but only in LHD cars). I liked the high geared steering which encourages absolute precision when placing the car in corners, and I'm sure given the chance of owning one new, and not driving Midge Ure's precious example, I could enjoy provoking oversteer in the classic *wischen* method. But no doubt the consequence would be flick rolling on to its roof.

The interior is simple and elegant, all beautifully finished, and like all great designs feels good to hold. The steering action may be a little dead but is very smooth. The drone and rumble of the flat-four after the initial novelty become tiresome, lacking the variety of tone and flexibility of the Alfa. I'm sure it is superbly engineered and very rugged, but I have never taken to flat configuration designs. The 356 reminded me too much of a friend's Beetle convertible when I was at art school, and I can't forgive any sports car that.

So which coupé would I ultimately choose? Well, if I ever owned a Porsche 356 it would have to be a Speedster, and much as I admire the 'designer' lines of the Erwin Komenda masterstroke, and the quality of Porsche's finish, I would not have been tempted into AFN's showrooms in the early sixties. The 356 is a very acquired taste, and for me it is just *too* idiosyncratic. I would soon tire of the 'floppy' gearchange, the tin toy looks, and the constant drone of the flat-four engine on the *autobahnen*.

The Elite wins in the styling stakes with the purity of Kirwan-Taylor's coupé conception. Along with the Citroën DS, it should stand alongside Henry Moore's sculpture in the Tate Gallery. I admire Chapman's wonderful rational engineering throughout the car, and marvel at the Elite's handling, but as an everyday road car it is just too basic. If I owned one it would have to be for taking historic racing. On the tracks I could really come to terms with its sensational roadholding, and progressive character on the limit.

If I could only own one of the coupés, it would have to be the Alfa. Nuccio Bertone's seductive styling of the Giulietta is, for me, absolutely perfect. Those beautiful lines, allied to that classic twin-cam engine and endearing handling, present a package that encapsulates everything I would wish from a road car. It is the only engine I could happily sit behind for a long continental trip, hopefully including the Stelvio pass, and my favourite restaurant in Milan. But what are my hopes of finding a virgin chassis saved from the ravages of a rust attack? Too few I'm afraid, but maybe one day a rich Italian girlfriend will tell me her uncle has one moth-balled away near Lake Como. If only …

Our eternal thanks are due to Miles Wilkins of Club Elite and Fibre Glass Services (telephone 0243 63320 and you'll discover there's nothing he doesn't know about most Lotus models), Alan Turner of the Giulietta Register, and Midge Ure of Ultravox and Band Aid fame for providing their cars for this feature.

SPECIFICATION	ALFA ROMEO GIULIETTA SPRINT	PORSCHE 356C	LOTUS ELITE
Engine	In-line four	Flat-four, air-cooled	In-line four
Bore/stroke	78 × 82	82.5 × 74mm	76.2 × 66.6mm
Capacity	1570cc	1582cc	1216cc
Valves	Twin ohc	Pushrod ohv	Single ohv
Compression	9.7:1	8.5:1	10:1
Power	80bhp @ 6000rpm	75bhp @ 5000rpm	71bhp @ 6100rpm
Torque	72lbs ft @ 3500rpm	86lbs ft @ 3700rpm	77lbs ft @ 3750rpm
Transmission	Four-speed	Four-speed	Four-speed
Final drive	8:41	4.43:1	4.22:1
Suspension front	Ind by coil springs and wishbones	Ind by transverse torsion bars, swing axles	Ind by coil springs and wishbones
Suspension rear	Live axle, coil springs, radius arms and A-bracket	Ind by transverse torsion bars, swing axles	Ind by coil springs, Chapman struts
Steering	Worm and sector	Worm and peg	Rack and pinion
Brakes	Drums	Discs	Discs
DIMENSIONS			
Length	12ft 7.5in	13ft 2in	12ft 2in
Width	5ft 0.5in	5ft 6in	4ft 11.25in
Height	4ft 4in	4ft 4in	3ft 10.5in
Weight	1935lbs	2060lbs	1455lbs
PERFORMANCE			
Max speed	111.8mph	107mph	118mph
0-60mph	10.5secs	13.5secs	11.0secs
Standing ¼ mile	17.4secs	18.8secs	17.5secs
Fuel consumption	38mpg	35mpg	35mpg
PRODUCTION			
Years built	1962-68	1963-65	1959-63
Numbers built	42,889	16,668	988
Price new	£2118	£2493	£2586